Theory of holors

Theory of holors

A generalization of tensors

PARRY MOON
Massachusetts Institute of Technology

DOMINA EBERLE SPENCER
University of Connecticut

The right of the University of Cambridge to print and sell all manner of books was granted by Henry VIII in 1534. The University has printed and published continuously since 1584.

CAMBRIDGE UNIVERSITY PRESS
Cambridge
London New York New Rochelle
Melbourne Sydney

CAMBRIDGE UNIVERSITY PRESS
Cambridge, New York, Melbourne, Madrid, Cape Town, Singapore, São Paulo

Cambridge University Press
The Edinburgh Building, Cambridge CB2 2RU, UK

Published in the United States of America by Cambridge University Press, New York

www.cambridge.org
Information on this title: www.cambridge.org/9780521245852

First published 1986
This digitally printed first paperback version 2005

A catalogue record for this publication is available from the British Library

Library of Congress Cataloguing in Publication data

Moon, Parry Hiram, 1898–
Theory of holors.
Bibliography: p.
Includes index.
1. Calculus of tensors. I. Spencer, Domina Eberle,
1920– II. Title. III. Title: Holors.
QA433.M66 1986 515′.63 85–6657

ISBN-13 978-0-521-24585-2 hardback
ISBN-10 0-521-24585-0 hardback

ISBN-13 978-0-521-01900-2 paperback
ISBN-10 0-521-01900-1 paperback

To

DIRK J. STRUIK

our teacher and our friend

Contents

Preface

The word *holor* indicates a mathematical entity that is made up of one or more independent quantities. Examples of holors are complex numbers, vectors, matrices, tensors, and other hypernumbers. The very fruitful idea of representing a collection of quantities by a single symbol apparently originated in 1673, when John Wallis represented a *complex number* as a point in 2-space. More complicated holors were developed by Hamilton (1843), Grassmann (1844), Gibbs (1881), Ricci (1884), Heaviside (1892), Levi-Civita (1901), Einstein (1916), and many others.

An unfortunate aspect of the subject is that each kind of holor was developed more or less independently, with its own nomenclature, its own theory, and its own textbooks. On the contrary, we propose to use a *single notation* that applies to all holors, be they tensors or nontensors. This step results in an immense simplification. Special symbols and notations, such as boldface type, ∇, grad, div, curl, are abandoned. Ordinary type is used, but with subscripts and superscripts. Such index notation is, of course, common in tensor theory but has not ordinarily supplanted the old notation of vector analysis and matrix theory. The first feature of our book, therefore, is the use of index notation throughout.

A second feature deals with coordinate transformation. A tensor is a holor that transforms in a particularly simple way. But many holors are not tensors. Indeed, applications can be cited where transformation has no physical significance. And even in applications where transformations are physically meaningful, the holor may not transform as a tensor. For instance, the ordinary derivative of a tensor is generally not a tensor.

Our book differs from most others in its generality. Not only does it cover nontensors having no transformation equation, but it also includes holors that transform in more complicated ways than ordinary tensors.

This opens up an exciting field of new possibilities. In devising a holor for a new physical application, we are no longer limited to a few conventional types of holor but can tailor our system to fit the physics of the subject.

The authors look forward to the golden day when university curricula will not separate vector and tensor analysis into different subjects but will cover all holors in a uniform treatment. The present book constitutes a step in this direction. The material has been taught at MIT and at the University of Connecticut.

Finally, it is our pleasant duty to thank Dr. Struik for his kindness in reading the historical introduction and Cambridge University Press for its cheerful cooperation.

P.M.
D.E.S.

Nomenclature

Nilvalent holor (Chapter 1): S.

Univalent holors (Chapter 1): $v^i = (v^1, v^2, \ldots, v^n)$, $v_i = (v_1, v_2, \ldots, v_n)$.

Bivalent holors (Chapter 1):

$$W^{ij} = \begin{bmatrix} W^{11} & W^{12} & \cdots & W^{1n} \\ W^{21} & W^{22} & \cdots & W^{2n} \\ \cdot & \cdot & \cdots & \cdot \\ W^{n1} & W^{n2} & \cdots & W^{nn} \end{bmatrix}, \qquad W^i_j = \begin{bmatrix} W^1_1 & W^1_2 & \cdots & W^1_n \\ W^2_1 & W^2_2 & \cdots & W^2_n \\ \cdot & \cdot & \cdots & \cdot \\ W^n_1 & W^n_2 & \cdots & W^n_n \end{bmatrix},$$

$$W_{ij} = \begin{bmatrix} W_{11} & W_{12} & \cdots & W_{1n} \\ W_{21} & W_{22} & \cdots & W_{2n} \\ \cdot & \cdot & \cdots & \cdot \\ W_{n1} & W_{n2} & \cdots & W_{nn} \end{bmatrix}.$$

Trivalent holors (Chapter 1): W^{ijk}, W^{ij}_k, W^i_{jk}, W_{ijk}.

Symmetric part of holor u^{ij} (Chapter 1): $u^{(ij)} = \frac{1}{2}[u^{ij} + u^{ji}]$.

Antisymmetric part of holor u^{ij} (Chapter 1): $u^{[ij]} = \frac{1}{2}[u^{ij} - u^{ji}]$.

Kronecker delta (Chapter 1):

$$\delta^i_j = \begin{bmatrix} 1 & 0 & 0 & \cdots & 0 \\ 0 & 1 & 0 & \cdots & 0 \\ \cdot & \cdot & \cdot & \cdots & \cdot \\ 0 & 0 & 0 & \cdots & 1 \end{bmatrix}.$$

Generalized Kronecker deltas (Chapter 1):

$$\delta^{ij}_{kl} = \begin{bmatrix} \delta^i_k & \delta^i_l \\ \delta^j_k & \delta^j_l \end{bmatrix}, \qquad \delta^{ijk}_{lmn} = \begin{bmatrix} \delta^i_l & \delta^i_m & \delta^i_n \\ \delta^j_l & \delta^j_m & \delta^j_n \\ \delta^k_l & \delta^k_m & \delta^k_n \end{bmatrix}.$$

Alternators (Chapter 1): δ^{12}_{ij}, δ^{ij}_{12}, δ^{123}_{ijk}, $\delta^{ijk}_{123}, \ldots$.

Determinant (Chapter 1): $|a^{ij}|$, $|a^i_j|$, $|a_{ij}|$.

Gamma products (Chapter 2): $\gamma^i_j v^j = w^i$, $w^{kl} = \gamma^{kl}_{ij} v^{ij}$, $\gamma^j_i u^i v_j = S$,

$\gamma^{kj}_i u^i v_j = w^k$, $\gamma^{klj}_i u^i v_j = w^{kl}$, etc.

Scalar (Chapter 4): $S' = S$.

Contravariant univalent tensor (Chapter 4):

$$v^{i'} = \frac{\partial x^{i'}}{\partial x^i} v^i.$$

Covariant univalent tensor (Chapter 4):

$$u_{i'} = \frac{\partial x^i}{\partial x^{i'}} u_i.$$

Contravariant bivalent tensor (Chapter 4):

$$w^{i'j'} = \frac{\partial x^{i'}}{\partial x^i} \frac{\partial x^{j'}}{\partial x^j} w^{ij}.$$

Covariant bivalent tensor (Chapter 4):

$$w_{i'j'} = \frac{\partial x^i}{\partial x^{i'}} \frac{\partial x^j}{\partial x^{j'}} w_{ij}.$$

Mixed bivalent tensor (Chapter 4):

$$w^{i'}_{j'} = \frac{\partial x^{i'}}{\partial x^i} \frac{\partial x^j}{\partial x^{j'}} w^i_j.$$

Contravariant univalent akinetor (Chapter 5):

$$\tilde{v}^{i'} = \sigma \frac{\partial x^{i'}}{\partial x^i} \tilde{v}^i.$$

Contravariant univalent akinetor of weight w (Chapter 5):

$$\tilde{v}^{i'} = \Delta^w \frac{\partial x^{i'}}{\partial x^i} \tilde{v}^i.$$

Covariant univalent akinetor (Chapter 5):

$$\tilde{u}_{i'} = \sigma \frac{\partial x^i}{\partial x^{i'}} \tilde{u}_i.$$

Covariant univalent akinetor of weight w (Chapter 5):

$$\tilde{u}_{i'} = \Delta^w \frac{\partial x i}{\partial x i'} \tilde{u}_i.$$

Pseudoarea (Chapter 5): $d\tilde{\mathfrak{A}}_i = \delta^{123}_{ijk}\, dv^j\, dw^k$.

Pseudovolume (Chapter 5): $d\tilde{\mathfrak{V}} = \delta^{123}_{ijk}\, du^i\, dv^j\, dw^k$.

Gradient of a scalar (Chapters 4, 5):

$$\frac{\partial \phi}{\partial x^i}.$$

Pseudodivergence (Chapter 5):

$$\frac{\partial \tilde{v}^i}{\partial x^i}.$$

Pseudocurl (Chapter 5):

$$\tilde{W}^i = \delta^{ijk}_{123} \frac{\partial u_k}{\partial x^j}.$$

Pseudodivergence of a bivalent contravariant alternating akinetor of weight +1 (Chapter 5):

$$\frac{\partial \tilde{F}^{ij}}{\partial x^i}.$$

Linear connection (Chapter 7):

$$\Gamma^i_{jk} = \frac{\partial x^i}{\partial x^{i'}} \frac{\partial x^{j'}}{\partial x^j} \frac{\partial x^{k'}}{\partial x^k} \Gamma^{i'}_{j'k'} + \frac{\partial x^i}{\partial x^{i'}} \frac{\partial^2 x^{i'}}{\partial x^j \partial x^k}.$$

Torsion tensor (Chapter 7): $S^i_{jk} = \frac{1}{2}(\Gamma^i_{jk} - \Gamma^i_{kj})$.

Covariant derivatives (Chapter 7):

$$\overset{1}{\nabla}_j v^k = \frac{\partial v^k}{\partial x^j} + v^l \Gamma^k_{jl}, \qquad \overset{2}{\nabla}_j v^k = \frac{\partial v^k}{\partial x^j} + v^l \Gamma^k_{lj}.$$

Riemann–Christoffel tensor of first kind (Chapter 8):

$$R^k_{ijl} = \frac{\partial \Gamma^k_{jl}}{\partial x^i} - \frac{\partial \Gamma^k_{il}}{\partial x^j} + \Gamma^m_{jl} \Gamma^k_{im} - \Gamma^m_{il} \Gamma^k_{jm}.$$

Riemann–Christoffel tensor of second kind (Chapter 8):

$$T^k_{ijl} = \frac{\partial \Gamma^k_{lj}}{\partial x^i} - \frac{\partial \Gamma^k_{li}}{\partial x^j} + \Gamma^m_{lj}\Gamma^k_{mi} - \Gamma^m_{li}\Gamma^k_{mj}.$$

Ricci tensors of first and second kind (Chapter 8):

$$R_{ij} = R^l_{ijl} = \frac{\partial \Gamma^l_{jl}}{\partial x^i} - \frac{\partial \Gamma^l_{il}}{\partial x^j}, \qquad T_{ij} = T^l_{ijl} = \frac{\partial \Gamma^l_{lj}}{\partial x^i} - \frac{\partial \Gamma^l_{li}}{\partial x^j}.$$

Ricci tensor of third kind (Chapter 10): $\mathcal{R}_{ij} = R^l_{lij}$.

Absolute differentials (Chapter 9):

$$\underset{1}{\delta} v^i = dv^i + \Gamma^i_{jk} v^k \, dx^j, \qquad \underset{2}{\delta} v^i = dv^i + \Gamma^i_{kj} v^k \, dx^j$$

Riemannian metric (Chapter 10): $(ds)^2 = g_{ij}\, dx^i\, dx^j$.

Linear connection in metric space (Chapter 10):

$$\Gamma^i_{jk} = \frac{1}{2} g^{il} \left[\frac{\partial g_{lj}}{\partial x^k} + \frac{\partial g_{lk}}{\partial x^j} - \frac{\partial g_{jk}}{\partial x^l} \right].$$

Area (Chapter 10): $d\mathcal{A}_i = g^{1/2} \delta^{123}_{ijk}\, dv^j\, dw^k$.

Volume (Chapter 10): $d\mathcal{V} = g^{1/2} \delta^{123}_{ijk}\, du^i\, dv^j\, dw^k$.

Components of vector v^i (Chapter 10):

$$(v)_1 = \sqrt{g_{11}}\, v^1, \qquad (v)_2 = \sqrt{g_{22}}\, v^2, \qquad (v)_3 = \sqrt{g_{33}}\, v^3.$$

Magnitude of vector v^i (Chapter 10): $v = (g_{ij} v^i v^j)^{1/2}$.

Angle between vectors u^i and v^j (Chapter 10):

$$\cos\theta = \frac{g_{ij} u^i v^j}{uv}.$$

Scalar product of vectors (Chapter 10): $S = g_{ij} u^i v^j = \mathbf{u} \cdot \mathbf{v}$.

Vector product of vectors (Chapter 10):

$$w^k = g^{1/2} g^{kl} \delta^{123}_{lij} u^i v^j, \qquad \mathbf{w} = \mathbf{u} \times \mathbf{v}.$$

Gradient of scalar (Chapter 10):

$$\operatorname{grad}_i \phi = \frac{\partial \phi}{\partial x^i}, \qquad \operatorname{grad} \phi.$$

Divergence of vector (Chapter 10):

$$\operatorname{div} \mathbf{v} = g^{-1/2} \frac{\partial}{\partial x^i} (g^{1/2} v^i).$$

Curl of vector (Chapter 10):

$$W^i = g^{-1/2}\delta_{123}^{ijk}\frac{\partial}{\partial x^j}(g_{kl}v^l), \qquad \text{curl } \mathbf{v}.$$

Scalar Laplacian (Chapter 10): $\nabla^2\phi = \text{div grad } \phi$.

Vector Laplacian (Chapter 10): ✡$\mathbf{v} = \text{grad div } \mathbf{v} - \text{curl curl } \mathbf{v}$.

Euclidean metric (Chapter 11):

$$(ds)^2 = g_{ij}\, dx^i\, dx^j$$

where

$$g_{ij} = \sum_{i'=1'}^{n'} \left(\frac{\partial x^{i'}}{\partial x^i}\right)^2.$$

0
Historical introduction

The birth of the holor concept may be set at 1673, when John Wallis suggested the geometric representation of complex numbers by points in a plane.[1] A complex number is an entity consisting of two parts. It is usually written as

$$z = x + iy,$$

where x and y are real numbers and i emphasizes the fact that x and y are independent quantities. A more modern notation writes a two-element holor as an ordered number pair,

$$z = (x, y)$$

or, in index notation,

$$x^i = (x^1, x^2).$$

This constitutes the first step in a fascinating mathematical development that now includes vectors, matrices, tensors, and other holors.

0.01. Quaternions

Mathematicians expected that the extension from the complex number, with $n = 2$ to $n = 3$, would be child's play; but considerable time elapsed before they found that no such extension is possible without violating a rule of ordinary algebra.[2] After wrestling with the problem for 15 years, Sir William Rowan Hamilton in 1843 found a holor with four elements,

$$x^i = (x^1, x^2, x^3, x^4),$$

which he called a *quaternion*.[3] The ordinary rules of algebra hold for quaternions except that multiplication is noncommutative. Hamilton believed that quaternions constituted his greatest contribution to mathematics and that most of physics would eventually be written in quaternion form. Tait[4] devoted himself to that end. But physicists were not convinced. For instance, Maxwell[5] in his celebrated treatise (1873) mentioned quaternions several times but was careful not to use them.

0.02. Linear associative algebras

The work of Hamilton should certainly have suggested the possibility of a host of new algebras in which associative or commutative properties are abandoned. This possibility, however, was not recognized until 1870, when Benjamin Peirce[6] invented a whole new branch of mathematics – the study of *linear associative algebras.* The subject became popular, and hundreds of algebras* were developed with their associated holors.[7] One would expect that a few of these algebras would have had interesting practical applications, but such does not seem to be the case. The reason for this disappointing result will now be explained.

Consider holors such as

$$x^i = (x^1, x^2, \ldots, x^n),$$
$$y^i = (y^1, y^2, \ldots, y^n),$$

and two operations, which we shall call addition and multiplication. The universally accepted definition of addition is

$$x^i + y^i = (x^1 + y^1, x^2 + y^2, \ldots, x^n + y^n).$$

It appears that nothing is to be gained by changing the definition of addition. But multiplication is another story! We are at liberty to select a multiplication table, and each table determines a new algebra with its own peculiar properties.

A linear associative algebra is a closed system: a sum or product is always a holor of the same system. For instance, the product of two quaternions is always a quaternion, the product of two matrices is always a matrix. This is a delightful property, closely associated with groups, rings, and other aspects of modern algebra. But it is not the kind of behavior that is needed for most physical applications. Experience shows that the product of two vectors, for instance, must sometimes be a vector, sometimes a scalar, sometimes a matrix. And this variety cannot

* This includes algebras of Dedekind, Frobenius, Scheffers, Peirce, Kronecker, Weierstrass, Dickson, Sylvester, and Cartan.

be obtained with any of the linear associative algebras. Here is the reason that this branch of mathematics, once pursued with such enthusiasm, has proved to be of limited value as regards physical applications.

0.03. Matrices

We have been considering *univalent* holors, such as

$$x^i = (x^1, x^2, \ldots, x^n).$$

Also important are *bivalent* holors (matrices), for instance,

$$A^{ij} = \begin{bmatrix} A^{11} & A^{12} & \cdots & A^{1n} \\ A^{21} & A^{22} & \cdots & A^{2n} \\ \cdot & \cdot & \cdots & \cdot \\ A^{n1} & A^{n2} & \cdots & A^{nn} \end{bmatrix}.$$

The theory of matrices* was initiated by Arthur Cayley[8] in 1855. The only product ordinarily used is the Cayley product, which in index notation is

$$A^{ij} B_{jk} = C^i_k.$$

This is an associative but noncommutative product. Thus, matrix algebra[9] may be regarded as one of the linear associative algebras.

0.04. Grassmann

In the same year (1844) that Hamilton published his first paper on quaternions, a much more general treatment of holors was published by Herman Grassmann.[10] Not only did *Die Ausdehnungslehre* deal with n-space but, unlike the linear associative algebras, it was not limited to a single product per algebra. Thus it opened up new possibilities, particularly in geometric and physical applications of holors. Hamilton's comment was typical[11]: "I have recently been reading...more than a hundred pages of Grassmann's *Ausdehnungslehre,* with great admiration and interest.... But it is curious to see how narrowly, yet how completely, Grassmann failed to hit off the quaternions." As if Grassmann, with his immensely more general treatment, would be particularly interested in the special case of the quaternion!

* The name "matrices" was first used by J. J. Sylvester, *Collected math. papers,* Vol. 1 (Cambridge University Press, Cambridge, 1904), p. 145.

0.05. Vector analysis

To other mathematicians and physicists at this time, the *Ausdehnungslehre* seemed quite incomprehensible; and quaternions, based on Hamilton's 800-page explanation, seemed hardly simpler. It was not until 40 years later that Williard Gibbs[12] (and independently, Oliver Heaviside)[13] worked out a special case of quaternions called vector analysis. This was based largely on Hamilton's pioneer work and even used Hamilton's terms *scalar* and *vector*, but it applied only to 3-space and was usually expressed in rectangular coordinates. It did have two distinct products, however, and it proved to be just what was needed in a large number of physical applications. As a consequence, vector analysis has been universally accepted and is widely used today.[14]

0.06. Invariance

If one reads an early treatment of vectors, such as that of Gibbs or Coffin,[15] he is impressed by the exclusive use of rectangular Cartesian coordinates. Even such concepts as divergence, curl, and scalar and vector Laplacians[16] are usually defined in rectangular coordinates. In a way, this procedure makes for simplicity. But it ignores the subject of invariance. Even with Gibbs, a vector was more than a set of numbers representing rectangular coordinates: it was a geometric entity that could be expressed equally well in other coordinate systems.[12] The merates changed when the coordinates were changed, but the geometric object maintained its identity.

Invariants were studied by Cayley[17] beginning in 1841. The subject received such enthusiastic support during the latter half of the nineteenth century that Sylvester[18] remarked in 1864, "As all roads lead to Rome, so I find in my own case at least that all algebraic inquiries, sooner or later, end at the Capitol of modern algebra over whose shining portal is inscribed the *Theory of Invariants.*" The subject is treated in a voluminous literature.[19]

An important application of invariant theory is differential geometry. Riemann[20] outlined an approach in his *Habilitationsschrift* (Göttingen, 1854). Felix Klein,[21] in his Erlanger program, stated a principle that affected the development of geometry for many years. Christoffel, Beltrami, Bianchi, and others made valuable contributions. Ricci[22] developed a new notation, which he called "the absolute differential calculus" but which is now called tensor* theory. Ricci wrote many papers, including a celebrated one with Levi-Civita[23] (1901).

Physicists in general were not interested in invariants and had never heard of Ricci's work. It was not until 1916 that Albert Einstein[24] applied tensors in formulating his general relativity. Because of the extraordinary popular appeal of this theory, tensors became famous overnight. Hundreds of books appeared on relativity and several on tensors.[25] And even the old subject of vectors took on new life when presented in tensor form.[14]

0.07. Tensors

Index notation was designed particularly to handle behavior under coordinate transformation. But the basic notation is applicable even when no coordinate transformations are involved. Thus, index notation can be employed advantageously to unify a great field of holors of various valences and dimensionalities, even where the question of invariance does not necessarily enter. We have seen how vector analysis originated with little thought for coordinate transformation. The same may be said for quaternions, matrices, and the holors treated in linear associative algebras.

Since 1916, however, the feeling seems to be prevalent that the only hypernumbers of any value are tensors. On the contrary, many valuable applications of holors do not need a consideration of coordinate transformation. The simplest application of holors specifies a collection of discrete objects – for instance, machine screws. Each merate gives the number of screws of a particular type. The complete holor designates the inventory of a particular tool room (as regards screws in stock). Transformation of coordinates is meaningless here. Another example has merates that specify the loop currents in a given l-loop electric network. Again, transformation does not enter. And there are examples in which transformation is possible but not of much interest. Thus, we have a host of practical applications of holors where the holor is definitely not a tensor. And we have, of course, a host of other applications where the holor is a tensor.

* The name *tensor* is often said to be the invention of Woldemar Voigt,[26] who used it to represent a bivector in crystal physics. The word, however, is much older, having been applied in connection with quaternions by Hamilton. See *The Mathematical papers of Sir William Rowan Hamilton* (Cambridge University Press), p. 237.

0.08. Holor notation

Heaviside[13] suggested boldfaced type to distinguish vectors from scalars, and this convention has been widely accepted. Some means of distinguishing other holors is definitely needed. In particular, a distinctive symbol should be established for matrices, though no such standard seems to have been fixed. Schouten[27] invented a complicated notation for holors in 1914. Struik[28] developed another ingenious system in 1922.

A quite different nomenclature for holors developed from invariance theory. A quadratic form is usually written as

$$A = \sum_{1}^{n} a_{ij} x^i x^j,$$

where $x^1, x^2, \ldots, x^n$ are real numbers, x^i is a vector in n-space, and a_{ij} is a matrix. Such expressions were used by Riemann in his *Habilitationsschrift* (1854) and were incorporated into Ricci's work. Thus, tensor analysis was printed in ordinary type, and the kind of quantity was distinguished by the number and position of the indices. The scheme was further developed by Levi-Civita, Schouten and Struik,[29] Veblen, Weyl, Cartan, Eisenhart, T. Y. Thomas, and many others, so that it may now be considered as a universal notation for all holors.*

Einstein introduced the summation convention (1916), which has become an important characteristic of modern index notation. Whereas Ricci always wrote a summation with the customary Σ, as

$$\Sigma_{rs}\, a_{rs}\, dx_r\, dx_s.$$

Einstein omitted the summation sign and wrote

$$a_{rs}\, dx^r\, dx^s.$$

He says,[24] "A glance at the equations of this paragraph shows that there is always a summation with respect to the indices which occur twice under a sign of summation. . .and only with respect to indices which occur twice. It is therefore possible, without loss of clearness, to omit the sign of summation. In its place we introduce the convention: If an index occurs twice in one term of an expression, it is always to be summed unless the contrary is expressly stated." This apparently trivial change had an astonishing effect on the simplicity and power of index notation.†

* Note that both Schouten and Struik abandoned their "direct" representations after 1922 and employed the standard tensor notation.

† The quotation ignores the important distinction between subscripts and superscripts. A more modern treatment is given in Chapter 2.

0.09. Derivatives

Derivatives of holors are frequently encountered. For example, a contravariant vector transforms as

$$v^{i'} = \frac{\partial x^{i'}}{\partial x^{i}} v^{i}$$

and a covariant vector as

$$u_{i'} = \frac{\partial x^{i}}{\partial x^{i'}} u_{i}.$$

For a metric space, Riemann (1854) wrote

$$(ds)^2 = g_{ij}\, dx^i\, dx^j.$$

He also used geodesics defined by the second-order differential equation,

$$\frac{d^2x^i}{ds^2} + \left\{ \begin{matrix} i \\ jk \end{matrix} \right\} \frac{dx^j}{ds}\frac{dx^k}{ds} = 0.$$

Christoffel[30] (1869) introduced the Christoffel symbols which were defined in terms of the metric coefficients g_{ij}.

In 1917, Levi-Civita[31] extended the idea of parallelism to a Riemannian n-space. The covariant derivative appeared, and the limitations of Riemann and Christoffel were removed to give a more general treatment. The linear connection Γ^i_{jk} is defined as a holor whose transformation equation is[30]

$$\Gamma^i_{jk} = \frac{\partial x^i}{\partial x^{i'}} \frac{\partial x^{j'}}{\partial x^j} \frac{\partial x^{k'}}{\partial x^k} \Gamma^{i'}_{j'k'} + \frac{\partial^2 x^{i'}}{\partial x^j \partial x^k} \frac{\partial x^i}{\partial x^{i'}}.$$

The linear connection may be considered as the sum of a symmetric and an antisymmetric part:

$$\Gamma^i_{jk} = S^i_{jk} + \Omega^i_{jk}.$$

The covariant derivative employs a symbolism used by Schouten,[32]

$$\nabla_j v^i = \frac{\partial v^i}{\partial x^j} + \Gamma^i_{jk} v^k.$$

Also noteworthy are the Riemann–Christoffel tensor,

$$R^k_{ijl} = \frac{\partial \Gamma^k_{jl}}{\partial x^i} - \frac{\partial \Gamma^k_{il}}{\partial x^j} + \Gamma^m_{jl}\Gamma^k_{im} - \Gamma^m_{il}\Gamma^k_{jm},$$

and the Ricci tensors,

$$R_{ij} = R^{l}_{ijl} \quad \text{and} \quad R_{ij} = R^{l}_{lij}.$$

These curvature tensors are very important in Einstein's general relativity.

0.10. Outline of the book

Our book begins with three chapters on holor notation and holor algebra. At this stage, we are interested in index notation but not at all in transformation properties. Thus, we consider addition of holors as well as uncontracted and contracted multiplication.

Part II introduces coordinate transformations and defines tensors, akinetors, and geometric objects.

Part III introduces holor calculus in a very general form in which the linear connection is entirely arbitrary except for its transformation equation. Note that the keystone of tensor calculus is a linear connection holor that is itself not a tensor.

Part IV depends on the introduction of a metric. It thus includes the important subjects of Riemannian spaces and the special case of Euclidean space.

I

Holors

1
Index notation

In ordinary algebra, we represent a real number by a symbol such as x. In complex algebra, we deal with pairs of real numbers but still write the ordered pair as a single symbol:

$$z = x + iy = (x, y).$$

In vector algebra, we deal with number triples, such as

$$v^i = (v^1, v^2, v^3).$$

Evidently, the process can be extended indefinitely. Thus n independent quantities can be considered as a single hypernumber,

$$v^i = (v^1, v^2, \ldots, v^n).$$

Two-dimensional and three-dimensional arrays of numbers constitute more complicated entities.

Definition I: A holor is a mathematical entity built up of one or more independent merates, where merate refers to the number $v^1, v^2, \ldots$.

The word *holor** comes from the Greek ὅλος, a whole (as in the English word *holistic,* pertaining to an entity).[1] The elements are called *merates* (Greek μέρος, a part). Merates may be real or complex numbers or they

* Holors could be called "hypernumbers," except that we wish to include the special case of $N = 0$ (the scalar), which is certainly not a hypernumber. On the other hand, holors are often called "tensors." But this is incorrect, in general, for the definition of a tensor includes a specific dependence on coordinate transformation. To achieve sufficient generality, therefore, it seems best to coin a new word such as *holor*.

A similar ambiguity occurs in the names of the elements making up the holor. We could use the word "components" to designate elements regarded as vectors, and "coordinates" to designate elements regarded as scalars. But the terms are used so loosely in practice that this suggestion does not seem fruitful. We therefore introduce the new word *merates*.

may be more complicated quantities such as matrices. In this book, however, we shall limit ourselves (unless the contrary is specifically stated) to merates that are real numbers.

1.01. Variety of holors

Holors may be classified with respect to*
(1) plethos n and
(2) valence N.
One of the most common forms of holor consists of n merates arranged in a row, as

$$v^i = (v^1, v^2, \dots, v^n).$$

The plethos of this holor is n. If v^i represents an ordinary complex number, the plethos is 2; if v^i represents a vector in 3-space, the plethos is 3; if v^i is an event in space–time, the plethos is 4. In all these cases, the valence is said to be 1, corresponding to the number of indices attached to the base letter.

More complicated holors are often necessary. An array like

$$A_{ij} = \begin{bmatrix} A_{11} & A_{12} & A_{13} \\ A_{21} & A_{22} & A_{23} \\ A_{31} & A_{32} & A_{33} \end{bmatrix}$$

has three merates in each row and three in each column. Thus, its plethos, both for index i and index j, is 3. But it differs from the previous holors in having a two-dimensional array of merates instead of a one-dimensional array. Hence its valence is said to be 2.

The reader is acquainted with the distinction between a matrix and a determinant. A matrix is a bivalent holor – a rectangular array of merates, the changing of any one of which gives a new matrix. A determinant, on the other hand, is a nilvalent holor – an ordinary number whose value is obtained by the usual process of determinant expansion. Associated with each square matrix v^{ij} is a determinant $|v^{ij}|$:

* The term *valence* was introduced by Schouten and Struik (Einführung in die neuren Methoden der Differentialgeometri, 1935, Bd. I, p. 7): "Statt 'Valenz' wurde oft 'Grad' oder 'Ordnung' benutzt. Die beiden letzten Ausdrücke haben schon so viele andere Beduetungen, dass sie leicht zu Verwirrung führen. Der hier gewählte Ausdruck, der in allen Sprachen übertragbar ist, hat den Vorteil an die chemische Valenz zu erinnern, deren Eigenschaften tatsächlich eine grosse Ähnlichkeit mit unserer Valenz aufweisen."
The word *plethos* comes from the Greek πλῆθος, indicating "dimensionality."

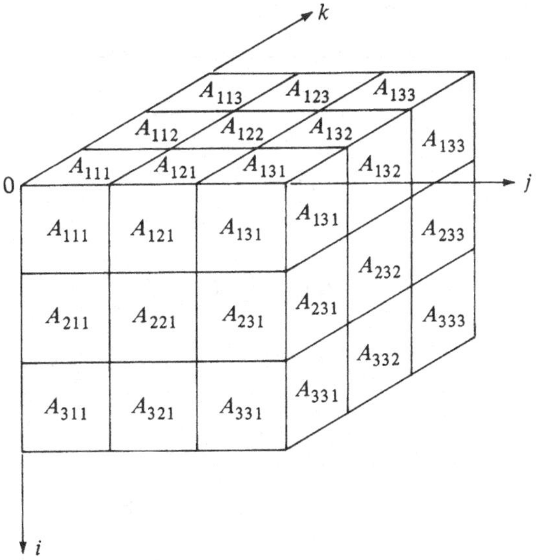

Figure 1.01. Three-dimensional matrix A_{ijk}.

$$|v^{ij}| = \begin{vmatrix} v^{11} & v^{12} & \cdots & v^{1n} \\ v^{21} & v^{22} & \cdots & v^{2n} \\ \cdot & \cdot & \cdots & \cdot \\ v^{n1} & v^{n2} & \cdots & v^{nn} \end{vmatrix}$$

$$= v^{11}M_{11} + v^{21}M_{21} + \cdots + v^{n1}M_{n1} = S,$$

where $\{M_{ij}\}$ is the cofactor of the merate $\{v^{ij}\}$. Evidently an infinite number of distinct matrices v^{ij} may be associated with the same value of the determinant $|v^{ij}| = S$.

Merates can be arranged also in a cubical array (Figure 1.01), giving A_{ijk}, a holor of valence 3. Evidently three indices are necessary in this case, each block of the diagram representing a merate $\{A_{ijk}\}$. Ordinarily, the plethos is the same for all indices, as in this case i, j, k range from 1 to 3. But trivalent holors are possible with different plethos for the three indices.

Valence is not limited to 3 but may be extended to any integer N. Similarly, plethos may have any integer value n. For the special case of $N = 0$, we have an ordinary number written without index,

Table 1.01. *Holors*

Name	Valence	Symbol	Representation
Nilvalent holor	0	S	S
Univalent holor	1	u^i, v^i	$u^i = (u^1, u^2, \ldots, u^n)$
Bivalent holor	2	U^{ij}, V_{ij}	$U^{ij} = \begin{bmatrix} U^{11} & U^{12} & \cdots & U^{1n} \\ U^{21} & U^{22} & \cdots & U^{2n} \\ \cdot & \cdot & \cdots & \cdot \\ U^{n1} & U^{n2} & \cdots & U^{nn} \end{bmatrix}$
Trivalent holor	3	U^{ijk}, V_{ijk}	$U^{ij1} = \begin{bmatrix} U^{111} & U^{121} & \cdots & U^{1n1} \\ U^{211} & U^{221} & \cdots & U^{2n1} \\ \cdot & \cdot & \cdots & \cdot \\ U^{n11} & U^{n21} & \cdots & U^{nn1} \end{bmatrix}$
			$U^{ij2} = \begin{bmatrix} U^{112} & U^{122} & \cdots & U^{1n2} \\ U^{212} & U^{222} & \cdots & U^{2n2} \\ \cdot & \cdot & \cdots & \cdot \\ U^{n12} & U^{n22} & \cdots & U^{nn2} \end{bmatrix}$
			etc.

as S. To summarize, holors include a great variety of entities, as indicated in Table 1.01.

1.02. Notation

In the past, holors have been represented by various symbols and have been printed in various kinds of type. A given notation is usually applicable to only one form of holor, such as a vector or a matrix. But in a general study of holors, we need a single notation that applies to all values of valence and all values of plethos. Such a notation is index notation, using both subscripts and superscripts.

For instance, a point in 3-space may be specified by giving three numbers representing its rectangular coordinates x, y, z. In index notation, the point may be called

$$x^{i'} = (x^{1'}, x^{2'}, x^{3'}),$$

where $x^{1'} = x$, $x^{2'} = y$, and $x^{3'} = z$. In a different coordinate system, the same point is represented by a different set of three numbers. It may be written

$$x^i = (x^1, x^2, x^3).$$

To emphasize the fact that we are dealing with the same geometric entity in the two coordinate systems, we use the same base letter x. To indicate that the coordinates are different in the two systems, we use different indices i and i'. It is customary to employ primes for rectangular coordinates, though this convention need not be used.

A univalent holor may be written with a superscript or with a subscript, though x^i and x_i are generally not equivalent. The distinction is concerned with the way in which the quantity behaves under coordinate transformation. This subject will be considered in later chapters. At present, we use both subscripts and superscripts and distinguish carefully between them.

We must also distinguish between superscripts as indices and as powers. The latter will be denoted by a superscript outside parentheses. Thus, the third merate of u^i is written u^3; but if this merate is squared, we have $(u^3)^2$. Also there is often a need for labels, which will be written as subscripts outside parentheses. For instance, there might be a need to distinguish a field vector v^i in region 1 from the same vector in region 2. One would then write $(v^i)_1$ and $(v^i)_2$. Unless some such convention is employed to discriminate among indices, powers, and labels, equations may be highly ambiguous.

We must also discriminate between a complete holor, represented by v^i or W^{ij}, and an arbitrary merate of such a holor. A base letter with indices always represents the holor: The same symbol in curly brackets represents an arbitrary merate of the holor. Thus, v^i is the complete holor; $\{v^i\}$ is any one of its merates. Of course, if we refer to a specific merate, as v^3, no brackets are required since no ambiguity is involved.

In writing a holor in terms of its merates, we may have, for instance,

$$v^i = (v^1, v^2, v^3, \ldots, v^n).$$

Strictly speaking, the nth merate should be written $\{v^n\}$ to emphasize the fact that it is a merate, not a holor. For simplicity, however, we shall omit the brackets in such cases, since the meaning seems obvious.

1.03. Conventions

The reader will find that index notation gives the ultimate in compactness and allows quick and easy manipulation that makes even the symbolism of vector analysis look clumsy. Ultimately, however, one must expand these condensed holor equations. In such expansions, it is advantageous to establish definite conventions as to the arrangements of the merates.

Convention I: Merates are arranged so that numerical indices are in ascending order from left to right and from top to bottom.

Thus, a univalent holor is written

$$u^i = (u^1, u^2, \dots, u^n),$$

never

$$u^i = (u^n, u^{n-1}, \dots, u^2, u^1) \quad \text{or} \quad u^i = (u^2, u^1, u^n, \dots, u^3).$$

A bivalent holor is written

$$W^{ij} = \begin{bmatrix} W^{11} & W^{12} & \cdots & W^{1n} \\ W^{21} & W^{22} & \cdots & W^{2n} \\ \cdot & \cdot & \cdots & \cdot \\ W^{n1} & W^{n2} & \cdots & W^{nn} \end{bmatrix} \begin{matrix} \leftarrow \text{1st row} \\ \leftarrow \text{2nd row} \\ . \\ \leftarrow n\text{th row} \end{matrix}$$

$$\begin{matrix} \uparrow & \uparrow & & \uparrow \\ \text{1st} & \text{2nd} & & n\text{th} \\ \text{column} & \text{column} & & \text{column} \end{matrix}$$

If we wish, we may think of i and j axes with indices increasing in unit steps in the positive directions of these axes.

For a trivalent holor V^{ijk}, the axes must form a left-hand system so that the array will appear as an ordinary matrix as viewed from the top or the side. Usually, it is convenient to write a trivalent holor in terms of ordinary matrices. If we take vertical sections through the box of Figure 1.01, we obtain the complete specification of V^{ijk} in terms of n matrices:

$$V^{ij1} = \begin{bmatrix} V^{111} & V^{121} & \cdots & V^{1n1} \\ V^{211} & V^{221} & \cdots & V^{2n1} \\ \cdot & \cdot & \cdots & \cdot \\ V^{n11} & V^{n21} & \cdots & V^{nn1} \end{bmatrix},$$

$$V^{ij2} = \begin{bmatrix} V^{112} & V^{122} & \cdots & V^{1n2} \\ V^{212} & V^{222} & \cdots & V^{2n2} \\ \cdot & \cdot & \cdots & \cdot \\ V^{n12} & V^{n22} & \cdots & V^{nn2} \end{bmatrix},$$

$$\cdots\cdots\cdots$$

$$V^{ijn} = \begin{bmatrix} V^{11n} & V^{12n} & \dots & V^{1nn} \\ V^{21n} & V^{22n} & \dots & V^{2nn} \\ \cdot & \cdot & \dots & \cdot \\ V^{n1n} & V^{n2n} & \dots & V^{nnn} \end{bmatrix}.$$

Equally valid are sections parallel to the *jk* plane or parallel to the *ik* plane. Holors of higher valence can be similarly expressed in terms of ordinary matrices.

With holors of valence 2 or higher, it is desirable also to establish a standard way of writing literal indices.

Convention II: For a bivalent holor with literal indices in alphabetical order,* the merates are arranged as in the usual matrix. For example, if $n=2$,

$$U^{ij} = \begin{bmatrix} U^{11} & U^{12} \\ U^{21} & U^{22} \end{bmatrix},$$

row (i) / column (j)

or

$$V_{pt} = \begin{bmatrix} V_{11} & V_{12} \\ V_{21} & V_{22} \end{bmatrix},$$

row (p) / column (t)

or

$$W^{l}_{m} = \begin{bmatrix} W^{1}_{1} & W^{1}_{2} \\ W^{2}_{1} & W^{2}_{2} \end{bmatrix}.$$

row (l) / column (m)

But if the literal indices are arranged antialphabetically, then the transpose matrix is indicated:

$$U^{ji} = \begin{bmatrix} U^{11} & U^{21} \\ U^{12} & U^{22} \end{bmatrix},$$

column (j) / row (i)

$$V_{tp} = \begin{bmatrix} V_{11} & V_{21} \\ V_{12} & V_{22} \end{bmatrix},$$

* Other alphabets may be employed for indices. For instance,

$$U^{\alpha\beta} = \begin{bmatrix} U^{11} & U^{12} \\ U^{21} & U^{22} \end{bmatrix}.$$

Or if primes are used, the normal order is taken as i', i or j'', j', j, and so on.

$$W^m_l = \begin{bmatrix} W^1_1 & W^2_1 \\ W^1_2 & W^2_2 \end{bmatrix},$$

Convention III: For mixed holors, we arbitrarily consider all superscripts before turning to subscripts. Thus, for $N=3$ and indices in normal order, there are only two possibilities:

$$W^i_{jk} \quad \text{and} \quad W^{ij}_k.$$

For $N=4$, we have

$$X^i_{jkl}, X^{ij}_{kl}, X^{ijk}_l.$$

1.04. The transpose

By the *transpose* of a given matrix is meant a new matrix obtained from the original by interchange of rows and columns.[2] Convention II allows the two to be easily distinguished. Thus, U^{ij} has rows and columns arranged in the usual manner; but U^{ji} still has rows represented by i, columns by j, so U^{ji} is the transpose* of U^{ij}.

Similarly, with subscripts, V_{ij} is written

$$V_{ij} = \begin{bmatrix} V_{11} & V_{12} \\ V_{21} & V_{22} \end{bmatrix},$$

row (i), column (j)

and its transpose is

* An alternative convention is often used in writing matrices: *The first index, regardless of its position in the alphabet, always refers to the row; the second index always refers to the column.* If the literal indices happen to be in alphabetical order, this convention gives exactly what we get by convention II. For instance,

$$u^{ij} = \begin{bmatrix} u^{11} & u^{12} \\ u^{21} & u^{22} \end{bmatrix}.$$

row (i), column (j)

But, since the alternative convention makes no use of alphabetical order, it writes

$$u^{ji} = \begin{bmatrix} u^{11} & u^{12} \\ u^{21} & u^{22} \end{bmatrix} = u^{ij}.$$

row (j), column (i)

This leaves no convenient way to distinguish a matrix from its transpose. Of course, one may write u^{ij} and $(u^{ij})_T$, but such notations seem rather clumsy and ad hoc. For this reason, we reject the alternative convention.

$$V_{ji} = \begin{bmatrix} V_{11} & V_{21} \\ V_{12} & V_{22} \end{bmatrix}.$$

column (j), row (i)

For a mixed bivalent holor,

$$W^i_j = \begin{bmatrix} W^1_1 & W^1_2 \\ W^2_1 & W^2_2 \end{bmatrix},$$

row (i), column (j)

the transpose is

$$W^j_i = \begin{bmatrix} W^1_1 & W^2_1 \\ W^1_2 & W^2_2 \end{bmatrix}.$$

column (j), row (i)

Note that the dots sometimes used in tensor books ($W^i_{\bullet j}$ and $W_i^{\bullet j}$) are redundant if we employ Convention II.

1.05. Symmetry and antisymmetry

Symmetric bivalent holors are often encountered:

$$\{u^{ij}\} = \{u^{ji}\}. \tag{1.01}$$

Here

$$u^{ij} = \begin{bmatrix} u^{11} & u^{12} & \cdots & u^{1n} \\ u^{12} & u^{22} & \cdots & u^{2n} \\ \cdot & \cdot & \cdots & \cdot \\ u^{1n} & u^{2n} & \cdots & u^{nn} \end{bmatrix}.$$

Index balance (Section 2.06) prevents us from writing

$$u^{ij} = u^{ji},$$

but symmetry is completely specified by Equation (1.01).

Antisymmetry is specified by the relation

$$\{u^{ij}\} = -\{u^{ji}\}, \tag{1.02}$$

the matrix then being

$$u^{ij} = \begin{bmatrix} 0 & u^{12} & u^{13} & \cdots & u^{1n} \\ -u^{12} & 0 & u^{23} & \cdots & u^{2n} \\ -u^{13} & -u^{23} & 0 & \cdots & u^{3n} \\ \cdot & \cdot & \cdot & \cdots & \cdot \\ -u^{1n} & -u^{2n} & \cdot & \cdots & 0 \end{bmatrix}.$$

It is easily shown that an arbitrary bivalent holor can always be represented as the sum of a symmetric holor and an antisymmetric holor. Given the arbitrary holor

$$W^{ij} = \begin{bmatrix} W^{11} & W^{12} \\ W^{21} & W^{22} \end{bmatrix}$$

and the symmetric and antisymmetric holors U^{ij} and V^{ij}:

$$U^{ij} = \begin{bmatrix} U^{11} & U^{12} \\ U^{12} & U^{22} \end{bmatrix}, \qquad V^{ij} = \begin{bmatrix} 0 & V^{12} \\ -V^{12} & 0 \end{bmatrix}.$$

Then

$$U^{ij} + V^{ij} = \begin{bmatrix} U^{11} & U^{12}+V^{12} \\ U^{12}-V^{12} & U^{22} \end{bmatrix}.$$

Irrespective of the merates $\{W^{ij}\}$, we can always represent them in terms of $\{U^{ij}\}$ and $\{V^{ij}\}$:

$$\{W^{ij}\} = \{U^{ij}\} + \{V^{ij}\}, \tag{1.03}$$

where

$$U^{12} = \tfrac{1}{2}(W^{12} + W^{21}), \qquad U^{11} = W^{11},$$
$$V^{12} = \tfrac{1}{2}(W^{12} - W^{21}), \qquad U^{22} = W^{22}.$$

Obviously, this theorem can be extended to any plethos n.

Symmetry and antisymmetry can be applied to holors of higher valence. For instance, the holor V^{ijk} may be symmetric in i, j,

$$\{V^{ijk}\} = \{V^{jik}\},$$

or in any pair of indices, or it may be completely symmetric,

$$\{V^{ijk}\} = \{V^{jki}\} = \{V^{kij}\} = \{V^{jik}\} = \{V^{ikj}\} = \{V^{kji}\}.$$

Antisymmetry is also possible.

A convenient way of indicating symmetry and antisymmetry* was suggested by Schouten.[3] A bivalent holor with arbitrary merates may be written

$$u^{ij} = \begin{bmatrix} u^{11} & u^{12} & \cdots & u^{1n} \\ u^{21} & u^{22} & \cdots & u^{2n} \\ \cdot & \cdot & \cdots & \cdot \\ u^{n1} & u^{n2} & \cdots & u^{nn} \end{bmatrix}. \tag{1.04}$$

A related antisymmetric holor $u^{[ij]}$ is defined by the merates:

* Apparently, this notation was originated by R. Bach, "Zur Weylschen Relativitätstheorie und der Weylschen Erweiterung des Krümmungsbegriffs," *Math. Zs.*, **9**, 1921, p. 110.

$$\{u^{[ij]}\} = \tfrac{1}{2}[\{u^{ij}\} - \{u^{ji}\}]. \tag{1.05}$$

Evidently this definition produces zeros on the principal diagonal and gives the new matrix

$$u^{[ij]} = \frac{1}{2}\begin{bmatrix} 0 & (u^{12}-u^{21}) & \cdots & (u^{1n}-u^{n1}) \\ -(u^{12}-u^{21}) & 0 & \cdots & (u^{2n}-u^{n2}) \\ \cdot & \cdot & \cdots & \cdot \\ -(u^{1n}-u^{n1}) & -(u^{2n}-u^{n2}) & \cdots & 0 \end{bmatrix}. \tag{1.06}$$

Similarly, the related symmetric matrix $u^{(ij)}$ has merates defined as

$$\{u^{(ij)}\} = \tfrac{1}{2}[\{u^{ij}\} + \{u^{ji}\}], \tag{1.07}$$

giving

$$u^{(ij)} = \frac{1}{2}\begin{bmatrix} 2u^{11} & u^{12}+u^{21} & \cdots & u^{1n}+u^{n1} \\ u^{21}+u^{12} & 2y^{22} & \cdots & u^{2n}+u^{n2} \\ \cdot & \cdot & \cdots & \cdot \\ u^{n1}+u^{1n} & u^{n2}+u^{2n} & \cdots & 2u^{nn} \end{bmatrix}. \tag{1.08}$$

Addition of merates gives

$$\begin{aligned} \{u^{[ij]}\} + \{u^{(ij)}\} &= \tfrac{1}{2}[\{u^{ij}\} - \{u^{ji}\} + \{u^{ij}\} + \{u^{ji}\}] \\ &= \{u^{ij}\}, \end{aligned}$$

so the sum of the matrices $u^{[ij]}$ and $u^{(ij)}$ gives the original matrix u^{ij}, Equation (1.04).

The same idea can be applied to holors of higher valence. For example, a trivalent holor W^{ijk} with arbitrary merates can be transformed into holors

$$W^{i[jk]},\ W^{[ij]k},\ W^{[i|j|k]},$$

which are antisymmetric on two indices, Equation (1.05). The vertical bars in the third designation indicates that j does not take part in the antisymmetry.

1.06. Diagonal holors

A bivalent holor is said to be *diagonal* if all merates are zero except those on the principal diagonal: $\{a^i_j\} = 0$ if $i \neq j$. An example is

$$a^i_j = \begin{bmatrix} a^1_1 & 0 & 0 & \cdots & 0 \\ 0 & a^2_2 & 0 & \cdots & 0 \\ \cdot & \cdot & \cdot & \cdots & \cdot \\ 0 & 0 & 0 & \cdots & a^n_n \end{bmatrix}. \qquad (1.09)$$

If all the nonzero merates of a diagonal matrix are the same, the matrix is said to be scalar:

$$a^i_j = \begin{bmatrix} a^1_1 & 0 & 0 & \cdots & 0 \\ 0 & a^1_1 & 0 & \cdots & 0 \\ \cdot & \cdot & \cdot & \cdots & \cdot \\ 0 & 0 & 0 & \cdots & a^1_1 \end{bmatrix}.$$

If a scalar matrix has $a^1_1 = 1$, the holor is said to be a *unit holor*. This is, in fact, the familiar Kronecker delta δ^i_j. The Kronecker delta is a bivalent holor

$$\delta^i_j = \begin{bmatrix} 1 & 0 & 0 & \cdots & 0 \\ 0 & 1 & 0 & \cdots & 0 \\ 0 & 0 & 1 & \cdots & 0 \\ \cdot & \cdot & \cdot & \cdots & \cdot \\ 0 & 0 & 0 & \cdots & 1 \end{bmatrix}, \qquad (1.10)$$

the merates being

$$\{\delta^i_j\} = 1 \quad \text{if } i = j; \qquad \{\delta^i_j\} = 0 \quad \text{if } i \neq j.$$

Evidently, these definitions can be extended to holors of higher valence. A holor $T^{ijk\cdots pq}$ is diagonal if all its merates are zero except those where $i = j = k = \cdots = q$.

1.07. Generalized Kronecker deltas

Murnaghan has shown how the Kronecker delta can be generalized[4] to give valuable results.[5] A generalized Kronecker delta of valence 4 is defined by the relation

$$\delta^{ij}_{kl} = \begin{vmatrix} \delta^i_k & \delta^i_l \\ \delta^j_k & \delta^j_l \end{vmatrix} = \delta^i_k \delta^j_l - \delta^i_l \delta^j_k. \qquad (1.11)$$

Similarly, a delta of valence 6 may be written

$$\delta^{ijk}_{lmn} = \begin{vmatrix} \delta^i_l & \delta^i_m & \delta^i_n \\ \delta^j_l & \delta^j_m & \delta^j_n \\ \delta^k_l & \delta^k_m & \delta^k_n \end{vmatrix}, \qquad (1.12)$$

and this procedure can be extended as far as needed:

$$\delta^{r_1 r_2 \cdots r_m}_{s_1 s_2 \cdots s_m} = \begin{vmatrix} \delta^{r_1}_{s_1} & \delta^{r_1}_{s_2} & \cdots & \delta^{r_1}_{s_m} \\ \delta^{r_2}_{s_1} & \delta^{r_2}_{s_2} & \cdots & \delta^{r_2}_{s_m} \\ \cdot & \cdot & \cdots & \cdot \\ \delta^{r_m}_{s_1} & \delta^{r_m}_{s_2} & \cdots & \delta^{r_m}_{s_m} \end{vmatrix}. \tag{1.13}$$

Consider a set of n integers $1, 2, \ldots, n$. From this set, we pick m numbers ($m \le n$) to form the superscript of δ; and we pick m numbers to form the subscript. Expansion of the determinant, Equation (1.13), gives the following rules for the merates of $\delta^{r_1 \cdots r_m}_{s_1 \cdots s_m}$:

(a) If the same integers do not appear in subscript and superscript, the merate is zero,

$$\{\delta^{r_1 \cdots r_m}_{s_1 \cdots s_m}\} = 0.$$

(b) If the same distinct integers appear in subscript and superscript, then

$$\{\delta^{r_1 \cdots r_m}_{s_1 \cdots s_m}\} = +1$$

if one set of integers is an even permutation of the other, and

$$\{\delta^{r_1 \cdots r_m}_{s_1 \cdots s_m}\} = -1$$

if one set of integers is an odd permutation of the other.

(c) If any two subscripts (or any two superscripts) are the same,

$$\{\delta^{r_1 \cdots r_m}_{s_1 \cdots s_m}\} = 0.$$

For example,

$$\delta^{11}_{12} = 0, \qquad \delta^{12}_{13} = 0, \qquad \delta^{23}_{12} = 0, \quad \text{etc.}$$

Also,

$$\delta^{12}_{12} = \delta^{21}_{21} = +1, \qquad \delta^{12}_{21} = \delta^{21}_{12} = -1, \qquad \delta^{13}_{31} = \delta^{31}_{13} = -1, \quad \text{etc.}$$

As a special case, we may fix the superscript as $12 \cdots m$:

$$\delta^{12 \cdots m}_{s_1 s_2 \cdots s_m},$$

which reduces the valence of the holor from $2m$ to m. Or we can fix the subscript, giving

$$\delta^{r_1 r_2 \cdots r_m}_{12 \cdots m}.$$

These holors are sometimes called *alternators*. For instance, if $m = 3$, we have a trivalent holor δ^{123}_{ijk}, which may be written in terms of three ordinary matrices:

$$\delta^{123}_{1jk} = \begin{bmatrix} 0 & 0 & 0 \\ 0 & 0 & 1 \\ 0 & -1 & 0 \end{bmatrix},$$

$$\delta^{123}_{2jk} = \begin{bmatrix} 0 & 0 & -1 \\ 0 & 0 & 0 \\ 1 & 0 & 0 \end{bmatrix},$$

$$\delta^{123}_{3jk} = \begin{bmatrix} 0 & 1 & 0 \\ -1 & 0 & 0 \\ 0 & 0 & 0 \end{bmatrix}.$$

A simple relation exists between alternators and the generalized Kronecker deltas.[4] Indeed,

$$\delta^{12\cdots m}_{s_1 s_2 \cdots s_m} \delta^{r_1 r_2 \cdots r_m}_{12\cdots m} = \delta^{r_1 r_2 \cdots r_m}_{s_1 s_2 \cdots s_m}. \tag{1.14}$$

For instance,

$$\delta^{12}_{kl} \delta^{ij}_{12} = \delta^{ij}_{kl}, \tag{1.14a}$$

as is easily shown by writing the merates for the two sides of the equation.

A distinction should be noted between the Kronecker deltas and the alternators. The Kronecker deltas

$$\delta^i_j, \delta^{ij}_{kl}, \delta^{ijk}_{lmp}, \ldots$$

have indices running from 1 to n, the dimensionality of the space. If the dimensionality is less than the number of subscripts, however, repeated indices appear and the delta becomes identically zero. Thus, in 3-space, the foregoing symbols are all valid, but

$$\delta^{ijkl}_{mpqr} \equiv 0.$$

For the alternators, also, the number of subscripts is ordinarily equal to n. Thus, in a 2-space, one uses δ^{ij}_{12} or δ^{12}_{kl}; and in a 3-space, one applies δ^{ijk}_{123} or δ^{123}_{klm}.

One of the applications of the Kronecker delta is to determinant theory.[5] Consider the square determinant

$$|a^i_j| = \begin{vmatrix} a^1_1 & a^1_2 & \cdots & a^1_n \\ a^2_1 & a^2_2 & \cdots & a^2_n \\ \cdot & \cdot & \cdots & \cdot \\ a^n_1 & a^n_2 & \cdots & a^n_n \end{vmatrix}.$$

The expansion consists of n terms, each term being the product of n factors. Expansion in terms of the elements of a row gives the sum of a series of terms of the form

$$\pm a_1^k a_2^l \cdots a_n^v,$$

where the superscripts are arranged in all possible ways. Evidently

$$|a_j^i| = \sum_{k=1}^{n} \sum_{l=1}^{n} \cdots \sum_{v=1}^{n} \delta_{k1\cdots v}^{12\cdots n} a_1^k a_2^l \cdots a_n^v. \tag{1.15}$$

Similarly, expansion in terms of elements of a column gives

$$|a_j^i| = \sum_{k=1}^{n} \sum_{l=1}^{n} \cdots \sum_{v=1}^{n} \delta_{12\cdots n}^{kl\cdots v} a_k^1 a_l^2 \cdots a_v^n. \tag{1.16}$$

For example, take the determinant

$$|a_j^i| = \begin{vmatrix} a_1^1 & a_2^1 & a_3^1 \\ a_1^2 & a_2^2 & a_3^2 \\ a_1^3 & a_2^3 & a_3^3 \end{vmatrix}.$$

According to Equation (1.15),

$$|a_j^i| = \sum_{k=1}^{3} \sum_{l=1}^{3} \sum_{m=1}^{3} \delta_{klm}^{123} a_1^k a_2^l a_3^m$$

$$= \delta_{123}^{123} a_1^1 a_2^2 a_3^3 + \delta_{231}^{123} a_1^2 a_2^3 a_3^1 + \delta_{312}^{123} a_1^3 a_2^1 a_3^2$$

$$+ \delta_{321}^{123} a_1^3 a_2^2 a_3^1 + \delta_{213}^{123} a_1^2 a_2^1 a_3^3 + \delta_{132}^{123} a_1^1 a_2^3 a_3^2$$

$$= a_1^1 a_2^2 a_3^3 + a_1^2 a_2^3 a_3^1 + a_1^3 a_2^1 a_3^2 - a_1^3 a_2^2 a_3^1 - a_1^2 a_2^1 a_3^3 - a_1^1 a_2^3 a_3^2,$$

which of course is the usual result.

A determinant can also be expressed[5] in terms of the generalized Kronecker delta $\delta_{\alpha\beta\cdots\omega}^{kl\cdots v}$:

$$|a_j^i| = \frac{1}{n!} \sum_{\alpha=1}^{n} \sum_{\beta=1}^{n} \cdots \sum_{\omega=1}^{n} \sum_{k=1}^{n} \sum_{l=1}^{n} \cdots \sum_{v=1}^{n} \delta_{\alpha\beta\cdots\omega}^{kl\cdots v} a_k^\alpha a_l^\beta \cdots a_v^\omega. \tag{1.17}$$

Or if the merates are written with subscripts,

$$|a_{ij}| = \frac{1}{n!} \sum_{k=1}^{n} \sum_{l=1}^{n} \cdots \sum_{v=1}^{n} \sum_{\alpha=1}^{n} \sum_{\beta=1}^{n} \cdots \sum_{\omega=1}^{n} \delta_{12\cdots n}^{kl\cdots v} \delta_{12\cdots n}^{\alpha\beta\cdots\omega} a_k^\alpha a_l^\beta \cdots a_v^\omega. \tag{1.18}$$

Similarly, with superscripts, we have

$$|a^{ij}| = \frac{1}{n!}\sum_{k=1}^{n}\sum_{l=1}^{n}\cdots\sum_{v=1}^{n}\sum_{\alpha=1}^{n}\sum_{\beta=1}^{n}\cdots\sum_{\omega=1}^{n}\delta_{kl\cdots v}^{12\cdots n}\delta_{\alpha\beta\cdots\omega}^{12\cdots n}a^{k\alpha}a^{l\beta}\cdots a^{v\omega}, \tag{1.19}$$

which of course is the usual result.

1.08. Summary

Chapter 1 considers the general subject of holors and their representation in index notation. Holors may be classified according to plethos n and valence N, as indicated in Table 1.01. A holor is written as a base letter with indices (subscripts or superscripts). Powers are written outside parentheses to distinguish them from ordinary superscripts. The square of v^3, for instance, is written $(v^3)^2$. Labels are also written outside parentheses. For example, the vector v^i at point P is distinguished from the vector at point Q by writing $(v^i)_P$ and $(v^i)_Q$. Symbols such as v^i or w_{ij} represent complete holors; while symbols in curly brackets indicate arbitrary merates, as $\{v^i\}$, $\{w_{ij}\}$.

In writing a holor in terms of its merates, we use the convention that for a bivalent holor, the alphabetically prior index always represents the row, the alphabetically later index always represents the column. This convention gives an easy distinction between a matrix and its transpose. If

$$\{u^{ij}\} = \{u^{ji}\}, \tag{1.01}$$

the matrix is said to be symmetric; if

$$\{u^{ij}\} = -\{u^{ji}\}, \tag{1.02}$$

the matrix is said to be antisymmetric.

The chapter concludes with a short treatment of the Kronecker delta,

$$\delta^i_j = \begin{bmatrix} 1 & 0 & 0 & \cdots & 0 \\ 0 & 1 & 0 & \cdots & 0 \\ 0 & 0 & 1 & \cdots & 0 \\ \cdot & \cdot & \cdot & \cdots & \cdot \\ 0 & 0 & 0 & \cdots & 1 \end{bmatrix} \tag{1.10}$$

and the generalized Kronecker delta, such as

$$\delta^{ij}_{kl} = \begin{vmatrix} \delta^i_k & \delta^i_l \\ \delta^j_k & \delta^j_l \end{vmatrix}. \tag{1.11}$$

These are holors whose merates are always zero or ± 1. They will appear frequently in subsequent chapters.

Problems

1.01 Specify valence and plethos for the following holors:

(a)
$$\{V^{ijk}\},$$

(b)
$$(u_1, u_2, u_3, u_4),$$

(c)
$$\begin{bmatrix} v^{11} & v^{12} & v^{13} \\ v^{21} & v^{22} & v^{23} \\ v^{31} & v^{32} & v^{33} \end{bmatrix},$$

(d)
$$\begin{bmatrix} x^1_1 & x^2_1 \\ x^1_2 & x^2_2 \end{bmatrix},$$

(e)
$$\begin{vmatrix} u_{11} & u_{12} & u_{13} & u_{14} \\ u_{21} & u_{22} & u_{23} & u_{24} \\ u_{31} & u_{32} & u_{33} & u_{34} \\ u_{41} & u_{42} & u_{43} & u_{44} \end{vmatrix}.$$

1.02 Give the valence of each of the following:

$$u_i, v^j_k, U^{ij}_{kl}, v_4, (u^7)^3, A_{ijk}, AX^2W_i, m, \{V^{ij}_l\}.$$

1.03 For plethos 3, write the following holors in terms of their merates:

$$u^i, u^j, u^{ij}, a_{kp}, Z^{ih}, A_{gkm}.$$

1.04 For plethos 2, write the following in terms of merates:

$$V^i, V^{ij}, V^{ijk}, V^{ijkl}, U^i_j, W^i_{jk}.$$

1.05 Express in terms of merates $(n=2)$:

$$u^i_j, u^i_k, u^i_{jk}, u^i_{kj}, u^j_{ik}, u^j_{ki}, u^k_{ij}, u^k_{ji}.$$

Which holors are equal?

1.06 Express in terms of merates in a 3-space:

$$V^{ps}, W^t_q, Y^{lm}_p.$$

1.07 Write the holor U^i_{kl} in expanded form for all permutations of the indices if $n=2$. How many of these holors are the same?

1.08 Write the holor U^{ij}_{kl} in terms of matrices. Repeat for U^{ik}_{jl}. Let $n=2$.

1.09 Write the holor U^{ij}_{pqr} in terms of matrices ($n=2$). How many distinct holors can be obtained by permutation of indices?

1.10
(a) Write the matrix a^i_j for $n=4$ if the matrix is symmetric.
(b) Repeat if the matrix is antisymmetric.
(c) Expand the determinants for these two matrices.

1.11 Express in matrix form, V^{ijk} for $n=2$:
(a) If $\{V^{ijk}\}=\{V^{jik}\}$.
(b) If $\{V^{ijk}\}=-\{V^{jik}\}$.
(c) If V^{ijk} is completely symmetric.

1.12 Write U^i_{jk} in terms of matrices if $n=3$ and the holor is symmetric in indices j and k.

1.13 Show that

$$\delta^{123}_{lmn}\delta^{ijk}_{123}=\delta^{ijk}_{lmn}.$$

1.14 Show that

$$\delta^{ijk}_{lmn}=\delta^{lmn}_{ijk}$$

(a) For 2-space.
(b) For 3-space.

1.15
(a) Show that δ^{ij}_{kl} is completely antisymmetric in the indices i, j, and k, l.
(b) For plethos 2, δ^{ij}_{kl} has only two merates that are $+1$:

$$\delta^{12}_{12}, \delta^{21}_{21}.$$

The suggestion is made, therefore, that

$$\delta^{ij}_{kl}=\delta^i_k\delta^j_l.$$

Discuss.

1.16 Consider a determinant $|a^i_j|$ with plethos 4.
(a) Expand this determinant by columns. Show how the expansion can be expressed in terms of the Kronecker deltas.
(b) Repeat (a), expanding by rows.

1.17
(a) Expand the determinant $|a^i_j|$ for $n=3$, using Equation (1.17).
(b) Expand the determinant $|a^{ij}|$ for $n=3$, using Equation (1.19).

1.18 Using Kronecker deltas, prove that any determinant having two identical columns is zero.

1.19 Using Kronecker deltas, obtain an expression for the expansion of any determinant in terms of cofactors.

1.20 Two determinants, $|A^i_j|$ and $|B^k_l|$ are to be multiplied.
(a) From Equation (1.15), obtain an expression for the product.
(b) Show that your result is correct for $n=2$.

2
Holor algebra

Chapter 1 has developed index notation and has given a certain amount of practice in the use of this notation. We now consider in greater detail the algebra of holors.

In the algebra of real numbers, the following relations apply:

Addition is commutative: $a+b=b+a$.

Addition is associative: $a+(b+c)=(a+b)+c$.

Multiplication is commutative: $ab=ba$.

Multiplication is associative: $a(bc)=(ab)c$.

Multiplication is distributive with respect to addition: $a(b+c)=ab+ac$.

In the eighteenth and nineteenth centuries, the naive belief was prevalent that mathematics was an absolute: something "present in the Divine Mind before Creation." The modern idea, on the other hand, is much more sophisticated: Mathematics is merely a man-made game whose rules can be changed at will. Thus, for holor manipulation, the rules of real algebra may be accepted or not, depending entirely on whether the results are what we need for a particular application.

The structure of the world being what it is, there seems to be little or no advantage in modifying the commutative and associative rules of addition. The distributive rule also seems to be universally accepted. With multiplication, however, considerable flexibility in holor algebra is required to meet the needs of modern applications. This chapter is devoted to a study of holor algebra and how it differs from the algebra of real numbers.

2.01. Equality of holors

For ordinary numbers, the equation

$$x=y$$

means that the two symbols are interchangeable; that one may be substituted for the other in any equation. With vectors, it is not sufficient to have the magnitudes equal. We know that the two vectors

$$u^i = (u^1, u^2, u^3) \quad \text{and} \quad v^j = (v^1, v^2, v^3)$$

are equal only if

$$u^1 = v^1, \quad u^2 = v^2, \quad \text{and} \quad u^3 = v^3.$$

Similarly, two matrices

$$u^{ij} = \begin{bmatrix} u^{11} & u^{12} & \cdots & u^{1n} \\ u^{21} & u^{22} & \cdots & u^{2n} \\ \cdot & \cdot & \cdots & \cdot \\ u^{n1} & u^{n2} & \cdots & u^{nn} \end{bmatrix} \quad \text{and} \quad v^{kl} = \begin{bmatrix} v^{11} & v^{12} & \cdots & v^{1n} \\ v^{21} & v^{22} & \cdots & v^{2n} \\ \cdot & \cdot & \cdots & \cdot \\ v^{n1} & v^{n2} & \cdots & v^{nn} \end{bmatrix}$$

are said to be equal[1] only if corresponding merates are equal:

$$\{u^{ij}\} = \{v^{ij}\}.$$

Extending these ideas to the general holor, we may state:

Definition II: Two holors are equal if and only if each corresponding pair of merates is equal.

For example, two holors A_k^{ij} and B_n^{lm} are equal if and only if corresponding merates are equal:

$$\{A_k^{ij}\} = \{B_k^{ij}\}. \tag{2.01}$$

Note that equality is possible only if corresponding merates exist. That is, the holors must have the same valence and the same plethos and indices must be arranged in the same way. To speak of the equality of u^i and v_i, or of A_k^{ij} and $B^i_{\ jk}$, or of u^i and v^j, where $i = 1, 2$ and $j = 1, 2, 3$, is meaningless. Definition II is, of course, arbitrary. But it is universally employed, and it seems to be the only possible definition of equality that has any significance.

Also needed is the definition of a null holor:

Definition III: A holor is zero if and only if each of its merates is zero.

For example, a univalent holor

$$u^i = (u^1, u^2, \ldots, u^n)$$

is zero if and only if

$$u^1 = u^2 = \cdots = u^n = 0.$$

Similarly, a bivalent holor

$$u_{ij} = \begin{bmatrix} u_{11} & u_{12} & \cdots & u_{1n} \\ u_{21} & u_{22} & \cdots & u_{2n} \\ \cdot & \cdot & \cdots & \cdot \\ u_{n1} & u_{n2} & \cdots & u_{nn} \end{bmatrix}$$

is zero if and only if

$$u_{11}=0, \quad u_{12}=0, \ldots, \quad \text{and } u_{nn}=0, \quad \text{or } \{u_{ij}\}=0.$$

It is hardly necessary to mention again the distinction between a matrix and a determinant. For instance, the matrix

$$\begin{bmatrix} 5 & 6 & 2 \\ 0 & 1 & 7 \\ 2 & 3 & 5 \end{bmatrix}$$

could be zero only if every element were zero. But the determinant of this matrix is zero:

$$\begin{vmatrix} 5 & 6 & 2 \\ 0 & 1 & 7 \\ 2 & 3 & 5 \end{vmatrix} = 0.$$

To emphasize the distinction between an ordinary zero used with a real number and a zero holor of higher valence, one may employ indices on the zero. Thus, a zero univalent holor may be written

$$v^i = 0^i = (0,0,0),$$

and a zero bivalent holor may be written

$$A_{ij} = 0_{ij} = \begin{bmatrix} 0 & 0 & 0 \\ 0 & 0 & 0 \\ 0 & 0 & 0 \end{bmatrix}.$$

2.02. Addition of holors

An important operation with holors is addition.

Definition IV: The sum of two holors $U^{ij\cdots n}$ and $V^{ij\cdots n}$ is a holor of the same valence and plethos, $W^{ij\cdots n}$; each merate of $W^{ij\cdots n}$ being obtained by addition of corresponding merates of $U^{ij\cdots n}$ and $V^{ij\cdots n}$:

$$\{W^{ij\cdots n}\} = \{U^{ij\cdots n}\} + \{V^{ij\cdots n}\}.$$

Addition is possible only if the holors have the same literal indices arranged in the same way. The sum then also has the same literal indices arranged in the same way.[2]

For example, take the univalent holors

$$u^i = (u^1, u^2, u^3) \quad \text{and} \quad v^i = (v^1, v^2, v^3).$$

The sum is

$$u^i + v^i = w^i = (u^1 + v^1, u^2 + v^2, u^3 + v^3).$$

Or consider the bivalent holors

$$U^{ij} = \begin{bmatrix} U^{11} & U^{12} & U^{13} \\ U^{21} & U^{22} & U^{23} \\ U^{31} & U^{32} & U^{33} \end{bmatrix} \quad \text{and} \quad V^{ij} = \begin{bmatrix} V^{11} & V^{12} & V^{13} \\ V^{21} & V^{22} & V^{23} \\ V^{31} & V^{32} & V^{33} \end{bmatrix}.$$

The sum is, by definition,

$$U^{ij} + V^{ij} = W^{ij} = \begin{bmatrix} W^{11} & W^{12} & W^{13} \\ W^{21} & W^{22} & W^{23} \\ W^{31} & W^{32} & W^{33} \end{bmatrix},$$

where any merate is

$$\{W^{ij}\} = \{U^{ij}\} + \{V^{ij}\}.$$

The same procedure applies to holors of higher valence. For instance,

$$U_{ijk} + V_{ijk} = W_{ijk},$$

where each merate of the trivalent holor W_{ijk} is obtained by adding corresponding merates of U_{ijk} and V_{ijk}:

$$\{W_{ijk}\} = \{U_{ijk}\} + \{V_{ijk}\}.$$

Note that, by Definition IV, the following expressions are meaningless:

$$u^i + v_i, \qquad u^i + v^j, \qquad u^i + v_{ij}, \qquad W^{ij} + W^{ik}.$$

As a matter of fact, the principle of index balance (Section 2.06) guards against the use of such nonsignificant combinations of symbols.

Subtraction introduces nothing new, since it is merely the addition of a negative quantity. For example,

$$u^i - v^i = (u^1 - v^1, u^2 - v^2, \ldots, u^n - v^n),$$

and

$$U^{ij}-V^{ij}=\begin{bmatrix} U^{11}-V^{11} & U^{12}-V^{12} & \cdots & U^{1n}-V^{1n} \\ U^{21}-V^{21} & U^{22}-V^{22} & \cdots & U^{2n}-V^{2n} \\ \cdot & \cdot & \cdots & \cdot \\ U^{n1}-V^{n1} & U^{n2}-V^{n2} & \cdots & U^{nn}-V^{nn} \end{bmatrix}.$$

The addition of two bivalent holors,

$$A_{ij}+B_{ij}=C_{ij}, \tag{2.02}$$

gives the matrix

$$C_{ij}=\begin{bmatrix} C_{11} & C_{12} & \cdots & C_{1n} \\ C_{21} & C_{22} & \cdots & C_{2n} \\ \cdot & \cdot & \cdots & \cdot \\ C_{n1} & C_{n2} & \cdots & C_{nn} \end{bmatrix},$$

where a typical merate is

$$\{C_{ij}\}=\{A_{ij}\}+\{B_{ij}\}. \tag{2.03}$$

For any values of i and j $(i, j=1,2,\ldots,n)$, we have a real number $\{C_{ij}\}$ obtained by adding real numbers, Equation (2.03). Since the addition of real numbers is commutative, Equation (2.03) may be written

$$\{C_{ij}\}=\{A_{ij}\}+\{B_{ij}\}=\{B_{ij}\}+\{A_{ij}\}.$$

Consequently, the holor equation may be written

$$A_{ij}+B_{ij}=B_{ij}+A_{ij}=C_{ij}.$$

Obviously, this procedure can be applied to any holors that allow addition; and in all cases, the sum is commutative. Thus, we can say that the addition of holors is always commutative regardless of valence or plethos.

Similarly, the addition of three holors gives another holor of the same valence and plethos, such as

$$X_{ij}+Y_{ij}+Z_{ij}=W_{ij}. \tag{2.04}$$

By definition of holor addition, each merate of the W_{ij} matrix is a real number, which follows the usual rules for addition of real merates:

$$\begin{aligned}\{W_{ij}\}&=[\{X_{ij}\}+\{Y_{ij}\}]+\{Z_{ij}\}\\ &=\{X_{ij}\}+[\{Y_{ij}\}+\{Z_{ij}\}].\end{aligned}$$

Since this is true for each merate, it is true for the complete holor, or

$$[X_{ij}+Y_{ij}]+Z_{ij}=X_{ij}+[Y_{ij}+Z_{ij}]=W_{ij}.$$

The same argument applies to holors of other valence, so addition of holors is always associative.

In real algebra, there is a unique element, zero, that satisfies the relation

$$a+0=a, \tag{2.05}$$

for any real number a. Similarly, for holors of any given valence and plethos, there is a zero holor (defined in Section 2.01) such that

$$\mathfrak{A}+\mathfrak{O}=\mathfrak{A}, \tag{2.06}$$

where $\mathfrak{A}$ is an arbitrary holor and $\mathfrak{O}$ is the zero holor having the same valence and plethos as $\mathfrak{A}$. For univalent holors of plethos n, Equation (2.06) becomes

$$A^i+0^i=A^i, \tag{2.06a}$$

or

$$(A^1,A^2,\ldots,A^n)+(0,0,\ldots,0)=(A^1,A^2,\ldots,A^n),$$

which is obviously true, according to the definitions of the zero holor and of addition. Similarly, for bivalent holors,

$$U^{ij}+0^{ij}=U^{ij} \tag{2.06b}$$

or

$$\begin{bmatrix} U^{11} & U^{12} & \cdots & U^{1n} \\ U^{21} & U^{22} & \cdots & U^{2n} \\ \cdot & \cdot & \cdots & \cdot \\ U^{n1} & U^{n2} & \cdots & U^{nn} \end{bmatrix} + \begin{bmatrix} 0 & 0 & \cdots & 0 \\ 0 & 0 & \cdots & 0 \\ \cdot & \cdot & \cdots & \cdot \\ 0 & 0 & \cdots & 0 \end{bmatrix} = \begin{bmatrix} U^{11} & U^{12} & \cdots & U^{1n} \\ U^{21} & U^{22} & \cdots & U^{2n} \\ \cdot & \cdot & \cdots & \cdot \\ U^{n1} & U^{n2} & \cdots & U^{nn} \end{bmatrix}.$$

Another detail of ordinary algebra is the cancellation rule for addition. If

$$a+b=a+c,$$

then

$$b=c.$$

Evidently, this rule applies to any holors that can be added. If

$$\mathfrak{A}+\mathfrak{B}=\mathfrak{A}+\mathfrak{C}, \tag{2.07}$$

then our definition of subtraction allows us to write

$$(\mathfrak{A}-\mathfrak{A})+\mathfrak{B}=\mathfrak{C},$$

or

$$\mathfrak{B} = \mathfrak{C}.$$

In other words, the familiar rules of addition and subtraction of real numbers apply to holors of any valence and plethos.

2.03. The uncontracted product

The product of two holors is written simply by placing the two symbols in juxtaposition. The valence of such an uncontracted product is equal to the sum of the valences of the factors.[3]

Definition V: The uncontracted product $W^{ij\cdots nst\cdots z}$ of two holors $U^{ij\cdots n}$ and $V^{st\cdots z}$ is defined as a holor whose valence is the sum of the valences of its factors, each merate of the product being equal to the product of merates:

$$\{W^{ij\cdots nst\cdots z}\} = \{U^{ij\cdots n}\}\{V^{st\cdots z}\}.$$

Thus, the uncontracted product of two univalent holors, u^i and v^j, is bivalent:

$$u^i v^j = w^{ij}. \qquad (2.08)$$

Note that indices are arranged in the same order on the two sides of the equation. The typical merate of the product is

$$\{w^{ij}\} = \{u^i v^j\} = \{u^i\}\{v^j\},$$

so

$$u^i v^j = \begin{bmatrix} w^{11} & w^{12} & \cdots & w^{1n} \\ w^{21} & w^{22} & \cdots & w^{2n} \\ \cdot & \cdot & \cdots & \cdot \\ w^{n1} & w^{n2} & \cdots & w^{nn} \end{bmatrix} = \begin{bmatrix} u^1 v^1 & u^1 v^2 & \cdots & u^1 v^n \\ u^2 v^1 & u^2 v^2 & \cdots & u^2 v^n \\ \cdot & \cdot & \cdots & \cdot \\ u^n v^1 & u^n v^2 & \cdots & u^n v^n \end{bmatrix}.$$

Similarly,

$$u_i v_j = W_{ij} \qquad (2.09)$$

and

$$u^i v_j = X^i_{\;j}, \qquad (2.10)$$

$$u_i v^j = Y_i^{\;j}. \qquad (2.11)$$

Note that the order of indices is preserved and that Conventions I and II are applied in writing the matrices.

The factors need not have the same valence. For instance, the product of a bivalent holor and a nilvalent holor is a bivalent holor. Thus,

$$SU^{ij} = V^{ij} \tag{2.12}$$

or

$$S\begin{bmatrix} U^{11} & U^{12} & \cdots & U^{1n} \\ U^{21} & U^{22} & \cdots & U^{2n} \\ \cdot & \cdot & \cdots & \cdot \\ U^{n1} & U^{n2} & \cdots & U^{nn} \end{bmatrix} = \begin{bmatrix} V^{11} & V^{12} & \cdots & V^{1n} \\ V^{21} & V^{22} & \cdots & V^{2n} \\ \cdot & \cdot & \cdots & \cdot \\ V^{n1} & V^{n2} & \cdots & V^{nn} \end{bmatrix},$$

where each merate of V^{ij} is given by

$$\{V^{ij}\} = \{SU^{ij}\}.$$

Similarly, the product of a bivalent and a univalent holor is trivalent. For example,

$$U^{ij}v^k = W^{ijk}, \tag{2.13}$$

where

$$W^{ij1} = \begin{bmatrix} U^{11}v^1 & U^{12}v^1 & \cdots & U^{1n}v^1 \\ U^{21}v^1 & U^{22}v^1 & \cdots & U^{2n}v^1 \\ \cdot & \cdot & \cdots & \cdot \\ U^{n1}v^1 & U^{n2}v^1 & \cdots & U^{nn}v^1 \end{bmatrix},$$

$$W^{ij2} = \begin{bmatrix} U^{11}v^2 & U^{12}v^2 & \cdots & U^{1n}v^2 \\ U^{21}v^2 & U^{22}v^2 & \cdots & U^{2n}v^2 \\ \cdot & \cdot & \cdots & \cdot \\ U^{n1}v^2 & U^{n2}v^2 & \cdots & U^{nn}v^2 \end{bmatrix}, \quad \text{etc.}$$

We have defined the sum of holors in terms of the sum of corresponding merates (Definition IV, Section 2.02). For example,

$$U^i + V^i = W^i \tag{2.14}$$

means that

$$\{U^i\} + \{V^i\} = \{W^i\}.$$

We have defined the uncontracted product of holors in terms of the product of merates. For example,

$$u^i v^j = w^{ij} \tag{2.08}$$

means that

$$\{u^i\}\{v^j\} = \{w^{ij}\}.$$

These equations for holors are identical with the equations for merates. Therefore, the algebra of holors, as regards these operations with real merates, is identical with the algebra of real numbers.

In particular, the product of real numbers is commutative,

$$\{u^i\}\{v^j\} = \{v^j\}\{u^i\},$$

so the uncontracted product of holors of any valence and any plethos is commutative. That this conclusion is in accordance with our conventions for the writing of matrices is easily shown. By Convention II, the product of Equation (2.08) is

$$u^i v^j = \underset{\text{row}\quad\text{column}}{w^{ij}} = \begin{bmatrix} w^{11} & w^{12} & \cdots & w^{1n} \\ w^{21} & w^{22} & \cdots & w^{2n} \\ \cdot & \cdot & \cdots & \cdot \\ w^{n1} & w^{n2} & \cdots & w^{nn} \end{bmatrix}$$

$$= \begin{bmatrix} u^1 v^1 & u^1 v^2 & \cdots & u^1 v^n \\ u^2 v^1 & u^2 v^2 & \cdots & u^2 v^n \\ \cdot & \cdot & \cdots & \cdot \\ u^n v^1 & u^n v^2 & \cdots & u^n v^n \end{bmatrix}.$$

Now commute the holors u^i and v^j to obtain a new matrix W^{ji}:

$$v^j u^i = W^{ji}. \tag{2.15}$$

The indices must be the same on both sides of the equation and must be arranged in the same order. But a different base letter is used to allow for possible noncommutivity. By Convention II, therefore,

$$v^j u^i = W^{ji} = \begin{bmatrix} W^{11} & W^{21} & \cdots & W^{n1} \\ W^{12} & W^{22} & \cdots & W^{n2} \\ \cdot & \cdot & \cdots & \cdot \\ W^{1n} & W^{2n} & \cdots & W^{nn} \end{bmatrix}$$

$$= \begin{bmatrix} v^1 u^1 & v^2 u^1 & \cdots & v^n u^1 \\ v^1 u^2 & v^2 u^2 & \cdots & v^n u^2 \\ \cdot & \cdot & \cdots & \cdot \\ v^1 u^n & v^2 u^n & \cdots & v^n u^n \end{bmatrix}.$$

The matrices for Equations (2.08) and (2.15) are seen to be identical, since the products of real numbers are commutative (e.g., $v^2u^1 = u^1v^2$). Thus, the order in which factors are written in an uncontracted holor product is completely unimportant.

Because of commutivity of the holor product, an uncontracted product of univalent holors is always a symmetric holor. Thus, the foregoing bivalent holors w^{ij} and W^{ji} are not arbitrary holors but are symmetric and should really be written $w^{(ij)}$ and $W^{(ji)}$. Index balance does not allow us to write

$$w^{ij} = W^{ji},$$

which indeed is not true in general. But there is no reason why we cannot write

$$w^{(ij)} = W^{(ji)}.$$

Similarly, the product of three univalent holors is a completely symmetric trivalent holor:

$$u^i v^j w^k = X^{(ijk)} = \frac{1}{3!}[u^i v^j w^k + v^j w^k u^i + w^k u^i v^j + w^k v^j u^i + v^j u^i w^k + u^i w^k v^j].$$

The uncontracted product of holors is not only commutative, it is also associative. For instance,

$$u^i(v^j w^k) = (u^i v^j)w^k. \qquad (2.16)$$

Also, the uncontracted product of holors is distributive with respect to addition:

$$u^i(v^j + w^j) = u^i v^j + u^i w^j. \qquad (2.17)$$

These conclusions apply to holors of any valence and any plethos because of the identity of equations for holors and for merates.

Theorem I. *The algebra of holors, as regards addition, subtraction, and uncontracted multiplication (of holors of any valence and plethos with real merates), is identical with the algebra of real numbers.*

Since the algebra of ordinary complex numbers is the same as the algebra of real numbers, the above theorem is also applicable when the merates are complex. In working with holors having more complicated merates such as submatrices, however, the rules of ordinary algebra may not be valid. Also, in dealing with the very important contracted product (Section 2.04), one must be careful because Theorem I may not apply.

2.04. The contracted product

We have considered the sum, difference, and uncontracted product of holors. Another important combination of holors is the sum of products:

$$u^1 v_1 + u^2 v_2 + \cdots + u^n v_n = \sum_{i=1}^{n} u^i v_i. \tag{2.18}$$

This combination occurs so frequently in applications of holors that Einstein[3] suggested that the summation sign be omitted whenever the two indices are the same. What could be more obvious, what could be more trivial, one says. Yet, as we shall see, this summation convention has been instrumental in making tensor analysis the powerful tool it is today.

Stated more precisely, the Einstein summation convention is:

Definition VI: Whenever the same literal index occurs as a superscript and as a subscript in the same term, summation is understood.

For instance,

$$u^i v_i = u^1 v_1 + u^2 v_2 + \cdots + u^n v_n = S$$

is a nilvalent holor. Similarly,

$$v_i U^{ij} = v_1 U^{1j} + v_2 U^{2j} + \cdots + v_n U^{nj} = V^j$$

is a univalent holor. Also,

$$A^{ij} w_{jk} = A^{i1} w_{1k} + A^{i2} w_{2k} + \cdots + A^{in} w_{nk} = W^i_k$$

is a bivalent holor.

The process of contraction consists in changing a free index so that an upper and a lower index are the same. Thus, the product $u^i v_j$ may be contracted by making $j = i$. Obviously, contraction results in a completely different holor having a valence 2 less than that of the uncontracted product. So $u^i v_j = w^i_j$ is a bivalent holor, but the contracted product $u^i v_i = S$ is a nilvalent holor. Similarly, $v_i U^{jk}$ is trivalent, but the contracted form $v_i U^{ik}$ is univalent. Multiple contraction is also possible, resulting in a reduction in valence of $2p$, where p is the number of contractions.

A repeated literal index occurring as subscript and superscript is called a *dummy* index. An unrepeated index is called a *free* index. In holor manipulation, it is often desirable to change a letter that is used as an index. Usually such a change can be effected just as in ordinary algebra. To make the operation as definite as possible, however, we shall state two theorems.

Theorem II. *A free literal index may be replaced by any other letter if* (a) *this change is made throughout an equation,* (b) *the new letter does not occur elsewhere in the equation, and* (c) *alphabetical order of indices is preserved.*

Requirement (c) is necessary because of Convention II regarding the alphabetical order of indices. Thus,

$$B^{i}_{kl}C^{lm}_{i} = D^{m}_{k}$$

has free indices k and m. This equation may be written equally well as

$$B^{i}_{kl}C^{lp}_{i} = D^{p}_{k}$$

or as

$$B^{i}_{jl}C^{ln}_{i} = D^{n}_{j}.$$

But

$$B^{i}_{pl}C^{lj}_{i} = D^{j}_{p}$$

gives the transpose of the original D matrix.

Theorem III. *A dummy index may be replaced by any other letter if this letter is not already used as an index in the equation.*

An equation

$$u_i A^{ik} = V^k$$

may be written equally well as[3]

$$u_h A^{hk} = V^k$$

or

$$u_j A^{jk} = V^k$$

if $h, i, j, k = 1, 2, \ldots, n$. But it must not be expressed as

$$u_k A^{kk} = V^k.$$

As an example of a change of dummy index, suppose that we have the expression

$$A^i W_{ik} = V_k \tag{2.19}$$

and wish to multiply it by a univalent holor u^i. The product

$$u^i A^i W_{ik} = A^i u^i W_{ik}$$

is ambiguous, since it might imply a summation $A^i W_{ik}$ or a summation $u^i W_{ik}$. The difficulty is eliminated by substituting another letter for the dummy index in the original expression:

$$A^j W_{jk} = V_k. \tag{2.19a}$$

This obviously gives the same summation as before but gives the unambiguous result

$$u^i A^j W_{jk} = X^i_{\ k}. \tag{2.20}$$

Note that summation applies only if the same literal index occurs on different levels, as

$$u^i v_i, A^i W_{ij}, \delta^{123}_{ijk}, du^i dv^j dw^k, \frac{\partial u^i}{\partial v^j} w^j.$$

There is no summation in

$$u^i v^i, \quad A^i W^i_{\ j}, \quad \text{or} \quad \frac{\partial u^i}{\partial v^j} w_j.$$

Likewise, there is no summation if the repeated index is a number: $u^1 v_1$ is a single merate.

It is easily demonstrated that the contracted product of holors is commutative, just as was the uncontracted product. For example, consider

$$u^{ij} v_{jk} = w^i_{\ k}. \tag{2.21}$$

By the summation convention,

$$\{w^i_{\ k}\} = \{u^{i1} v_{1k}\} + \{u^{i2} v_{2k}\} + \cdots + \{u^{in} v_{nk}\}.$$

Interchange of u^{ij} and v_{jk} gives

$$v_{jk} u^{ij} = W^i_{\ k} \tag{2.22}$$

or

$$\{W^i_{\ k}\} = \{v_{1k} u^{i1}\} + \{v_{2k} u^{i2}\} + \cdots + \{v_{nk} u^{in}\}.$$

Since the merates are the same and are arranged in the same way in the matrices (Conventions I and II), the contracted product is commutative. Associative and distributive properties are found similarly.

Theorem IV. *The product – uncontracted, singly contracted, or multiply contracted – of any two holors is commutative, associative, and distributive with respect to addition.*

For anyone familiar with matrices, this conclusion may seem startling. How can it be reconciled with the known fact that the product of two matrices is noncommutative? Stated more definitely:

(a) The product of two holors is always commutative (Theorem IV).

(b) A matrix is a form of holor.

(c) Therefore, the product of two matrices is commutative.

(d) But, according to matrix theory, the product of matrices is noncommutative.

The point is that in matrix theory there is only one product, which in index notation is a contracted product with summation on second and third indices. This inflexibility of the matrix product leads to the lack of commutivity.

Given two matrices,

$$A^{ij} = \begin{bmatrix} A^{11} & A^{12} \\ A^{21} & A^{22} \end{bmatrix} \quad \text{and} \quad B_{jk} = \begin{bmatrix} B_{11} & B_{12} \\ B_{21} & B_{22} \end{bmatrix},$$

the *matrix product* is obtained by moving horizontally in the first matrix and vertically in the second,[1] giving

$$\overrightarrow{\begin{bmatrix} A^{11} & A^{12} \\ A^{21} & A^{22} \end{bmatrix}} \downarrow \begin{bmatrix} B_{11} & B_{12} \\ B_{21} & B_{22} \end{bmatrix} = \begin{bmatrix} A^{11}B_{11}+A^{12}B_{21} & A^{11}B_{12}+A^{12}B_{22} \\ A^{21}B_{11}+A^{22}B_{21} & A^{21}B_{12}+A^{22}B_{22} \end{bmatrix}.$$

But this is exactly what is obtained as the contracted holor product

$$A^{ij}B_{jk} = \begin{bmatrix} A^{11}B_{11}+A^{12}B_{21} & A^{11}B_{12}+A^{11}B_{22} \\ A^{21}B_{11}+A^{22}B_{21} & A^{21}B_{12}+A^{22}B_{22} \end{bmatrix},$$

and we know that this product is commutative:

$$A^{ij}B_{jk} = B_{jk}A^{ij}. \tag{2.23}$$

But the right side of Equation (2.23) is no longer a matrix product, since it is now summed over the first and fourth indices.

2.05. Matrices

It is instructive to see exactly how index notation fits into the traditional picture of the matrix product.[4] The holor product of two bivalent holors must be simply contracted if the product is to be bivalent. Evidently there

Table 2.01. *Eight products*

$$A^{ij} = \mathfrak{A} = \begin{bmatrix} A^{11} & A^{12} \\ A^{21} & A^{22} \end{bmatrix} \qquad B_{jk} = \mathfrak{B} = \begin{bmatrix} B_{11} & B_{12} \\ B_{21} & B_{22} \end{bmatrix}$$

Index notation	Summation on indices	Holor product	Matrix product
$A^{ij}B_{jk}$	2, 3	$\begin{bmatrix} A^{11}B_{11}+A^{12}B_{21} & A^{11}B_{12}+A^{12}B_{22} \\ A^{21}B_{11}+A^{22}B_{21} & A^{21}B_{12}+A^{22}B_{22} \end{bmatrix}$	$\mathfrak{A}\mathfrak{B}$
$A^{ij}B_{kj}$	2, 4	$\begin{bmatrix} A^{11}B_{11}+A^{12}B_{12} & A^{11}B_{21}+A^{12}B_{22} \\ A^{21}B_{11}+A^{22}B_{12} & A^{21}B_{21}+A^{22}B_{22} \end{bmatrix}$	$\mathfrak{A}\mathfrak{B}_T$
$A^{ji}B_{jk}$	1, 3	$\begin{bmatrix} A^{11}B_{11}+A^{21}B_{21} & A^{11}B_{12}+A^{21}B_{22} \\ A^{12}B_{11}+A^{22}B_{21} & A^{12}B_{12}+A^{22}B_{22} \end{bmatrix}$	$\mathfrak{A}_T\mathfrak{B}$
$A^{ji}B_{kj}$	1, 4	$\begin{bmatrix} A^{11}B_{11}+A^{21}B_{12} & A^{11}B_{21}+A^{21}B_{22} \\ A^{12}B_{11}+A^{22}B_{12} & A^{12}B_{21}+A^{22}B_{22} \end{bmatrix}$	$\mathfrak{A}_T\mathfrak{B}_T$
$B_{jk}A^{ij}$	1, 4	$\begin{bmatrix} A^{11}B_{11}+A^{12}B_{21} & A^{11}B_{12}+A^{12}B_{22} \\ A^{21}B_{11}+A^{22}B_{21} & A^{21}B_{12}+A^{22}B_{22} \end{bmatrix}$	$(\mathfrak{B}_T\mathfrak{A}_T)_T$
$B_{kj}A^{ij}$	2, 4	$\begin{bmatrix} A^{11}B_{11}+A^{12}B_{12} & A^{11}B_{21}+A^{12}B_{22} \\ A^{21}B_{11}+A^{22}B_{12} & A^{21}B_{21}+A^{22}B_{22} \end{bmatrix}$	$(\mathfrak{B}\mathfrak{A}_T)_T$
$B_{jk}A^{ji}$	1, 3	$\begin{bmatrix} A^{11}B_{11}+A^{21}B_{21} & A^{11}B_{12}+A^{21}B_{22} \\ A^{12}B_{11}+A^{22}B_{21} & A^{12}B_{12}+A^{22}B_{22} \end{bmatrix}$	$(\mathfrak{B}_T\mathfrak{A})_T$
$B_{kj}A^{ji}$	2, 3	$\begin{bmatrix} A^{11}B_{11}+A^{21}B_{12} & A^{11}B_{21}+A^{21}B_{22} \\ A^{12}B_{11}+A^{22}B_{12} & A^{12}B_{21}+A^{22}B_{22} \end{bmatrix}$	$(\mathfrak{B}\mathfrak{A})_T$

are eight possible products of this kind, obtained by interchanging indices and by interchanging factors (Table 2.01). Because of commutivity, however, only four of the products are distinct.

The matrices of Table 2.01 may be written also in terms of matrices $\mathfrak{A}$ and $\mathfrak{B}$ and their transposes $\mathfrak{A}_T$ and $\mathfrak{B}_T$. Comparison of the fourth and eighth matrices shows that

$$(\mathfrak{B}\mathfrak{A})_T = \mathfrak{A}_T\mathfrak{B}_T.$$

This is an important theorem of matrix algebra: The transpose of a matrix product is equal to the matrix product of the transposes in reverse order.

Table 2.02. *Products of three bivalent holors*[a]

Holor product	Summation on indices	Matrix product
$A^{ij}B_{jk}C^{kl}$	23; 45	$\mathfrak{A}\mathfrak{B}\mathfrak{C}$
$A^{ij}B_{kj}C^{kl}$	24; 35	$\mathfrak{A}\mathfrak{B}_T\mathfrak{C}$
$A^{ij}B_{jk}C^{lk}$	23; 46	$\mathfrak{A}\mathfrak{B}\mathfrak{C}_T$
$A^{ij}B_{kj}C^{lk}$	24; 36	$\mathfrak{A}\mathfrak{B}_T\mathfrak{C}_T$
$A^{ji}B_{jk}C^{kl}$	13; 45	$\mathfrak{A}_T\mathfrak{B}\mathfrak{C}$
$A^{ji}B_{kj}C^{kl}$	14; 35	$\mathfrak{A}_T\mathfrak{B}_T\mathfrak{C}$
$A^{ji}B_{jk}C^{lk}$	13; 46	$\mathfrak{A}_T\mathfrak{B}\mathfrak{C}_T$
$A^{ji}B_{kj}C^{lk}$	14; 36	$\mathfrak{A}_T\mathfrak{B}_T\mathfrak{C}_T$

[a] Each holor product can have its factors arranged in six ways, giving a total of 48 products. But because of commutivity, there are only eight distinct holor products.

By using index notation, we eliminate the need for such theorems: Everything is taken care of by the indices. Besides its greater flexibility, the holor product is commutative and allows the freedom of arranging the factors of a product in any order. Thus, we eliminate the annoying requirement of matrix theory that factors be written in an inflexible order.

A similar table can be made for triple products (Table 2.02). Here there are eight distinct holor products, obtained by interchanging indices. We can also arrange the factors in six ways; but because of commutivity of the holor product, these provide nothing new. The table also includes the eight possibilities expressed as matrix products of the matrices and their transposes.

2.06. Index balance

A helpful aspect of our notation is provided by index balance. In a surprising number of cases, index balance suggests possible forms of equations, while other notations give no help whatsoever. For instance, if the physics of a problem shows that the product of a certain bivalent holor and a trivalent holor must be univalent, we write

$$A^{ij}B_{ijk}=V_k,$$

the double contraction leaving only the free index k, which appears on the right side. On the other hand, if a long series of mathematical operations finally results in the equation

$$A^{ij}B_{ijk}C^l_m=D_m,$$

we know immediately that an error has been made, since indices on the two sides of the equation do not balance.

The principle of index balance rests on our definition of addition; the sum of holors is a holor whose merates are sums of corresponding merates. For example,

$$A^{ij}+B^{ij}=C^{ij}.$$

Addition does not allow such expressions as

$$A^{ij}+B^{ji}=C^{jk}.$$

Thus, we have:

Theorem V. *Each term of a holor equation must have the same literal indices in the same positions and in the same order.*

In this strict sense, the theorem rules out such expressions as

$$u^{ij}=u^{ji}$$

to indicate a symmetric matrix or

$$u^iv^j=v^ju^i$$

to indicate a commutative product. Such equations are often found in tensor books. It seems best, however, to keep the theorem as stated and to indicate the special case of symmetry by writing $u^{(ij)}$ or by merely saying that certain merates are equal:

$$\{u^{ij}\}=\{u^{ji}\}.$$

Similarly, commutivity is indicated by making a statement about the merates:

$$\{u^iv^j\}=\{v^ju^i\}.$$

These are not holor equations, and thus Theorem V does not apply. An examination of all previous holor equations in this book shows that index balance has been used, though probably the reader has not noticed it.

What is the significance of index balance? Is it really necessary, or is it merely a superficial refinement that can be ignored if it happens to be inconvenient?

A simple example of an equation without index balance is

$$v^i = \phi,$$

which says that the ordered set of numbers $(v^1, v^2, \ldots, v^n)$ is equal to a single number. The only conceivable interpretation would be that one of the merates, say v^1, is equal to ϕ, the others being zero. But even if this were unambiguous, it would hold in only one coordinate system. Thus, the proposed equation has no significance.

Another simple example is

$$v^i = w^j,$$

where $i, j = 1, 2, \ldots, n$. Since i and j are chosen independently from the set $1, 2, \ldots, n$, this equation might mean that

$$v^1 = w^2, \quad v^2 = w^n, \quad \text{etc.}$$

or it might stand for any other correlation. Again, the equation has no definite significance. Obviously, the same difficulty occurs with holors of higher valence.

A lesser violation of index balance employs the same indices on both sides of the equation but in different order, such as

$$u^{ij} = w^{ji}.$$

This might be interpreted as a statement that one matrix is the transpose of the other. But such a statement is made more specifically by writing

$$\{u^{ij}\} = \{w^{ji}\}.$$

We conclude that index balance is a very important property of index notation. Indeed, index balance should be observed rigorously in all cases: A violation of Theorem V will generally result in meaningless equations.

2.07. Unit holors

In the field of real numbers, unity is defined by the relation

$$a\,1 = a, \tag{2.24}$$

Table 2.03. *Characteristics of unit holors*

Unit holor	Valence	Number of contractions (p)	Valid for holors with N equal to or greater than
1	0	0	0
δ_j^k	2	1	1
$\delta_i^k\delta_j^l$	4	2	2
$\delta_i^l\delta_j^m\delta_k^n$	6	3	3
$\delta_i^m\delta_j^n\delta_k^p\delta_l^q$	8	4	4
.	.	.	.

for an arbitrary number a. Evidently this same unity may be used for any holor $\mathfrak{A}$, since

$$\mathfrak{A}1=\mathfrak{A}. \tag{2.25}$$

It is interesting to see if other holors can be considered as unit holors by extending the definition in analogy with Equation (2.24). If we have a univalent holor v^j instead of the nilvalent a, we can write

$$v^j\delta_j^k=v^k \tag{2.26}$$

or, for $n=2$,

$$v^1\delta_1^1+v^2\delta_2^1=v^1, \qquad v^1\delta_1^2+v^2\delta_2^2=v^2,$$

where the merate $\{\delta_j^k\}$ of the Kronecker delta is zero unless $k=j$. Similarly, for $N=2$,

$$u^{ij}\delta_j^k=u^{ik} \tag{2.27}$$

or, for $n=2$,

$$u^{11}\delta_1^1+u^{12}\delta_2^1=u^{11}, \qquad u^{11}\delta_1^2+u^{12}\delta_2^2=u^{12},$$

$$u^{21}\delta_1^1+u^{22}\delta_2^1=u^{21}, \qquad u^{21}\delta_1^2+u^{22}\delta_2^2=u^{22}.$$

Evidently, the Kronecker delta δ_j^k acts as a unit holor for the singly contracted product with an arbitrary holor of any valence above zero.

Similarly, unit holors of higher valence can be introduced. Consider $\delta_i^k\delta_j^l$ as a unit holor. Then

$$u^{ij}\delta_i^k\delta_j^l=u^{kl} \tag{2.28}$$

or

$$u^{11}\delta_1^1\delta_1^1+u^{12}\delta_1^1\delta_2^1+u^{21}\delta_2^1\delta_1^1+u^{22}\delta_2^1\delta_2^1=u^{11},$$

$$u^{11}\delta_1^1\delta_1^2+u^{12}\delta_1^1\delta_2^2+u^{21}\delta_2^1\delta_1^2+u^{22}\delta_2^1\delta_2^2=u^{12}, \quad \text{etc.}$$

The quadrivalent unit holor $\delta_i^k\delta_j^l$ can be used with any holor of valence 2 or higher and for any plethos.

This set of unit holors can be extended indefinitely. The valence of a unit holor must always be an even number (as with a product of Kronecker deltas); otherwise index balance cannot be obtained. A list of unit holors is found in Table 2.03.

2.08. The inverse

In the algebra of holors, there is no division. The one exception is division by a nilvalent holor, which of course is feasible just as in real algebra. In general, instead of dividing by a holor, one multiplies by the *inverse* of this holor.

With real numbers, one often encounters an expression such as

$$uv=w. \tag{2.29}$$

If u and w are known, then v is obtained by division:

$$v=w/u.$$

But v could be obtained by multiplying both sides of the original expression by u^{-1}, the inverse of u:

$$u^{-1}uv=u^{-1}w.$$

Since $u^{-1}u=1$, we have again

$$v=u^{-1}w=w/u.$$

The idea is easily extended to bivalent holors. Then a product corresponding to Equation (2.29) is

$$u^{ij}v_{jk}=w_k^i. \tag{2.30}$$

The definition of the inverse U_{hi} of u^{ij} is

$$U_{hi}u^{ij}=\delta_h^j. \tag{2.31}$$

Multiplying both sides of Equation (2.30) by U_{hi} gives

$$U_{hi}u^{ij}v_{jk} = U_{hi}w^i_k,$$

or, by Equation (2.31),

$$\delta^j_h v_{jk} = v_{hk} = U_{hi}w^i_k. \tag{2.32}$$

The order of indices is important if one is to distinguish between the final result v_{hk} and its transpose. Probably the simplest procedure is to write all holors with indices in alphabetical order. Thus,

(1) $$\underset{\rightarrow}{u^{ij}}\,\underset{\rightarrow}{v_{jk}} = w^i_k\downarrow, \quad \text{not} \quad \underset{\leftarrow}{u^{lj}}\,\underset{\rightarrow}{v_{jk}} = w^l_k\uparrow.$$

Also,

(2) $$\underset{\rightarrow}{U_{hi}}\,\underset{\rightarrow}{u^{ij}} = \delta^j_h\uparrow, \quad \text{not} \quad \underset{\leftarrow}{U_{li}}\,\underset{\rightarrow}{u^{ij}} = \delta^j_l\downarrow,$$

and

(3) $$(U_{hi}u^{ij})v_{jk} = \delta^j_h v_{jk} = U_{hi}w^i_k$$

or

$$\underset{\rightarrow}{v_{hk}} = \underset{\rightarrow}{U_{hi}}\,w^i_k\downarrow, \quad \text{not} \quad \underset{\rightarrow}{v_{jk}} = \underset{\leftarrow}{U_{ji}}\,w^i_k\downarrow.$$

Other formulations are, of course, possible, but they may require some thought before one is sure about the arrangement of the final matrix.

We now consider the evaluation of the inverse. For $n=2$, Equation (2.31) is written

$$U_{11}u^{11} + U_{12}u^{21} = \delta^1_1 = 1,$$
$$U_{11}u^{12} + U_{12}u^{22} = \delta^2_1 = 0,$$
$$U_{21}u^{11} + U_{22}u^{21} = \delta^1_2 = 0,$$
$$U_{21}u^{12} + U_{22}u^{22} = \delta^2_2 = 1.$$

If the determinant $|u^{ij}| \neq 0$, these four equations suffice to give a unique inverse U_{hi}; merates U_{11} and U_{12} being obtained by solution of the first pair of equations, and merates U_{21} and U_{22} being obtained from the second pair.

For arbitrary u^{ij}, the system matrix for the first pair is

$$\begin{bmatrix} u^{11} & u^{21} \\ u^{12} & u^{22} \end{bmatrix}$$

and the augmented matrix is

$$\begin{bmatrix} u^{11} & u^{21} & -1 \\ u^{12} & u^{22} & 0 \end{bmatrix}.$$

In general, both are of rank 2, and a solution exists if the determinant is not zero.[5]

The determinant is

$$\Delta = \begin{vmatrix} u^{11} & u^{21} \\ u^{12} & u^{22} \end{vmatrix} = u^{11}u^{22} - u^{12}u^{21}.$$

By Cramer's rule,

$$U_{11} = \frac{1}{\Delta}\begin{vmatrix} 1 & u^{21} \\ 0 & u^{22} \end{vmatrix} = \frac{u^{22}}{\Delta},$$

$$U_{12} = \frac{1}{\Delta}\begin{vmatrix} u^{11} & 1 \\ u^{12} & 0 \end{vmatrix} = -\frac{u^{12}}{\Delta}.$$

The other pair of equations is handled in the same way, giving the inverse

$$U_{hi} = \frac{1}{\Delta}\begin{bmatrix} u^{22} & -u^{12} \\ -u^{21} & u^{11} \end{bmatrix}.$$

The inverse is evaluated similarly for any plethos n. Equation (2.31) applies:

$$U_{hi}u^{ij} = \delta_h^j,$$

with

$$u^{ij} = \begin{bmatrix} u^{11} & u^{12} & \cdots & u^{1n} \\ u^{21} & u^{22} & \cdots & u^{2n} \\ \cdot & \cdot & \cdots & \cdot \\ u^{n1} & u^{n2} & \cdots & u^{nn} \end{bmatrix}$$

and determinant

$$\Delta = \begin{bmatrix} u^{11} & u^{12} & \cdots & u^{1n} \\ u^{21} & u^{22} & \cdots & u^{2n} \\ \cdot & \cdot & \cdots & \cdot \\ u^{n1} & u^{n2} & \cdots & u^{nn} \end{bmatrix}.$$

The inverse is, according to Conventions I and II,

$$U_{hi} = \begin{bmatrix} U_{11} & U_{12} & \cdots & U_{1n} \\ U_{21} & U_{22} & \cdots & U_{2n} \\ \cdot & \cdot & \cdots & \cdot \\ U_{n1} & U_{n2} & \cdots & U_{nn} \end{bmatrix}.$$

Any merate $\{U_{hi}\}$ is obtained by dividing the cofactor of $\{u^{ih}\}$ by Δ. Note the interchange of indices, which should be familiar from ordinary algebra.[6]

For instance, if $n=3$,

$$u^{ij} = \begin{bmatrix} u^{11} & u^{12} & u^{13} \\ u^{21} & u^{22} & u^{23} \\ u^{31} & u^{32} & u^{33} \end{bmatrix}$$

and

$$\Delta = u^{11}u^{22}u^{33} + u^{12}u^{23}u^{31} + u^{13}u^{21}u^{32} \\ - u^{13}u^{22}u^{31} - u^{12}u^{21}u^{33} - u^{11}u^{23}u^{32}.$$

Then U_{11} is equal to the cofactor of u^{11} divided by Δ, or

$$U_{11} = \frac{1}{\Delta}\begin{vmatrix} u^{22} & u^{23} \\ u^{32} & u^{33} \end{vmatrix} = \frac{1}{\Delta}[u^{22}u^{33} - u^{23}u^{32}].$$

Similarly, U_{12} is equal to the cofactor of u^{21} divided by Δ:

$$U_{12} = -\frac{1}{\Delta}\begin{vmatrix} u^{12} & u^{13} \\ u^{32} & u^{33} \end{vmatrix} = -\frac{1}{\Delta}[u^{12}u^{33} - u^{13}u^{32}], \quad \text{etc.}$$

Let us note again the exact relation between index notation and matrices. Evidently, there are four possibilities for bivalent holors written with superscripts. According to our conventions (Chapter 1), the normal form of matrix is represented by a base letter with two indices arranged in alphabetical order, as

$$A^{ij} = \begin{bmatrix} A^{11} & A^{12} \\ A^{21} & A^{22} \end{bmatrix}.$$

The transpose has indices in anti-alphabetical order, as A^{ji}. The inverse, a_{hi}, satisfies the relation

$$a_{hi}A^{ij} = \delta_h^j.$$

The transpose of the inverse, which is also the inverse of the transpose, is written a_{ih}. These four forms are listed in Table 2.04.

For holors with subscripts, the procedure is exactly the same. In fact, the base letter could be the same for the normal form and for the inverse. So we could write the four forms as B_{ij}, B_{ji}, B^{hi}, and B^{ih} without ambiguity. The same applies to A^{ij}, A^{ji}, A_{hi}, and A_{ih}. But for the mixed

Table 2.04. *Bivalent holors*

Superscripts

$$A^{ij} = \begin{bmatrix} A^{11} & A^{12} \\ A^{21} & A^{22} \end{bmatrix} \qquad \text{normal form}$$

$$A^{ji} = \begin{bmatrix} A^{11} & A^{21} \\ A^{12} & A^{22} \end{bmatrix} \qquad \text{transpose}$$

$$a_{hi} A^{ij} = \delta_h^j$$

$$a_{hi} = \begin{bmatrix} a_{11} & a_{12} \\ a_{21} & a_{22} \end{bmatrix} = \frac{1}{\Delta} \begin{bmatrix} A^{22} & -A^{12} \\ -A^{21} & A^{11} \end{bmatrix} \qquad \text{inverse}$$

$$a_{ih} = \begin{bmatrix} a_{11} & a_{21} \\ a_{12} & a_{22} \end{bmatrix} = \frac{1}{\Delta} \begin{bmatrix} A^{22} & -A^{21} \\ -A^{12} & A^{11} \end{bmatrix} \qquad \text{transpose of inverse}$$

where

$$\Delta = \begin{vmatrix} A^{11} & A^{12} \\ A^{21} & A^{22} \end{vmatrix}$$

Subscripts

$$B_{ij} = \begin{bmatrix} B_{11} & B_{12} \\ B_{21} & B_{22} \end{bmatrix} \qquad \text{normal form}$$

$$B_{ji} = \begin{bmatrix} B_{11} & B_{21} \\ B_{12} & B_{22} \end{bmatrix} \qquad \text{transpose}$$

$$b^{hi} B_{ij} = \delta_j^h$$

$$b^{hi} = \begin{bmatrix} b^{11} & b^{12} \\ b^{21} & b^{22} \end{bmatrix} = \frac{1}{\Delta} \begin{bmatrix} B_{22} & -B_{12} \\ -B_{21} & B_{11} \end{bmatrix} \qquad \text{inverse}$$

$$b^{ih} = \begin{bmatrix} b^{11} & b^{21} \\ b^{12} & b^{22} \end{bmatrix} = \frac{1}{\Delta} \begin{bmatrix} B_{22} & -B_{21} \\ -B_{12} & B_{11} \end{bmatrix} \qquad \text{transpose of inverse}$$

where

$$\Delta = \begin{vmatrix} B_{11} & B_{12} \\ B_{21} & B_{22} \end{vmatrix}$$

Mixed

$$C_j^i = \begin{bmatrix} C_1^1 & C_2^1 \\ C_1^2 & C_2^2 \end{bmatrix} \qquad \text{normal form}$$

$$C_i^j = \begin{bmatrix} C_1^1 & C_1^2 \\ C_2^1 & C_2^2 \end{bmatrix} \qquad \text{transpose}$$

$$c_i^h C_j^i = \delta_j^h$$

Table 2.04 *(cont.)*

	$c^h_i = \begin{bmatrix} c^1_1 & c^1_2 \\ c^2_1 & c^2_2 \end{bmatrix} = \dfrac{1}{\Delta}\begin{bmatrix} C^2_2 & -C^1_2 \\ -C^2_1 & C^1_1 \end{bmatrix}$	inverse
	$c^i_h = \begin{bmatrix} c^1_1 & c^2_1 \\ c^1_2 & c^2_2 \end{bmatrix} = \dfrac{1}{\Delta}\begin{bmatrix} C^2_2 & -C^2_1 \\ -C^1_2 & C^1_1 \end{bmatrix}$	transpose of inverse
where	$\Delta = \begin{vmatrix} C^1_1 & C^1_2 \\ C^2_1 & C^2_2 \end{vmatrix}$	

form of bivalent holor, this flexibility is no longer possible. The normal form is

$$C^i_j = \begin{bmatrix} C^1_1 & C^1_2 \\ C^2_1 & C^2_2 \end{bmatrix},$$

and the transpose is

$$C^j_i = \begin{bmatrix} C^1_1 & C^2_1 \\ C^1_2 & C^2_2 \end{bmatrix},$$

leaving no way of distinguishing the inverse. Dots could be used, but the simplest solution of the difficulty is to employ different base letters for the normal form and for the inverse, as is done in Table 2.04. Different symbols for a quantity and its inverse are commonly used in physics and engineering. The inverse of electrical impedance Z, for instance, is admittance Y. Such a convention seems advisable also with tensors.

One of the principal applications of the inverse occurs in the solution of a set of linear equations. Given

$$\begin{aligned} y_1 &= a_{11}x^1 + a_{22}x^2 + \cdots + a_{1n}x^n, \\ y_2 &= a_{21}x^1 + a_{22}x^2 + \cdots + a_{2n}x^n, \\ &\vdots \\ y_n &= a_{n1}x^1 + a_{n2}x^2 + \cdots + a_{nn}x^n, \end{aligned}$$

which is written

$$y_i = a_{ij}x^j. \tag{2.33}$$

The coefficient matrix is

$$a_{ij} = \begin{bmatrix} a_{11} & a_{12} & \cdots & a_{1n} \\ a_{21} & a_{22} & \cdots & a_{2n} \\ \cdot & \cdot & \cdots & \cdot \\ a_{n1} & a_{n2} & \cdots & a_{nn} \end{bmatrix}, \qquad (2.34)$$

and its determinant is

$$\Delta = \begin{vmatrix} a_{11} & a_{12} & \cdots & a_{1n} \\ a_{21} & a_{22} & \cdots & a_{2n} \\ \cdot & \cdot & \cdots & \cdot \\ a_{n1} & a_{n2} & \cdots & a_{nn} \end{vmatrix}.$$

The inverse A^{hi} is defined by the relation

$$A^{hi} a_{ij} = \delta^h_j. \qquad (2.35)$$

Multiplication of both sides of Equation (2.33) by A^{hi} gives

$$A^{hi} y_i = A^{hi} a_{ij} x^j = \delta^h_j x^j$$

or

$$x^h = A^{hi} y_i. \qquad (2.36)$$

The inverse is written

$$A^{hi} = \begin{bmatrix} A^{11} & A^{12} & \cdots & A^{1n} \\ A^{21} & A^{22} & \cdots & A^{2n} \\ \cdot & \cdot & \cdots & \cdot \\ A^{n1} & A^{n2} & \cdots & A^{nn} \end{bmatrix}, \qquad (2.37)$$

where each merate $\{A^{hi}\}$ is obtained by dividing the merate $\{a_{ih}\}$ of Equation (2.34) by Δ.

2.09. General conditions for an inverse

That an inverse is not always possible is shown if we take holors whose valence is not 2. Consider a univalent holor u^i and a defining equation

$$U_i u^i = 1. \qquad (2.38)$$

Expanded for plethos n, Equation (2.38) gives the single equation

$$U_1 u^1 + U_2 u^2 + \cdots + U_n u^n = 1.$$

Here we have one equation for the determination of n unknowns, $U_1, U_2, \ldots, U_n$. Obviously, no unique inverse defined by Equation (2.38) is possible.

As another equation that gives index balance, one could try the uncontracted product

$$U_j u^i = \delta^i_j. \tag{2.39}$$

For $n=2$, this gives

$$U_1 u^1 = \delta^1_1 = 1, \qquad U_2 u^1 = \delta^1_2 = 0,$$
$$U_1 u^2 = \delta^2_1 = 0, \qquad U_2 u^2 = \delta^2_2 = 1,$$

or four equations with two unknowns. For arbitrary u^i, the equations are inconsistent.

The foregoing examples indicate that inverses cannot be ascribed to holors in general. Evidently an inverse exists only when the expanded equations for the inverse have a unique solution for the merates. In analogy with real numbers, the defining equation for the inverse is

$$U_{fg\ldots} u^{ij\cdots} = \mathfrak{I}, \tag{2.40}$$

where $\mathfrak{I}$ is one of the unit holors considered in Section 2.07, and the product on the left is uncontracted or is contracted one or more times. Let p = number of contractions, q = valence of $\mathfrak{I}$, N = valence of $u^{ij\cdots}$ and valence of $U_{fg\ldots}$, and n = plethos. Then the valence of the product is $2N-2p$, which, for index balance, must be equal to q. Thus,

$$N - p = \tfrac{1}{2} q. \tag{2.41}$$

We must now consider the number of linear equations obtained by expanding Equation (2.40). Evidently the number of equations is equal to the number of merates of $\mathfrak{I}$, or $(n)^q$. The number of unknowns in the set of linear equations is equal to the number of merates of $U_{fg\ldots}$, which is $(n)^N$. For a unique solution of the set of linear equations, these two numbers must be equal, or

$$(n)^q = (n)^N.$$

Thus, the condition for a unique inverse is

$$q = N, \tag{2.42}$$

or, by Equation (2.41),

$$2N - 2p = N$$

or

$$p = \tfrac{1}{2}N. \tag{2.43}$$

The number of equations is $(n)^N$. Note that Equation (2.43) cannot be satisfied unless N is even: no inverse exists for a holor of odd valence.

For the nilvalent holor u, we have $N = 0$, $n = 1$, and $q = 0$. Thus, Equation (2.43) is satisfied with $p = 0$. The next possibility is $N = 2$, which requires a singly contracted product ($p = 1$) and a bivalent unit holor δ^i_j, as noted in Section 2.08. Other inverses occur for holors of valence $4, 6, 8, \ldots$, as indicated below.

N	p	q	*Number of equations*
0	0	0	1
2	1	2	$(n)^2$
4	2	4	$(n)^4$
6	3	6	$(n)^6$
8	4	8	$(n)^8$
.	.	.	.

One likes to think that holor equations can be manipulated much like ordinary algebraic equations. The lack of an inverse, however, means that, except in special cases, there is no way of solving for one of the factors of a product. For example, take the equation

$$\mathfrak{U}\mathfrak{V} = \mathfrak{W},$$

where $\mathfrak{U}$ and $\mathfrak{V}$ are holors of arbitrary valence. In general, there is no way of solving explicitly for $\mathfrak{V}$.

2.10. Null products

Consider the null product of nilvalent holors,

$$uv = 0. \tag{2.44}$$

If $u \neq 0$, then $v = 0$. This simple result, however, is not always true for contracted products.

Consider first the *uncontracted null product*. With $N = 1$ and $n = 2$, for instance,

$$u^i v_j = 0 \tag{2.45}$$

or

$$u^1 v_1 = 0, \qquad u^2 v_1 = 0,$$
$$u^1 v_2 = 0, \qquad u^2 v_2 = 0.$$

If $u^i \neq 0$, there will be at least one merate, say u^2, that is not zero. This will force both v_1 and v_2 to be zero, so

$$v_j = 0,$$

just as with real numbers, Equation (2.44). Obviously, this conclusion applies to any uncontracted null product, irrespective of the valence and plethos of the factors.

Theorem VI. *For the uncontracted null product of any two holors, one of the holors must be a null holor.*

This conclusion may not be valid, however, if the product is contracted. Take, for example,

$$u^{ij} v_{jk} = 0, \tag{2.46}$$

which is equivalent to the two pairs of linear equations

$$u^{11} v_{11} + u^{12} v_{21} = 0, \qquad u^{11} v_{12} + u^{12} v_{22} = 0,$$
$$u^{21} v_{11} + u^{22} v_{21} = 0, \qquad u^{21} v_{12} + u^{22} v_{22} = 0.$$

If the merates of u^{ij} are linearly independent,[7] the only way to satisfy the set of equations is to make $v_{11} = v_{12} = v_{21} = v_{22} = 0$, or

$$v_{jk} = 0.$$

This is the case where the rank of the augmented matrix is not equal to the rank of the system matrix. The result is the same as for nilvalent holors:

Theorem VII. *For a contracted null product of any two holors, with the merates of one holor linearly independent, the other holor must be zero.*

There remains the contracted product where the merates of one holor are related. In the above example, for instance, the rank of the augmented and system matrices become the same if the determinant of the coefficients is zero:

$$\Delta = \begin{vmatrix} u^{11} & u^{12} \\ u^{21} & u^{22} \end{vmatrix} = u^{11}u^{22} - u^{12}u^{21} = 0.$$

This gives a relation among the merates of u^{ij} for which Equation (2.46) is satisfied without requiring that v_{jk} be zero. Similar results can be obtained for other contracted products.

2.11. Cancellation rule for products

With real numbers, if

$$AB = AC \tag{2.47}$$

and $A \neq 0$, then $B = C$. This is sometimes called the *cancellation rule* for multiplication. Is it applicable also to holors of higher valence?

Let $\mathfrak{A}$, $\mathfrak{B}$, and $\mathfrak{C}$ be holors of any valence and plethos, and let

$$\mathfrak{A}\mathfrak{B} = \mathfrak{A}\mathfrak{C}. \tag{2.48}$$

Subtraction is always possible (Section 2.02), so Equation (2.48) can be written

$$\mathfrak{A}(\mathfrak{B} - \mathfrak{C}) = 0.$$

Thus, we have a null product of holor $\mathfrak{A}$ and holor $\mathfrak{B} - \mathfrak{C}$. According to Theorem VI,

$$\mathfrak{B} - \mathfrak{C} = 0 \quad \text{or} \quad \mathfrak{C} = \mathfrak{B} \tag{2.49}$$

is always true if the product $\mathfrak{A}\mathfrak{B}$ is uncontracted. According to Theorem VII, this conclusion holds also for the contracted product if the merates of $\mathfrak{A}$ are linearly independent. If, however, a relation exists among the merates of $\mathfrak{A}$, then in general $\mathfrak{B} \neq \mathfrak{C}$ and the cancellation rule does not hold.

2.12. Summary

The algebra of holors may be summarized as follows:

(1) Equality of holors occurs only if corresponding merates are equal.

(2) A zero holor is one in which every merate is zero.

(3) Addition of holors is effected by adding corresponding merates.

(4) The uncontracted product is written by placing the symbols in juxtaposition. For example, the product

$$U^{ij} v^k = W^{ijk} \tag{2.13}$$

gives a trivalent holor, each of whose merates is

$$\{W^{ijk}\} = \{U^{ij}\}\{v^k\}.$$

Theorem I. *The algebra of holors, as regards addition, subtraction, and uncontracted multiplication, is identical with the algebra of real numbers.*

(5) *Contracted multiplication* employs the summation convention; for example,

$$u^i v_i = u^1 v_1 + u^2 v_2 + \cdots + u^n v_n. \tag{2.18}$$

Theorem II. *A free literal index may be replaced by any other letter if* (a) *this change is made throughout an equation,* (b) *the new letter does not occur elsewhere in the equation, and* (c) *alphabetical order of indices is preserved.*

Theorem III. *A dummy index may be replaced by any other letter if this letter is not already used as an index in the equation.*

Theorem IV. *The product of any two holors is commutative, associative, and distributive with respect to addition.*

(6) *Index balance* is explained in

Theorem V. *Each term of a holor equation must have the same literal indices in the same positions and in the same order.*

(7) Unit holors are $1, \delta^i_j, \delta^i_j \delta^k_l, \ldots$. Valence must be an even number, and the number of contractions, in Equation (2.28), for instance,

$$u^{ij} \delta^k_i \delta^l_j = u^{kl}, \tag{2.28}$$

must be equal to half the valence of the unit holor.

(8) The inverse of a bivalent holor u^{ij} is U_{hi}, defined by the relation

$$U_{hi} u^{ij} = \delta^j_h. \tag{2.31}$$

Inverses of holors of other valences are not generally possible, though a few cases occur.

(9) Null products are considered:

Theorem VI. *For the uncontracted null product of any two holors, one of the holors must be a null holor.*

Theorem VII. *For a contracted null product of any two holors, with the merates of one holor linearly independent, the other holor must be zero.*

In the general case of a contracted product, however, the fact that the product is zero does not ensure that either factor is zero.

Problems

2.01 Determine the valence of each of the following holors:

$$u^i v_j, u^i v_i, au^2, v_3, x^{ij} y_k, A_{ijk} x^k, BX^3 v^i, (Ca)^2 mX_{ij}.$$

2.02 For plethos 3, expand the following:

(a) $ax^i Z_i$.

(b) $U^{ij} u_i v_j$.

(c) $\dfrac{\partial x^i}{\partial x^{i'}} \dfrac{\partial x^j}{\partial x^{j'}} v^{i'j'}$.

(d) $A_i^{i'} A_j^{j'} A_k^{k'} u_{i'} v_{j'} w_{k'}$ (expand fully for one merate only).

(e) State the valence of each of the above products, and write each product as a single symbol with proper indices.

2.03 Given two bivalent holors

$$u_{ij} = \begin{bmatrix} u_{11} & u_{12} & u_{13} \\ u_{21} & u_{22} & u_{23} \\ u_{31} & u_{32} & u_{33} \end{bmatrix} \quad \text{and} \quad v^{jk} = \begin{bmatrix} v^{11} & v^{12} & v^{13} \\ v^{21} & v^{22} & v^{23} \\ v^{31} & v^{32} & v^{33} \end{bmatrix}.$$

(a) Write the product

$$w_i^k = u_{ij} v^{jk}$$

by the usual rules of matrix products.

(b) Compare your result of (a) with that obtained by direct expansion of $u_{ij} v^{jk}$.

(c) Expand $v^{jk} u_{ij}$. What relation does this product have to (a)?

2.04 Expand R_{kl}^{ij} and R_{ij}^{kl} for plethos 2.

2.05 A bivalent holor A_{ij} is arbitrary except that it is antisymmetric:

$$\{A_{ij}\} = -\{A_{ji}\}.$$

(a) For plethos 3, evaluate the expression

$$A_{ij}[u^i v^j + u^j v^i].$$

(b) If the expression (a) is equal to zero, what does this fact tell about the bracketed quantity?

2.06 A trivalent holor A^{ijk} is completely symmetric in i, j, k. Evaluate

$$A^{ijk}[u_i v_j w_k + u_j v_i w_k + u_i v_k w_j + u_k v_j w_i]$$

in simplest form
(a) In 2-space.
(b) In 3-space.

2.07 If

$$U^{ij}V_{jk} = \begin{bmatrix} 1 & 0 & 0 \\ 0 & 1 & 0 \\ 0 & 0 & 1 \end{bmatrix},$$

determine V_{jk} in terms of U^{ij} for $n=3$.

2.08 Prove that if

$$U^{ij}V_{jk}=0$$

for arbitrary V_{jk}, then

$$U^{ij}=0$$

(a) In 2-space.
(b) In 3-space.

2.09 A holor U^i_j is to act like unity, so

$$U^i_j x^j_k = x^i_k$$

for an arbitrary bivalent holor x^i_k in 3-space. Expand the above equation and express U^i_j in the form of a matrix.

2.10 An arbitrary univalent holor u^j lies in the x^1x^2 plane. Consider the product in 3-space:

$$\gamma_{ijk}u^j v^k = 0,$$

where $\gamma_{123}=\gamma_{231}=\gamma_{312}=+1, \gamma_{213}=\gamma_{132}=\gamma_{321}=-1$, all other gammas zero.
(a) Investigate the properties of u^j if v^k is arbitrary.
(b) Repeat if $v^k=(1,1,1)$.
(c) Repeat if $v^k=(1,1,0)$.

2.11 For plethos 3, evaluate the following:

$$\delta^i_i, \delta^i_j \delta^j_i, \delta^i_j \delta^j_k \delta^k_i.$$

2.12 The inverse B_k^j of a bivalent holor A_j^i is defined by the equation

$$A_j^i B_k^j = \delta_k^i.$$

(a) For plethos 3, expand this equation as a set of simultaneous algebraic equations.
(b) Solve these equations for B_j^i in terms of A_j^i.
(c) Repeat (a) and (b) for plethos 4.

2.13 Given a bivalent holor A_{ij} and its inverse B^{jk}:
(a) Write the equation

$$A_{ij} B^{jk} = \delta_i^k$$

in matrix form for plethos 4.
(b) Write the simultaneous equations corresponding to the above relation.
(c) Obtain an expression for the inverse B^{jk} and evaluate the merates $\{B^{jk}\}$.

2.14 The inverse of a bivalent holor A_{ij} is obtained from the relation

$$A_{ij} B^{jk} = \delta_i^k.$$

Can this same process be employed to define the inverse of an arbitrary univalent holor u_i by writing

$$u_i v^j = \delta_i^j?$$

Expand the equation for plethos 3 and discuss the value of this equation in obtaining v^j as the inverse of u_i.

2.15 Consider the possibility of defining the inverse of an arbitrary trivalent holor A^{ijk} by the equation

$$A^{ijk} B_{jkl} = \delta_l^i.$$

Is the inverse B_{jkl} uniquely determined?

2.16 The product $u^i v_j$ may be written as the matrix

$$u^i v_j = \begin{bmatrix} u^1 v_1 & u^1 v_2 & u^1 v_3 \\ u^2 v_1 & u^2 v_2 & u^2 v_3 \\ u^3 v_1 & u^3 v_2 & u^3 v_3 \end{bmatrix}.$$

If we take the special case of $j = i$, only terms on the principal diagonal are allowable and thus

$$u^i v_i = \begin{bmatrix} u^1 v_1 & 0 & 0 \\ 0 & u^2 v_2 & 0 \\ 0 & 0 & u^3 v_3 \end{bmatrix}.$$

Is this correct? Explain.

2.17 In Problem 2.14, an attempt was made to obtain an inverse of u_i by assuming the relation

$$u_i v^j = \delta_i^j.$$

Another possibility that has index balance is

$$u_i v^i = 1.$$

Here u_i is an arbitrary univalent holor and v^i is supposed to be its inverse. Show that this relation defines (or does not define) a unique inverse v^i.

2.18 In the equation

$$A_{ij} u^j = 0,$$

suppose that A_{ij} is an arbitrary holor with plethos 3. Prove that u^j is, or is not, a null holor.

2.19 In the equation

$$A_{ij} u^j = 0,$$

let A_{ij} be a symmetric but otherwise arbitrary holor of plethos 3. Prove that u^j is, or is not, a null holor.

2.20 In the equation

$$A_{ij} u^j = 0,$$

let A_{ij} be an antisymmetric but otherwise arbitrary holor of plethos 3. Prove that u^j is, or is not, a null holor.

2.21 Consider a null holor (plethos n),

$$A_{ijkl} u^{lm} = 0.$$

If A_{ijkl} is an arbitrary quadrivalent holor, prove that u^{lm} is zero or is not zero.

2.22 In the example immediately following Theorem II,

$$B^i_{kl} C^{lm}_i = D^m_k, \qquad B^i_{pl} C^{lj}_i = D^j_p.$$

Expand these equations and show that they are not identical.

2.23 Using index notation, prove that the product of two transpose matrices $U_T V_T$ is the transpose of the product VU.

3
Gamma products

The previous chapter has included a discussion of uncontracted and contracted holor products. These products give a fairly broad range of possibilities, though not enough to cover all physical applications. Accordingly, in this chapter we consider how increased flexibility can be obtained by use of the γ-products.[1]

Suppose that we are dealing with a given holor but would prefer to have the indices arranged differently. This can be accomplished by multiplying the original holor by another holor γ, tailored to give the desired result.

Or one may wish to associate with the given holor a simpler holor having a lower valence. In general relativity, for instance, Einstein used the Riemann–Christoffel tensor R^{i}_{jkl}, a quadrivalent quantity giving considerable information about curvature of the 4-space. Contraction gives a related holor of valence 2. This holor necessarily gives less information than the original holor; but it may be adequate for some purposes and is much easier to manipulate than the original. Even greater flexibility is obtained by multiplying the original holor by γ^{kl}, whose merates can be chosen at will. Thus valence can be changed by a γ-product.

A third possibility is to obtain a related symmetric or a related antisymmetric holor by multiplying a given holor by the proper γ-holor. Thus, γ-products are useful in converting a given holor into a related holor as follows:

(a) altering positions of indices,
(b) altering valence, or
(c) obtaining a symmetric or an antisymmetric holor.

Still another application of γ-products consists in altering the characteristics of the product of two holors.[2] For example, the ordinary product of two univalent holors is either bivalent or nilvalent, depending on whether it is uncontracted or contracted. These are the only possibilities, according to Chapter 2. But physical applications frequently require a univalent product, as in the cross product of vector analysis. This can be obtained by including a γ, as in

$$w_h = \gamma_{hij} u^i v^j.$$

By choosing a proper γ, one can in this way obtain a product that has almost any desired characteristics.

3.01. Change of index positions

As noted in the previous section, the positions of indices can be altered by a γ-product. For instance, if one is dealing with a holor v^j and wishes to obtain a related univalent holor with subscript instead of superscript, he multiplies by a bivalent holor γ_{ij}:

$$\gamma_{ij} v^j = w_i. \tag{3.01}$$

The merates of γ_{ij} can be chosen at will to obtain the particular w_i that is needed. A possible choice in 2-space is

$$\gamma_{ij} = \delta^{12}_{ij}.$$

Then

$$\delta^{12}_{1j} v^j = \delta^{12}_{12} v^2 = v^2, \qquad \delta^{12}_{2j} v^j = \delta^{12}_{21} v^1 = -v^1$$

or

$$w_i = (w_1, w_2) = (v^2, -v^1).$$

Similarly, given a holor v_j, one can obtain a related holor w^i:

$$\gamma^{ij} v_j = w^i. \tag{3.02}$$

Or with a bivalent holor v^{jk}, γ-products give

$$\gamma_{ij} v^{jk} = w_i^k \tag{3.03}$$

and

$$\gamma_{hijk} v^{jk} = w_{hi}. \tag{3.04}$$

A variety of such changes is listed in Table 3.01.

Table 3.01. *Change of index positions*

Valence	Original holor	General case	Special case
1	v^j	$\gamma_{ij} v^j = w_i$	$\delta_{ij}^{12} v^j = w_i$
	v_j	$\gamma^{ij} v_j = w^i$	$\delta_{12}^{ij} v_j = w^i$
2	v^{jk}	$\gamma_{hijk} v^{jk} = w_{hi}$	
	v^{jk}	$\gamma_{ij} v^{jk} = w_i^k$	$\delta_{ij}^{12} v^{jk} = w_i^k$
	v_{jk}	$\gamma^{hijk} v_{jk} = w^{hi}$	
	v_{jk}	$\gamma^{ij} v_{jk} = w_k^i$	$\delta_{12}^{ij} v_{jk} = w_k^i$
	v_k^j	$\gamma^{ik} v_k^j = w^{ij}$	$\delta_{12}^{ik} v_k^j = w^{ij}$
	v_k^j	$\gamma_{ij} v_k^j = w_{ik}$	$\delta_{ij}^{12} v_k^j = w_{ik}$
3	v^{jkl}	$\gamma_{ij} v^{jkl} = w_i^{kl}$	$\delta_{ij}^{12} v^{jkl} = w_i^{kl}$
	v^{jkl}	$\gamma_{hijk} v^{jkl} = w_{hi}^l$	
	v_{jkl}	$\gamma^{ij} v_{jkl} = w_{kl}^i$	$\delta_{12}^{ij} v_{jkl} = w_{kl}^i$
	v_{jkl}	$\gamma^{hijk} v_{jkl} = w_l^{hi}$	
	v_{kl}^j	$\gamma^{hikl} v_{kl}^j = w^{hij}$	
	v_{kl}^j	$\gamma_{ij} v_{kl}^j = w_{ikl}$	$\delta_{ij}^{12} v_{kl}^j = w_{ikl}$
	v_{kl}^j	$\gamma_i^k v_{kl}^j = w_{il}^j$	
	v_{kl}^j	$\gamma^{ik} v_{kl}^j = w_l^{ij}$	$\delta_{12}^{ik} v_{kl}^j = w_l^{ij}$
	v_l^{jk}	$\gamma^{il} v_l^{jk} = w^{ijk}$	$\delta_{12}^{il} v_l^{jk} = w^{ijk}$
	v_l^{jk}	$\gamma_{hijk} v_l^{jk} = w_{hil}$	
	v_l^{jk}	$\gamma_{ij} v_l^{jk} = w_{il}^k$	$\delta_{ij}^{12} v_l^{jk} = w_{il}^k$
	v_l^{jk}	$\gamma_i^l v_l^{jk} = w_i^{jk}$	

3.02. Change of valence

A change of valence is easily accomplished by a γ-product. The purpose is usually to obtain a related holor that is simpler than the original. Thus, the change ordinarily involves a reduction in valence rather than an increase. For example, obtain a nilvalent holor associated with the univalent holor v^j:

$$v^j = (v^1, v^2, v^3).$$

Valence is reduced to zero by multiplying by a univalent γ:

$$\gamma_j v^j = S, \qquad (3.05)$$

and the merates of γ_j can be chosen at will. Suppose that we take

$$\gamma_j = \tfrac{1}{3}(1,1,1).$$

Then

$$S = \gamma_1 v^1 + \gamma_2 v^2 + \gamma_3 v^3 = \tfrac{1}{3}(v^1 + v^2 + v^3) = v_{av}. \tag{3.05a}$$

Thus, S is the average of the three merates of v^j.

To associate a nilvalent holor S with a bivalent holor u^{ij}, we may write

$$S = \gamma_{ij} u^{ij}. \tag{3.06}$$

Thus, in 2-space,

$$S = \gamma_{11} u^{11} + \gamma_{12} u^{12} + \gamma_{21} u^{21} + \gamma_{22} u^{22},$$

where the γ's may be chosen at will. For instance, if γ_{ij} be taken as the unit matrix, then

$$S = u^{11} + u^{22}.$$

Similarly, for an $n \times n$ matrix,

$$S = u^{11} + u^{22} + \cdots + u^{nn}, \tag{3.07}$$

which is the measure of u^{ij}, called the *trace*.[3]

A bivalent holor can also be reduced to a univalent holor, as in

$$w^k = \gamma_j u^{jk} \tag{3.08}$$

or

$$w_i = \gamma_{ijk} u^{jk}. \tag{3.09}$$

A special case of Equation (3.09) for 3-space is

$$w_i = \delta_{ijk}^{123} u^{jk}, \tag{3.09a}$$

or

$$w_1 = \delta_{123}^{123} u^{23} + \delta_{132}^{123} u^{32} = u^{23} - u^{32},$$
$$w_2 = u^{31} - u^{13},$$
$$w_3 = u^{12} - u^{21}.$$

In general, the related holor w_i contains less information than the original holor u^{jk}. In 3-space, the matrix u^{jk} has nine merates while w_i has only three. But for the special case of an antisymmetric matrix u^{jk}, there are only three independent merates in the matrix. This is an interesting case where the reduction of valence from 2 to 1 causes no loss of information. Instead of

$$u^{jk} = \begin{bmatrix} 0 & u^{12} & u^{13} \\ -u^{12} & 0 & u^{23} \\ -u^{13} & -u^{23} & 0 \end{bmatrix},$$

we have, by Equation (3.09a), the related univalent holor

$$w_i = (u^{23} - u^{32}, u^{31} - u^{13}, u^{12} - u^{21})$$

or

$$w_i = 2(u^{23}, u^{13}, u^{12}). \tag{3.09b}$$

Starting with any valence N_1, we can obtain a related holor having any valence N_2 by use of a γ-product. Some examples are listed in Table 3.02. Let

N_1 = valence of original holor,

N_2 = valence desired for related holor,

N_3 = number of indices for γ,

p = number of contractions in product.

Then

$$N_1 + N_3 - 2p = N_2. \tag{3.10}$$

For example, if $N_1 = 3$, $N_2 = 1$, and $p = 2$, Equation (3.10) shows that the valence of γ must be 2, or

$$\gamma_{jk} u^{jkl} = w^l. \tag{3.11}$$

3.03. Symmetric and antisymmetric matrices

A third application of the γ-product transforms a normal matrix into a symmetric or an antisymmetric one. Valence remains the same, and position of indices may be the same. Suppose, for instance, that we have a matrix v^{ij} with arbitrary merates and wish to obtain a related antisymmetric matrix. Consider

$$w^{kl} = \gamma_{ij}^{kl} v^{ij}, \tag{3.12}$$

with

$$\{w^{kl}\} = -\{w^{lk}\}.$$

Table 3.02. *Change of valence*

N_1 (Valence of H)	N_2 (Valence of w)	N_3 (Valence of γ)	p (Number of contractions)
1	0	1	1
	1	2	1
	2	1	0
	2	3	1
2	0	2	2
	1	1	1
	1	3	2
	2	2	1
	2	4	2
	3	1	0
	3	3	1
3	0	3	3
	1	2	2
	1	4	3
	2	1	1
	2	3	2
	3	2	1
	3	4	2
	4	1	0
	4	3	1
4	0	4	4
	1	3	3
	2	2	2
	2	4	3
	3	1	1
	3	3	2
	4	2	1
	4	4	2
	5	1	0
	5	3	1

Evidently, the antisymmetry of w^{kl} requires antisymmetry of γ_{ij}^{kl} in the upper indices:

$$\{\gamma_{ij}^{kl}\} = -\{\gamma_{ij}^{lk}\}.$$

For 3-space, let

$$\gamma_{ij}^{12} = -\gamma_{ij}^{21} = \frac{1}{2}\begin{bmatrix} 0 & 1 & 0 \\ -1 & 0 & 0 \\ 0 & 0 & 0 \end{bmatrix},$$

$$\gamma_{ij}^{13} = -\gamma_{ij}^{31} = \frac{1}{2}\begin{bmatrix} 0 & 0 & 1 \\ 0 & 0 & 0 \\ -1 & 0 & 0 \end{bmatrix},$$

$$\gamma_{ij}^{23} = -\gamma_{ij}^{32} = \frac{1}{2}\begin{bmatrix} 0 & 0 & 0 \\ 0 & 0 & 1 \\ 0 & -1 & 0 \end{bmatrix}.$$

Applying these matrices to Equation (3.12), we obtain

$$w^{12} = v^{12} - v^{21}, \qquad w^{13} = v^{13} - v^{31}, \qquad w^{23} = v^{23} - v^{32},$$

so the new antisymmetric matrix is

$$w^{kl} = \frac{1}{2}\begin{bmatrix} 0 & (v^{12} - v^{21}) & (v^{13} - v^{31}) \\ -(v^{12} - v^{21}) & 0 & (v^{23} - v^{32}) \\ -(v^{13} - v^{31}) & -(v^{23} - v^{32}) & 0 \end{bmatrix}. \qquad (3.13)$$

This result is identical (except for a factor of 2) with that obtained from Equation (3.12) with

$$\gamma_{ij}^{kl} = \delta_{ij}^{kl}.$$

It is also the same as the antisymmetric formulation $v^{[ij]}$ of Schouten[4] (see Section 1.06):

$$\{v^{[ij]}\} = \tfrac{1}{2}[\{v^{ij}\} - \{v^{ji}\}].$$

Now suppose that w^{kl} is to be symmetric instead of antisymmetric. As before,

$$w^{kl} = \gamma_{ij}^{kl} v^{ij}, \qquad (3.12)$$

but now

$$\{w^{kl}\} = \{w^{lk}\} \quad \text{and} \quad \{\gamma_{ij}^{kl}\} = \{\gamma_{ij}^{lk}\}.$$

Evidently symmetry is obtained by taking the γ-matrices

$$\gamma_{ij}^{11} = \begin{bmatrix} 1 & 0 & 0 \\ 0 & 0 & 0 \\ 0 & 0 & 0 \end{bmatrix}, \qquad \gamma_{ij}^{22} = \begin{bmatrix} 0 & 0 & 0 \\ 0 & 1 & 0 \\ 0 & 0 & 0 \end{bmatrix}, \qquad \gamma_{ij}^{33} = \begin{bmatrix} 0 & 0 & 0 \\ 0 & 0 & 0 \\ 0 & 0 & 1 \end{bmatrix},$$

$$\gamma_{ij}^{12}=\gamma_{ij}^{21}=\frac{1}{2}\begin{bmatrix}0&1&0\\1&0&0\\0&0&0\end{bmatrix},$$

$$\gamma_{ij}^{13}=\gamma_{ij}^{31}=\frac{1}{2}\begin{bmatrix}0&0&1\\0&0&0\\1&0&0\end{bmatrix},$$

$$\gamma_{ij}^{23}=\gamma_{ij}^{32}=\frac{1}{2}\begin{bmatrix}0&0&0\\0&0&1\\0&1&0\end{bmatrix}.$$

Then

$$w^{ii}=v^{ii},$$
$$w^{12}=\tfrac{1}{2}(v^{12}+v^{21}),$$
$$w^{13}=\tfrac{1}{2}(v^{13}+v^{31}),$$
$$w^{23}=\tfrac{1}{2}(v^{23}+v^{32}),$$

and the symmetric matrix is

$$w^{kl}=\begin{bmatrix}v^{11}&\tfrac{1}{2}(v^{12}+v^{21})&\tfrac{1}{2}(v^{13}+v^{31})\\\tfrac{1}{2}(v^{12}+v^{21})&v^{22}&\tfrac{1}{2}(v^{23}+v^{32})\\\tfrac{1}{2}(v^{13}+v^{31})&\tfrac{1}{2}(v^{23}+v^{32})&v^{33}\end{bmatrix}. \qquad (3.14)$$

This matrix is the same as $v^{(ij)}$ obtained in Section 1.06:

$$\{v^{(ij)}\}=\tfrac{1}{2}[\{v^{ij}\}+\{v^{ji}\}].$$

Note that the sum of Equations (3.13) and (3.14) is the original matrix v^{ij}. This process is, of course, not limited to bivalent holors with plethos 3 but may be extended to higher values of valence and plethos.

3.04. Products of univalent holors

Consider the product of two holors, each of which is univalent. According to Chapter 2, two possibilities exist: Either the product is bivalent,

$$u^i v_j = w^i_j,$$

or it is nilvalent,

$$u^i v_i = S.$$

These are the only products of univalent quantities that are given in most treatments of holors. Yet experience shows that in many cases one needs a product that is univalent.

The products introduced in Chapter 2 are too restrictive. A univalent product can be obtained, however, by introducing an arbitrary trivalent holor γ_i^{hj}. Then

$$\gamma_i^{hj} u^i v_j = w^h.$$

One may, in fact, introduce a new symbolism[1] and write

$$u^i \odot v_j = \gamma_i^j u^i v_j = S, \tag{3.15}$$

$$u^i \otimes v_j = \gamma_i^{hj} u^i v_j = w^h, \tag{3.16}$$

$$u^i \bigcirc v_j = \gamma_i^{klj} u^i v_j = W^{kl}, \tag{3.17}$$

depending on whether a product of valence 0, 1, or 2 is wanted. Flexibility is obtained by making the γ's arbitrary.

The symbolism on the left, with the signs $\odot$, $\otimes$, and $\bigcirc$, indicates that the three are different kinds of γ-products of the same holors u^i and v_j. This symbolism was suggested by the very successful nomenclature of vector analysis. It has the disadvantage, however, that index balance is not preserved. So if one prefers, the new shorthand notation may be ignored, the γ's written as in Equation (3.15)–(3.17), and index balance preserved. In either case, the product is not completely specified unless all the merates of the γ's are given.

As an example, take the γ-product

$$u^i \odot v_j = \gamma_i^j u^i v_j = S$$

with

$$\gamma_i^j = \delta_i^j = \begin{bmatrix} 1 & 0 & 0 & \cdots & 0 \\ 0 & 1 & 0 & \cdots & 0 \\ \cdot & \cdot & \cdot & \cdots & \cdot \\ 0 & 0 & 0 & \cdots & 1 \end{bmatrix}.$$

Therefore,

$$u^i \odot v_j = \delta_i^j u^i v_j = u^i v_i = u^1 v_1 + u^2 v_2 + \cdots + u^n v_n.$$

In vector analysis,[2] this particular product, in rectangular coordinates and 3-space, is called the dot product or scalar product and is written

$$\mathbf{u} \cdot \mathbf{v} = u^1 v_1 + u^2 v_2 + u^3 v_3.$$

On the other hand, a different choice of γ_i^j will give a different product, though always a nilvalent one. For instance, let $n = 2$ and

$$\gamma_i^j = \begin{bmatrix} 0 & 1 \\ -1 & 0 \end{bmatrix}.$$

Then

$$u^i \odot v_j = \gamma_1^1 u^1 v_1 + \gamma_1^2 u^1 v_2 + \gamma_2^1 u^2 v_1 + \gamma_2^2 u^2 v_2$$
$$= u^2 v_1 - u^1 v_2.$$

As an example of a univalent product with plethos 2, consider

$$u^i \otimes v^j = \gamma_{ij}^h u^i v^j = w^h,$$

with

$$\gamma_{ij}^1 = \begin{bmatrix} 1 & 0 \\ 0 & -1 \end{bmatrix}, \qquad \gamma_{ij}^2 = \begin{bmatrix} 0 & 1 \\ 1 & 0 \end{bmatrix}.$$

Expanding, we obtain

$$w^1 = \gamma_{11}^1 u^1 v^1 + \gamma_{12}^1 u^1 v^2 + \gamma_{21}^1 u^2 v^1 + \gamma_{22}^1 u^2 v^2$$
$$= u^1 v^1 - u^2 v^2,$$
$$w^2 = u^1 v^2 + u^2 v^1.$$

But this is merely the product of two complex numbers

$$u^i = u^1 + iu^2, \qquad v^j = v^1 + iv^2,$$

or

$$u^i \otimes v^j = (u^1 v^1 - u^2 v^2) + i(u^1 v^2 + u^2 v^1).$$

The products of Equations (3.15)–(3.17) may of course be extended to other arrangements of indices. If both holors have superscripts,

$$u^i \odot v^j = \gamma_{ij} u^i v^j = S, \tag{3.15a}$$

$$u^i \otimes v^j = \gamma_{ij}^h u^i v^j = w^h, \tag{3.16a}$$

$$u^i \bigcirc v^j = \gamma_{ij}^{kl} u^i v^j = W^{kl}. \tag{3.17a}$$

If both holors have subscripts,

$$u_i \odot v_j = \gamma^{ij} u_i v_j = S, \tag{3.15b}$$

$$u_i \otimes v_j = \gamma_h^{ij} u_i v_j = w_h, \tag{3.16b}$$

$$u_i \bigcirc v_j = \gamma_{kl}^{ij} u_i v_j = W_{kl}. \tag{3.17b}$$

A γ-product of two univalent holors may be commutative or noncommutative, associative or nonassociative, depending on the choice of the γ's. Let us be a little more specific. The symbols u^i and v^j are always commutable in expressions having index balance, so

$$\gamma_{ij}^h u^i v^j = \gamma_{ij}^h v^j u^i = w^h.$$

But such results do not necessarily apply to the shorthand expressions. In general,

$$u^i \otimes v^j \neq v^j \otimes u^i.$$

All products, however, are distributive with respect to addition. Take, for example,

$$u^i \otimes (v_j + w_j) = \gamma_i^{hj} u^i (v_j + w_j).$$

Expansion gives

$$\begin{aligned} u^i \otimes (v_j + w_j) = {} & \gamma_1^{h1} u^1 (v_1 + w_1) + \gamma_2^{h1} u^2 (v_1 + w_1) + \cdots \\ & + \gamma_n^{h1} u^n (v_1 + w_1) + \gamma_1^{h2} u^1 (v_2 + w_2) + \gamma_2^{h2} u^2 (v_2 + w_2) + \cdots \\ & + \gamma_n^{h2} u^n (v_2 + w_2) + \cdots + \gamma_n^{hn} u^n (v_n + w_n). \end{aligned}$$

Since u^1, v_1, w_1, and so on, are ordinary real numbers, the distributive relation holds for them, and

$$\begin{aligned} u^i \otimes (v_j + w_j) &= (\gamma_1^{h1} u^1 v_1 + \gamma_2^{h1} u^2 v_1 + \cdots + \gamma_n^{hn} u^n v_n) \\ &\quad + (\gamma_1^{h1} u^1 w_1 + \gamma_2^{h1} u^2 w_1 + \cdots + \gamma_n^{hn} u^n w_n) \\ &= u^i \otimes v_j + u^i \otimes w_j. \end{aligned} \tag{3.18}$$

Obviously, the same reasoning applies to the other products. Thus, the distributive relation holds for γ-products.

3.05. Nilvalent products

Consider the nilvalent γ-product of two univalent holors:

$$u^i \odot v^j = \gamma_{ij} u^i v^j = S. \tag{3.15a}$$

If γ_{ij} is so chosen that it is symmetric,

$$\{\gamma_{ij}\}=\{\gamma_{ji}\},$$

then the product is evidently commutative:

$$u^i \odot v^j = v^j \odot u^i. \tag{3.19}$$

Otherwise, the product is noncommutative.

Theorem VIII. *The necessary and sufficient condition that a nilvalent γ-product of two univalent holors be commutative is that the γ-matrix be symmetric.* Since this product is not a univalent holor, the question of associativity does not arise.

3.06. Univalent products

Consider the univalent γ-product of two univalent holors[2]:

$$u^i \otimes v^j = \gamma_{hij} u^i v^j = w_h. \tag{3.20}$$

For plethos 3, we may write the trivalent holor γ_{hij} as

$$\gamma_{1ij} = \begin{bmatrix} \gamma_{111} & \gamma_{112} & \gamma_{113} \\ \gamma_{121} & \gamma_{122} & \gamma_{123} \\ \gamma_{131} & \gamma_{132} & \gamma_{133} \end{bmatrix},$$

$$\gamma_{2ij} = \begin{bmatrix} \gamma_{211} & \gamma_{212} & \gamma_{213} \\ \gamma_{221} & \gamma_{222} & \gamma_{223} \\ \gamma_{231} & \gamma_{232} & \gamma_{233} \end{bmatrix},$$

$$\gamma_{3ij} = \begin{bmatrix} \gamma_{311} & \gamma_{312} & \gamma_{313} \\ \gamma_{321} & \gamma_{322} & \gamma_{323} \\ \gamma_{331} & \gamma_{332} & \gamma_{333} \end{bmatrix}.$$

For commutative multiplication,

$$\begin{aligned} w_h &= u^i \otimes v^j = \gamma_{hij} u^i v^j \\ &= v^j \otimes u^i = \gamma_{hji} v^j u^i = \gamma_{hji} u^i v^j. \end{aligned}$$

Evidently this relation is satisfied if and only if γ_{hji} is symmetric in i, j. For plethos 3, this means that each of the above matrices must be symmetric. This conclusion can obviously be extended to any plethos.

Theorem IX. *The necessary and sufficient condition that the univalent γ-product of two univalent holors be commutative is that the γ-holor be symmetric in the indices of the univalent holors.*

There is also the question of associativity. If the univalent γ-product is to be associative,

$$(u^i \otimes v^j) \otimes s^l = u^i \otimes (v^j \otimes s^l). \tag{3.21}$$

Let

$$u^i \otimes v^j = \gamma_{ij}^{h} u^i v^j, \qquad v^j \otimes s^l = \gamma_{jl}^{g} v^j s^l.$$

Substitution into Equation (3.21) gives

$$\gamma_{hl}^{f} (\gamma_{ij}^{h} u^i v^j) s^l = \gamma_{ig}^{f} (\gamma_{jl}^{g} v^j s^l) u^i.$$

Because of commutivity in ordinary holor multiplication, this equation may be written

$$\gamma_{hl}^{f} \gamma_{ij}^{h} u^i v^j s^l = \gamma_{ig}^{f} \gamma_{jl}^{g} u^i v^j s^l.$$

Since $u^i v^j s^l$ is completely arbitrary, the coefficients must be equal, or

$$\gamma_{hl}^{f} \gamma_{ij}^{h} = \gamma_{ig}^{f} \gamma_{jl}^{g}. \tag{3.22}$$

Theorem X. *The necessary and sufficient condition that a univalent γ-product of univalent holors be associative is that* Equation (3.22) *be satisfied.*

Equation (3.22) represents $(n)^4$ relations among the γ's, where $n =$ plethos. All of these relations must be satisfied to obtain an associative product. As an example, consider the product

$$w^h = u^i \otimes v^j = \gamma_{ij}^{h} u^i v^j, \tag{3.23}$$

where γ_{ij}^{h} is specified by the matrices

$$\gamma_{ij}^{1} = \begin{bmatrix} 0 & 0 & 0 \\ 0 & 0 & 1 \\ 0 & -1 & 0 \end{bmatrix}, \qquad \gamma_{ij}^{2} = \begin{bmatrix} 0 & 0 & -1 \\ 0 & 0 & 0 \\ 1 & 0 & 0 \end{bmatrix}, \qquad \gamma_{ij}^{3} = \begin{bmatrix} 0 & 1 & 0 \\ -1 & 0 & 0 \\ 0 & 0 & 0 \end{bmatrix}.$$

Since γ_{ij}^{h} is not symmetric in i, j, the product is not commutative.

To show that it is also not associative, we need exhibit only one case in which Equation (3.22) is not satisfied. For instance, if $f = i = 1$ and $j = l = 3$, associativity requires that

$$\gamma_{13}^{1}\gamma_{13}^{1} + \gamma_{23}^{1}\gamma_{13}^{2} + \gamma_{33}^{1}\gamma_{13}^{3} = \gamma_{11}^{1}\gamma_{33}^{1} + \gamma_{12}^{1}\gamma_{33}^{2} + \gamma_{13}^{1}\gamma_{33}^{3}$$

or

$$-1 = \gamma_{23}^{1}\gamma_{13}^{2} = 0.$$

Equation (3.22) is not satisfied, so the product is nonassociative.

The example is, of course, a familiar one. Expansion of Equation (3.23) gives

$$w^h = \{(u^2v^3 - u^3v^2), (u^3v^1 - u^1v^3), (u^1v^2 - u^2v^1)\}, \tag{3.23a}$$

which is the cross product of vector analysis:

$$\mathbf{w} = \mathbf{u} \times \mathbf{v}.$$

It is well known that this product is noncommutative,

$$\mathbf{u} \times \mathbf{v} \neq \mathbf{v} \times \mathbf{u},$$

and nonassociative,

$$\mathbf{u} \times (\mathbf{v} \times \mathbf{s}) \neq (\mathbf{u} \times \mathbf{v}) \times \mathbf{s}.$$

3.07. Alternating currents

As an example of a univalent product, consider the determination of power in ac circuits.[1] The original method of handling ac circuits was to express voltage and current as functions of time. This led to very complicated expressions. Kennelly[5] realized that this complexity was unnecessary in the usual steady-state case with sinusoidal time variation. He found that by representing currents and voltages by complex numbers, he could reduce the equations of steady-state ac circuit theory to essentially the simplicity of the dc case. His complex number representation is widely used today.[6]

Two distinct products occur in circuit theory: the product of current and impedance, which gives voltage, and the product of current and voltage, which gives power. The former product works perfectly with complex numbers, but the latter requires *ad hoc* modification. The determination of power, therefore, has always constituted a logical flaw in the complex number representation of circuits.[7]

Consider now a new representation[1] by holors. Let voltage and current be specified by univalent holors of plethos 2.

$$V_i = (V_1, V_2), \qquad I^j = (I^1, I^2). \tag{3.24}$$

The relation between them is given by a generalized Ohm equation,

$$V_i = Z_{ij} I^j, \tag{3.25}$$

where the impedance Z_{ij} is a bivalent holor.

For pure resistance,

$$Z_{ij} = R\begin{bmatrix} 1 & 0 \\ 0 & 1 \end{bmatrix}.$$

For inductance and resistance,

$$Z_{ij} = \begin{bmatrix} R & \omega L \\ -\omega L & R \end{bmatrix}.$$

For capacitance and resistance,

$$Z_{ij} = \begin{bmatrix} R & -1/\omega C \\ 1/\omega C & R \end{bmatrix}.$$

In general,

$$Z_{ij} = \begin{bmatrix} R & X \\ -X & R \end{bmatrix}, \tag{3.26}$$

where the reactance X is

$$X = \omega L - 1/\omega C. \tag{3.27}$$

This new representation emphasizes the physical difference between impedance and current by making the former a bivalent holor.

Equation (3.25) may be solved for current by writing

$$I^j = Y^{ji} V_i, \tag{3.28}$$

where the admittance Y^{ji} is the inverse of the impedance:

$$Y^{ij} Z_{jk} = \delta^i_k. \tag{3.29}$$

Instantaneous power $P(t)$ may be written

$$P(t) = |V|\,|I|\,[\cos(\alpha - \beta) + \cos 2\omega t \cos(\alpha + \beta) + \sin 2\omega t \sin(\alpha + \beta)], \tag{3.30}$$

where α and β are phase angles of voltage and current with respect to an arbitrary reference. The time-invariant part, usually called average power, may be represented by P^0, while the two double-frequency parts may be represented by P^1 and P^2:

$$P^0 = |V|\,|I| \cos(\alpha - \beta), \qquad P^1 = |V|\,|I| \cos(\alpha + \beta),$$

$$P^2 = |V|\,|I| \sin(\alpha + \beta). \tag{3.31}$$

These three quantities are taken as merates of a new univalent holor P^{κ}:

$$P^{\kappa}=(P^0,P^1,P^2). \tag{3.32}$$

In analogy with dc power, this power holor should be equal to some kind of a product of voltage and current. Let

$$P^{\kappa}=V_i \otimes I^j=\gamma_j^{\kappa i} V_i I^j, \tag{3.33}$$

where $i,j=1,2$; $\kappa=0,1,2$.

It is found that the trivalent holor $\gamma_j^{\kappa i}$ is

$$\gamma_j^{0i}=\begin{bmatrix}1 & 0\\ 0 & 1\end{bmatrix}, \quad \gamma_j^{1i}=\begin{bmatrix}1 & 0\\ 0 & -1\end{bmatrix}, \quad \gamma_j^{2i}=\begin{bmatrix}0 & 1\\ 1 & 0\end{bmatrix}. \tag{3.34}$$

Thus, we have obtained a direct analogy between the equations for dc circuits,

$$V=RI \quad \text{and} \quad P=VI,$$

and the equations for ac circuits,

$$V_i=Z_{ij}I^j \quad \text{and} \quad P^{\kappa}=V_i \otimes I^j.$$

3.08. Bivalent products

In previous sections, we have considered the nilvalent and univalent γ-products, denoted by $\odot$ and $\otimes$. Now take a bivalent product of two univalent holors:

$$u^i \bigcirc v^j=\gamma_{ij}^{kl}u^i v^j=W^{kl}. \tag{3.35}$$

Here γ_{ij}^{kl} is an arbitrary quadrivalent holor. For plethos 2, this holor may be represented by four matrices of the form

$$\gamma_{ij}^{kl}=\begin{bmatrix}\gamma_{11}^{kl} & \gamma_{12}^{kl}\\ \gamma_{21}^{kl} & \gamma_{22}^{kl}\end{bmatrix},$$

and for plethos 3 by nine matrices of the form

$$\gamma_{ij}^{kl}=\begin{bmatrix}\gamma_{11}^{kl} & \gamma_{12}^{kl} & \gamma_{13}^{kl}\\ \gamma_{21}^{kl} & \gamma_{22}^{kl} & \gamma_{23}^{kl}\\ \gamma_{31}^{kl} & \gamma_{32}^{kl} & \gamma_{33}^{kl}\end{bmatrix}.$$

Though the bivalent product is distributive with respect to addition, it may or may not be commutative and associative. Evidently the product is commutative if and only if

$$\{\gamma_{ij}^{kl}\}=\{\gamma_{ji}^{kl}\}.$$

Theorem XI. *The necessary and sufficient condition that the bivalent product of two univalent holors,* Equation (3.35), *be commutative is that* γ_{ij}^{kl} *be symmetric in* i, j.

A simple example of the bivalent γ-product is obtained by letting γ_{ij}^{kl} be a Kronecker delta,

$$\gamma_{ij}^{kl} = \delta_{ij}^{kl}. \tag{3.36}$$

Then, for plethos 2,

$$\gamma_{12}^{kl} = \delta_{12}^{kl} = \begin{bmatrix} 0 & 1 \\ -1 & 0 \end{bmatrix},$$

$$\gamma_{21}^{kl} = \delta_{21}^{kl} = \begin{bmatrix} 0 & -1 \\ 1 & 0 \end{bmatrix},$$

$$\gamma_{11}^{kl} = \delta_{22}^{kl} = 0.$$

Thus, Equation (3.35) becomes

$$\begin{aligned} u^i \bigcirc v^j &= \gamma_{11}^{kl} u^1 v^1 + \gamma_{12}^{kl} u^1 v^2 + \gamma_{21}^{kl} u^2 v^1 + \gamma_{22}^{kl} u^2 v^2 \\ &= W^{kl}, \end{aligned}$$

or

$$W^{kl} = (u^1 v^2 - u^2 v^1) \begin{bmatrix} 0 & 1 \\ -1 & 0 \end{bmatrix}. \tag{3.37}$$

Since γ_{ij}^{kl} is not symmetric in i, j, the product is not commutative. The product is not univalent, so the question of associativity does not arise.

3.09. Other products

We have considered a variety of γ-products of univalent holors, including the three types designated by $\odot$, $\otimes$, and $\bigcirc$ in Section 3.04. Products of higher valence can, of course, be obtained in the same way. A trivalent holor, for instance, can be produced as a γ-product of two univalent holors:

$$\gamma_{ij}^{klm} u^i v^j = W^{klm}, \tag{3.38}$$

and holors of even higher valence can be obtained similarly.

Also, multiplier and multiplicand need not be univalent. Products of a univalent holor and a bivalent holor may be written

$$u^i \odot V^{jk} = \gamma_{ijk} u^i V^{jk} = S, \tag{3.39}$$

$$u^i \otimes V^{jk} = \gamma_{ijk}^{l} u^i V^{jk} = W^l, \tag{3.40}$$

$$u^i \bigcirc V^{jk} = \gamma_{ijk}^{lm} u^i V^{jk} = W^{lm}. \tag{3.41}$$

The symbols $\odot$, $\otimes$, and $\bigcirc$ indicate a product that is nilvalent, univalent, and bivalent, respectively, as in Section 3.04.

For example, take the ordinary matrix product of a univalent and a bivalent holor. If all indices are superscripts,

$$W^l = u^i \otimes V^{jk} = \gamma_{ijk}^l u^i V^{jk}.$$

For $n=2$,

$$W^l = (u^1 V^{11} + u^2 V^{21}, u^1 V^{12} + u^2 V^{22})$$

from matrix theory.[8] This result is obtained from Equation (3.40) if γ_{ijk}^l is written

$$\begin{aligned} \gamma_{1jk}^1 &= \begin{bmatrix} 1 & 0 \\ 0 & 0 \end{bmatrix}, \qquad \gamma_{2jk}^1 = \begin{bmatrix} 0 & 0 \\ 1 & 0 \end{bmatrix}, \\ \gamma_{1jk}^2 &= \begin{bmatrix} 0 & 1 \\ 0 & 0 \end{bmatrix}, \qquad \gamma_{2jk}^2 = \begin{bmatrix} 0 & 0 \\ 0 & 1 \end{bmatrix}. \end{aligned} \tag{3.42}$$

Similar results for higher plethos are easily formulated.

Also important are the γ-products of bivalent holors. Keeping the same meaning for the symbols $\odot$, $\otimes$, and $\bigcirc$, we write

$$U^{ij} \odot V^{kl} = \gamma_{ijkl} U^{ij} V^{kl} = S, \tag{3.43}$$

$$U^{ij} \otimes V^{kl} = \gamma_{ijkl}^m U^{ij} V^{kl} = w^m, \tag{3.44}$$

$$U^{ij} \bigcirc V^{kl} = \gamma_{ijkl}^{mn} U^{ij} V^{kl} = W^{mn}. \tag{3.45}$$

For the particular case where the product of two bivalent holors is an ordinary matrix product,[8]

$$\begin{aligned} W^{mn} &= U^{ij} \bigcirc V^{kl} = \gamma_{ijkl}^{mn} U^{ij} V^{kl} \\ &= \begin{bmatrix} U^{11}V^{11} + U^{12}V^{21} & U^{11}V^{12} + U^{12}V^{22} \\ U^{21}V^{11} + U^{22}V^{21} & U^{21}V^{12} + U^{22}V^{22} \end{bmatrix}. \end{aligned} \tag{3.46}$$

This result is obtained if

$$\gamma_{1111}^{11} = \gamma_{1221}^{11} = \gamma_{1112}^{12} = \gamma_{1222}^{12} = \gamma_{2111}^{21} = \gamma_{2221}^{21} = \gamma_{2112}^{22} = \gamma_{2222}^{22} = +1,$$

all others being zero.

3.10. Summary

This chapter has sketched some of the possibilities of γ-products. Given a single holor, we may multiply it by the proper γ-holor to achieve the following:
(a) change the positions of indices,
(b) change the valence, or
(c) obtain a symmetric or an antisymmetric holor.

As an example of (a), suppose that one desires a holor w^{ij} with two superscripts instead of a mixed holor v^j_k. Obviously,

$$\gamma^{ik} v^j_k = w^{ij}.$$

As an example of (b), a nilvalent holor S may be associated with a bivalent holor u^{ij}:

$$\gamma_{ij} u^{ij} = S.$$

For (c), the antisymmetric $u^{[ij]}$ and the symmetric $u^{(ij)}$ are easily expressible as γ-products.

The new holors obtained in this way are distinct from the original holors but are related to them. Perhaps the most useful of these operations is the simplification obtained by reduction in valence (b). The merates of the γ-holor are arbitrary and may be chosen to give a wide variety of characteristics to fit various physical applications.

The remainder of the chapter is devoted to γ-products with more than one given holor. The three types are:
(1) Nilvalent product:

$$u^i \odot v^j = \gamma_{ij} u^i v^j = S. \tag{3.15a}$$

(2) Univalent product:

$$u^i \otimes v^j = \gamma^h_{ij} u^i v^j = w^h. \tag{3.16a}$$

(3) Bivalent product:

$$u^i \bigcirc v^j = \gamma^{kl}_{ij} u^i v^j = W^{kl}. \tag{3.17a}$$

Similar results can be obtained with factors of higher valence; and the γ's can be tailored to fit any desired geometric or physical situation.

Problems

3.01 Discuss the associative property of $Au^i v^j$. Write the products $(Au^i)v^j$ and $A(u^i v^j)$ and compare.

3.02

(a) Given the product $u^i \otimes v^j$ with plethos n, determine the necessary and sufficient condition that this product be commutative.

(b) If $n=3$ and

$$\gamma_{ij}^k = \begin{bmatrix} 0 & \gamma_{12}^k & \gamma_{13}^k \\ -\gamma_{12}^k & 0 & \gamma_{23}^k \\ -\gamma_{13}^k & -\gamma_{23}^k & 0 \end{bmatrix},$$

where $\gamma_{12}^k \neq \gamma_{13}^k \neq \gamma_{23}^k$, write $u^i \otimes v^j = x^k$ and $v^k \otimes u^i = y^k$ and compare.

(c) Is the algebra (b) associative?

3.03 In a new application of holors for $n=2$, the quantities are to behave like ordinary complex numbers except that multiplication must be

$$u^i \otimes v^j = (u^1 v^1, u^2 v^2).$$

(a) Obtain the necessary γ_{ij}^k matrix.

(b) Is the algebra associative and is it commutative?

3.04 Cayley[9] classified algebras of holors ($n=2$) under the following seven classes:

$$\text{(I)} \quad u^i \otimes v^j = (u^1 v^1, u^2 v^2),$$

$$\text{(II)} \quad u^i \otimes v^j = (u^1 v^1, u^1 v^2 + u^2 v^1),$$

$$\text{(III)} \quad u^i \otimes v^j = (u^1 v^1, 0),$$

$$\text{(IV)} \quad u^i \otimes v^j = (u^2 v^2, 0),$$

$$\text{(V)} \quad u^i \otimes v^j = (u^1 v^1, u^2 v^1),$$

$$\text{(VI)} \quad u^i \otimes v^j = (u^1 v^1, u^1 v^2),$$

$$\text{(VII)} \quad u^i \otimes v^j = (0, 0).$$

(a) Write the matrices γ_{ij}^k for these seven algebras.

(b) Which algebras are commutative?

(c) Which algebras are associative?

(d) Which allow a zero product when neither u^i nor v^i is zero?

(e) These are elementary products. Other products can be obtained by linear combinations such as

$$\gamma_{ij}^k = (A)_1 (\gamma_{ij}^k)_1 + (A)_2 (\gamma_{ij}^k)_2,$$

where $(A)_1$ and $(A)_2$ are constants and $(\gamma_{ij}^k)_1$ and $(\gamma_{ij}^k)_2$ are obtained from the elementary Cayley products. Show how ordinary complex algebra is obtained in this way.

3.05 E. Study[10] classified algebras of univalent holors with $n=3$ as

(I) $u^i \otimes v^j = (u^1v^1, u^2v^2, u^3v^3)$,

(II) $u^i \otimes v^j = (u^1v^1, u^1v^2+u^2v^1, u^3v^3)$,

(III) $u^i \otimes v^j = (u^1v^1, u^1v^2+u^2v^1, u^2v^2+u^1v^3+u^3v^1)$,

(IV) $u^i \otimes v^j = (u^1v^1+u^2v^2, u^1v^2+u^2v^1, u^1v^3+u^2v^3+u^3v^1-u^3v^2)$,

(V) $u^i \otimes v^j = (u^1v^1, u^1v^2+u^2v^1, u^1v^3+u^3v^1)$.

Repeat (a)–(d) of Problem 3.04 but for Study's algebras.

(e) Determine whether the vector product of ordinary vector algebra can be obtained by a linear combination of Study's products.

3.06 A possible algebra for $n=4$ employs the product

$$u^i \otimes v^j = w^k,$$

with

$$\gamma_{ij}^{1} = \begin{bmatrix} 1 & 0 & 0 & 0 \\ 0 & 0 & 0 & 0 \\ 0 & 0 & 0 & 0 \\ 0 & 0 & 0 & 0 \end{bmatrix}, \qquad \gamma_{ij}^{2} = \begin{bmatrix} 0 & 1 & 0 & 0 \\ 1 & 0 & 0 & 0 \\ 0 & 0 & 0 & 0 \\ 0 & 0 & 0 & 0 \end{bmatrix},$$

$$\gamma_{ij}^{3} = \begin{bmatrix} 0 & 0 & 0 & 0 \\ 0 & 0 & 0 & 0 \\ 0 & 0 & 1 & 0 \\ 0 & 0 & 0 & 0 \end{bmatrix}, \qquad \gamma_{ij}^{4} = \begin{bmatrix} 0 & 0 & 0 & 0 \\ 0 & 0 & 0 & 0 \\ 0 & 0 & 0 & 0 \\ 0 & 0 & 0 & 1 \end{bmatrix}.$$

(a) Write the product w^k in terms of its merates.

(b) Determine if the algebra is commutative and associative.

3.07 The so-called reactive power Q in an ac circuit is defined as

$$Q = |V|\,|I| \sin(\beta - \alpha).$$

We wish to express this quantity as the nilvalent product,

$$Q = V_i \odot I^j.$$

Determine the γ-matrix.

3.08 Given the equations

$$S = \gamma_{ij} v^i v^j, \qquad v^i = \gamma^{ik} v_k, \qquad \gamma_{ij}\gamma^{jk} = \delta_i^k,$$

(a) Express S in terms of v's with subscripts instead of superscripts.

(b) Expand both expressions for S in 3-space.

3.09 Expand $u^i \odot v^j$ and $v^j \odot u^i$ for plethos 3, and determine the necessary and sufficient condition for equality of these two expressions.

3.10 Repeat Problem 3.09 for $u_i \odot v^j$ and $v^j \odot u_i$. Is the product $u_i \odot v^j$ commutative for

$$\gamma^i_j = \begin{bmatrix} a & c \\ d & b \end{bmatrix}?$$

3.11 Write all possible products

$$\mathfrak{u} \otimes \mathfrak{v} = \mathfrak{w},$$

where u, v, and w may have either subscripts or superscripts:

$$\mathfrak{u} = u^i \quad \text{or} \quad u_i, \qquad \mathfrak{v} = v^j \quad \text{or} \quad v_j, \qquad \mathfrak{w} = w^k \quad \text{or} \quad w_k.$$

Determine conditions for a commutative product in each case.

3.12 If

$$u_i \otimes v^j = \gamma^{ki}_j u_i v^j = w^k,$$

determine the necessary and sufficient condition that the product

$$u_i \otimes v^j \otimes s^l = w^m$$

be associative.

3.13 For a series circuit containing R, L, C, the average power dissipated is

$$P^0 = \gamma^{0i}_j V_i I^j,$$

where

$$V_i = Z_{ij} I^j.$$

(a) Express the average power in terms of current.
(b) Obtain expressions for P^1 and P^2 in terms of current.

II

Transformations

4

Tensors

Previous chapters have developed the subject of holor algebra. We are now in a position to introduce the concept of a tensor and to enunciate the rules of tensor algebra. If we are interested in only one coordinate system, say rectangular coordinates, and never use any other, then the tensor idea is not required. But if we are concerned with some kind of geometric object that remains invariant with respect to coordinate transformation, tensors form the appropriate mathematical tool. The whole subject of vector analysis, for example, deals with a geometric object – the directed line segment or arrow – that keeps its integrity irrespective of how one changes the coordinate system. This vector is actually a univalent tensor; and vector analysis is a branch of tensor analysis.[1]

In physics, too, we deal with quantities that are invariant with respect to coordinate transformations. Area, volume, temperature, potential, force, velocity, electric field strength, and so on, are physical entities whose existence is obviously quite independent of what coordinate system we arbitrarily select to designate points in the underlying Euclidean space. With force, for instance, the merates are different in rectangular and spherical coordinates; but the force itself is above these petty questions of choice of coordinates. Thus, the tensor concept seems to be ideally suited to physical applications. The idea of invariance is implicit in some of the older disciplines, such as vector analysis. In tensor theory, however, the idea is brought out into the open, resulting in a subject of great usefulness and generality.

4.01. Linear transformations

A familiar coordinate transformation is indicated in Figure 4.01. The orthogonal Cartesian axes in Euclidean 2-space are labeled $x^{1'}, x^{2'}$. These

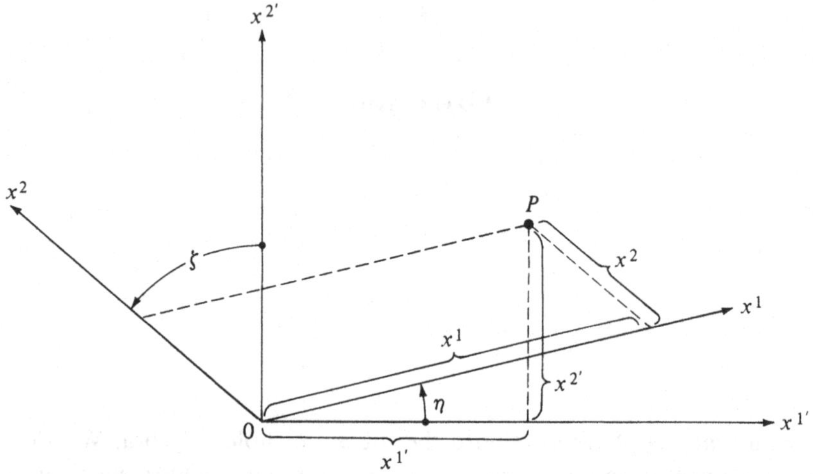

Figure 4.01. Transformation of coordinates in a plane. A point P may be expressed in terms of orthogonal Cartesian coordinates $(x^{1'},x^{2'})$ or in terms of skew coordinates (x^1,x^2). The transformation is

$$x^{i'}=A_i^{i'}x^i.$$

axes may be rotated about the origin through angles η and ζ, respectively, to form a nonorthogonal coordinate system x^1,x^2. Evidently the transformation is

$$\begin{aligned} x^{1'} &= x^1\cos\eta - x^2\sin\zeta,\\ x^{2'} &= x^1\sin\eta + x^2\cos\zeta. \end{aligned} \tag{4.01}$$

Since the transformation equations are linear in the x's, the transformation is said to be *linear* or *affine*. Using the summation convention, we write Equation (4.01) as

$$x^{i'}=A_i^{i'}x^i, \tag{4.01a}$$

where the transformation matrix is

$$A_i^{i'}=\begin{bmatrix}\cos\eta & -\sin\zeta\\ \sin\eta & \cos\zeta\end{bmatrix}. \tag{4.02}$$

Thus, the determinant of the transformation is

$$\begin{aligned} \Delta^{-1}=|A_i^{i'}| &= \begin{vmatrix}\cos\eta & -\sin\zeta\\ \sin\eta & \cos\zeta\end{vmatrix}\\ &= \cos\eta\cos\zeta+\sin\eta\sin\zeta=\cos(\eta-\zeta). \end{aligned} \tag{4.03}$$

If $|A_i^{i'}| \neq 0$, Equation (4.01a) can be inverted, giving

$$x^i = B_{i'}^i x^{i'}, \tag{4.04}$$

where

$$B_{i'}^i A_j^{i'} = \delta_j^i \tag{4.05}$$

or

$$B_{i'}^i = \frac{1}{|A_i^{i'}|}\begin{bmatrix} \cos\zeta & \sin\zeta \\ -\sin\eta & \cos\eta \end{bmatrix}. \tag{4.06}$$

If orthogonality of the axes is preserved, $\zeta = \eta$ and Equation (4.01) becomes

$$\begin{aligned} x^{1'} &= x^1 \cos\eta - x^2 \sin\eta, \\ x^{2'} &= x^1 \sin\eta + x^2 \cos\eta. \end{aligned} \tag{4.01b}$$

Equation (4.01a) is not confined to 2-space but applies equally well to any linear transformation with fixed origin in n-space. Here x^i and $x^{i'}$ are univalent holors with $i = 1, 2, \ldots, n$, $i' = 1', 2', \ldots, n'$; and the transformation matrix is

$$A_i^{i'} = \begin{bmatrix} A_1^{1'} & A_2^{1'} & \cdots & A_n^{1'} \\ A_1^{2'} & A_2^{2'} & \cdots & A_n^{2'} \\ \cdot & \cdot & \cdots & \cdot \\ A_1^{n'} & A_2^{n'} & \cdots & A_n^{n'} \end{bmatrix}, \tag{4.07}$$

where the A's are arbitrary constants independent of the x's. Linear transformations are not confined to rotations but may include arbitrary changes of scale for the different axes.[1] With n-space and $\Delta^{-1} \neq 0$, Equations (4.04) and (4.05) continue to apply, as in 2-space.

4.02. Invariants

An important consideration is to find invariants[2] of a given coordinate transformation: that is, to find entities that are unchanged when the coordinates are transformed. Obviously, such invariants tend to be of greater significance than quantities that are different in each coordinate system. In Figure 4.01, for instance, the point P is an invariant geometric object. Exactly why is it invariant? Evidently because it was taken as a fixed point in developing the transformation equation, (4.01). Had P been allowed to move as the axes were rotated, Equation (4.01) would no longer be valid.

Another invariant for Euclidean 2-space with orthogonal Cartesian systems ($\zeta=\eta$) is

$$(S)^2=g_{ij}x^ix^j, \tag{4.08}$$

where

$$g_{ij}=\begin{bmatrix}1 & 0\\ 0 & 1\end{bmatrix}.$$

Here S is the invariant distance from the origin to the fixed point P:

$$(S)^2=(x^1)^2+(x^2)^2. \tag{4.08a}$$

Obviously S is an invariant for the transformation of Equation (4.01) with $\zeta=\eta$, but it is not an invariant for the general affine transformation, which may include a translation and where the scales may be different along the two axes.

Three fixed points determine a triangle, which is thus an invariant geometric object under the transformation of Equation (4.01). Grassmann[3] considered the determinant

$$\mathfrak{A}=\frac{1}{2}\begin{bmatrix}(x^1)_1 & (x^2)_1 & 1\\ (x^1)_2 & (x^2)_2 & 1\\ (x^1)_3 & (x^2)_3 & 1\end{bmatrix}, \tag{4.09}$$

where the labels outside the parentheses refer to the three fixed points. The determinant may be written

$$\mathfrak{A}=\tfrac{1}{2}[(x^1)_1(x^2)_2+(x^2)_1(x^1)_3+(x^1)_2(x^2)_3 \\ -(x^2)_1(x^1)_2-(x^1)_1(x^2)_3-(x^2)_2(x^1)_3].$$

Substitution of Equation (4.01a) gives

$$\mathfrak{A}=\tfrac{1}{2}[A^1_{1'}A^2_{2'}-A^1_{2'}A^2_{1'}][(x^{1'})_1(x^{2'})_2+(x^{2'})_1(x^{1'})_3+(x^{1'})_2(x^{2'})_3 \\ -(x^{2'})_1(x^{1'})_2-(x^{1'})_1(x^{2'})_3-(x^{2'})_2(x^{1'})_3]$$

$$=[A^1_{1'}A^2_{2'}-A^1_{2'}A^2_{1'}]\mathfrak{A}'. \tag{4.09a}$$

Thus, $\mathfrak{A}$ is not an invariant for the general linear transformation. For rotations, Equation (4.01), the above equation becomes

$$\mathfrak{A}=\cos(\eta-\zeta)\mathfrak{A}'.$$

Even here, we do not have an invariant. But if both systems are orthogonal ($\zeta=\eta$),

$$\mathfrak{A} = \mathfrak{A}',$$

$\mathfrak{A}$ is an invariant. Indeed, it represents the area of the triangle. This area is obviously independent of the angle η at which the orthogonal coordinates are placed. If we allow the general linear transformation (which includes a possible change of scale for one of the axes), however, the concept of area loses its meaning and $\mathfrak{A}$ is not an absolute invariant.

Still another example of a geometric object is the vector v^i. This fixed vector may be written

$$v^i = (v^1, v^2)$$

in the unprimed system or

$$v^{i'} = (v^{1'}, v^{2'})$$

in the primed system. The two are related by the equations

$$v^{1'} = v^1 \cos\eta - v^2 \sin\eta,$$
$$v^{2'} = v^1 \sin\eta + v^2 \cos\eta,$$

or

$$\begin{aligned} v^1 &= v^{1'} \cos\eta + v^{2'} \sin\eta, \\ v^2 &= -v^{1'} \sin\eta + v^{2'} \cos\eta. \end{aligned} \tag{4.10}$$

These equations show how the merates of the geometric object change under rotation of orthogonal axes in Euclidean 2-space. The geometric object is invariant because it was taken as a fixed arrow in obtaining Equation (4.10), but the merates of the object change as the coordinate system is rotated.

The above applies to a vector associated with a fixed point P. If v^i is a field vector, it will be a function of position; but at an arbitrary point it will behave according to Equation (4.10). Suppose, for instance, that the field is at 45° from the $x^{1'}$ axis and varies exponentially with $x^{1'}$:

$$v^{1'} = Ve^{-\alpha x^{1'}}, \qquad v^{2'} = Ve^{-\alpha x^{1'}} = v^{1'}.$$

Then in the unprimed system,

$$v^1 = Ve^{-\alpha(x^1 \cos\eta - x^2 \sin\eta)}(\cos\eta + \sin\eta),$$
$$v^2 = Ve^{-\alpha(x^1 \cos\eta - x^2 \sin\eta)}(\cos\eta - \sin\eta).$$

Thus, the merates of the vector are functions of η, but the vector itself is an invariant geometric object. In particular, if $\eta=45°$, $v^2=0$; though in the primed system, $v^{2'}=v^{1'}$.

It is hoped that these elementary examples will give a rough idea of what is meant by invariance and by geometric object. Evidently these terms depend on the particular coordinate transformation: An invariant under one transformation will in general not be an invariant under another transformation. The subject will now be developed in greater detail.

4.03. Groups

A transformation of coordinates provides a bridge by which we cross from one designation of an arbitrary point P to another designation of the same point. If a transformation T_1 takes us from the designation x^i to the designation $x^{i'}$, and if T_2 takes us from $x^{i'}$ to $x^{i''}$, then there is a transformation T_3 that goes directly from x^i to $x^{i''}$. This is an example of the group property.

A group[4] is defined as a set of elements and one operation (written $\bigcirc$) such that:

(1) If a and b are elements of the set, then $a\bigcirc b$ is an element of the set.

(2) If a, b, and c are elements of the set,

$$a\bigcirc(b\bigcirc c)=(a\bigcirc b)\bigcirc c.$$

(3) The set contains an identity element I, such that for any element a,

$$I\bigcirc a=a\bigcirc I=a.$$

(4) An inverse exists for each element a, such that

$$a^{-1}\bigcirc a=a\bigcirc a^{-1}=I.$$

Evidently the rotations of Equation (4.01b) form a group. The elements of the group are the various coordinate systems; the operation is a rotation. The group has an infinite number of elements, corresponding to all possible values of the angle η. Any transformation $a\bigcirc b$ gives another element of the group, (1). A rotation through angle α followed by a rotation through angle β gives the same coordinate system as a rotation through β followed by a rotation through α, (2). There is an identity transformation, where $\eta=0$, (3); and there is an inverse, corresponding to a negative η, (4).

This idea of a group of transformations has been a fruitful one. It was emphasized by Felix Klein[5] in his Erlanger program of 1872. The idea that a geometry is a study of invariance under a group of transformations has been of immense value. A set of transformations does not always form a group, however, as shown by Veblen[6] and others.

4.04. General transformations

The linear transformation is, of course, only a very special case of a more general functional transformation. Even the familiar transformation from rectangular to polar coordinates in a plane, or from rectangular to spherical coordinates in 3-space, is a nonlinear transformation. We may write the general transformation as

$$x^{i'} = F^{i'}(x^i), \tag{4.11}$$

where $i = 1, 2, \ldots, n$, $i' = 1', 2', \ldots, n'$, and $\{F^{i'}\}$ is an arbitrary function.

In the transformation between rectangular coordinates $(x^{1'}, x^{2'}, x^{3'})$ and spherical coordinates (x^1, x^2, x^3) in Euclidean 3-space, for instance, Equation (4.11) becomes

$$x^{1'} = x^1 \sin x^2 \cos x^3,$$

$$x^{2'} = x^1 \sin x^2 \sin x^3,$$

$$x^{3'} = x^1 \cos x^2.$$

The function $\{F^{i'}\}$ may be chosen at will, each choice representing a possible coordinate transformation. For convenience, however, it is desirable to impose some limitations on $\{F^{i'}\}$:

(a) $\{F^{i'}\}$ is a single-valued function,

(b) $\{F^{i'}\}$ is analytic, and

(c) the Jacobian of the transformation is different from zero at every point.

The first restriction means that to every set of values x^i corresponds a unique set $x^{i'}$. This eliminates such transformations as

$$x^{i'} = \pm\sqrt{x^i}\,.$$

The second restriction means that the chosen functions possess derivatives of all orders. The third restriction ensures that each transformation shall have a unique inverse. The Jacobian is

$$\Delta = \left|\frac{\partial x^i}{\partial x^{i'}}\right| = \begin{vmatrix} \dfrac{\partial x^1}{\partial x^{1'}} & \dfrac{\partial x^1}{\partial x^{2'}} & \cdots & \dfrac{\partial x^1}{\partial x^{n'}} \\ \dfrac{\partial x^2}{\partial x^{1'}} & \dfrac{\partial x^2}{\partial x^{2'}} & \cdots & \dfrac{\partial x^2}{\partial x^{n'}} \\ \cdot & \cdot & \cdots & \cdot \\ \dfrac{\partial x^n}{\partial x^{1'}} & \dfrac{\partial x^n}{\partial x^{2'}} & \cdots & \dfrac{\partial x^n}{\partial x^{n'}} \end{vmatrix} \neq 0. \tag{4.12}$$

The necessary and sufficient condition that a functional transformation

$$x^{i'} = F^{i'}(x^i) \tag{4.11}$$

have a unique inverse

$$x^i = G^i(x^{i'})$$

is that the Jacobian $|\partial x^{i'}/\partial x^i| \neq 0$.

Note the convention of calling the Jacobian Δ if the unprimed coordinates are in the numerator, Δ^{-1} if the unprimed coordinates are in the denominator.

Manipulations of Equation (4.11) tend to be complicated because of the nonlinear functions involved. But the problem can be linearized if we are willing to deal with infinitesimal changes. Then Equation (4.11) leads to

$$dx^{i'} = \frac{\partial x^{i'}}{\partial x^1} dx^1 + \frac{\partial x^{i'}}{\partial x^2} dx^2 + \cdots + \frac{\partial x^{i'}}{\partial x^n} dx^n$$

or

$$dx^{i'} = \frac{\partial x^{i'}}{\partial x^i} dx^i. \tag{4.13}$$

The transformation matrix is

$$\frac{\partial x^{i'}}{\partial x^i} = \begin{bmatrix} \dfrac{\partial x^{1'}}{\partial x^1} & \dfrac{\partial x^{1'}}{\partial x^2} & \cdots & \dfrac{\partial x^{1'}}{\partial x^n} \\ \dfrac{\partial x^{2'}}{\partial x^1} & \dfrac{\partial x^{2'}}{\partial x^2} & \cdots & \dfrac{\partial x^{2'}}{\partial x^n} \\ \cdot & \cdot & \cdots & \cdot \\ \dfrac{\partial x^{n'}}{\partial x^1} & \dfrac{\partial x^{n'}}{\partial x^2} & \cdots & \dfrac{\partial x^{n'}}{\partial x^n} \end{bmatrix}, \tag{4.14}$$

whose determinant is

$$\Delta^{-1} = \left|\frac{\partial x^{i'}}{\partial x^{i}}\right| = \begin{vmatrix} \frac{\partial x^{1'}}{\partial x^{1}} & \frac{\partial x^{1'}}{\partial x^{2}} & \cdots & \frac{\partial x^{1'}}{\partial x^{n}} \\ \frac{\partial x^{2'}}{\partial x^{1}} & \frac{\partial x^{2'}}{\partial x^{2}} & \cdots & \frac{\partial x^{2'}}{\partial x^{n}} \\ \cdot & \cdot & \cdots & \cdot \\ \frac{\partial x^{n'}}{\partial x^{1}} & \frac{\partial x^{n'}}{\partial x^{2}} & \cdots & \frac{\partial x^{n'}}{\partial x^{n}} \end{vmatrix}.$$

If $\Delta^{-1} \neq 0$, the inverse of Equation (4.13) is

$$dx^{i} = \frac{\partial x^{i}}{\partial x^{i'}} dx^{i'}, \tag{4.15}$$

where the transformation matrix is

$$\frac{\partial x^{i}}{\partial x^{i'}} = \begin{bmatrix} \frac{\partial x^{1}}{\partial x^{1'}} & \frac{\partial x^{1}}{\partial x^{2'}} & \cdots & \frac{\partial x^{1}}{\partial x^{n'}} \\ \frac{\partial x^{2}}{\partial x^{1'}} & \frac{\partial x^{2}}{\partial x^{2'}} & \cdots & \frac{\partial x^{2}}{\partial x^{n'}} \\ \cdot & \cdot & \cdots & \cdot \\ \frac{\partial x^{n}}{\partial x^{1'}} & \frac{\partial x^{n}}{\partial x^{2'}} & \cdots & \frac{\partial x^{n}}{\partial x^{n'}} \end{bmatrix} \tag{4.16}$$

and

$$\Delta = \left|\frac{\partial x^{i}}{\partial x^{i'}}\right| = \begin{vmatrix} \frac{\partial x^{1}}{\partial x^{1'}} & \frac{\partial x^{1}}{\partial x^{2'}} & \cdots & \frac{\partial x^{1}}{\partial x^{n'}} \\ \frac{\partial x^{2}}{\partial x^{1'}} & \frac{\partial x^{2}}{\partial x^{2'}} & \cdots & \frac{\partial x^{2}}{\partial x^{n'}} \\ \cdot & \cdot & \cdots & \cdot \\ \frac{\partial x^{n}}{\partial x^{1'}} & \frac{\partial x^{n}}{\partial x^{2'}} & \cdots & \frac{\partial x^{n}}{\partial x^{n'}} \end{vmatrix}.$$

4.05. Geometric objects

The incentive for the development of tensor theory has been partly the needs of differential geometry, where geometric entities are of primary importance, and partly the needs of physics, where the concepts and equations should be divorced as much as possible from trivial aspects associated with particular coordinate systems. The latter aim has been achieved in a narrow sense by vector analysis. The tensor idea is a much broader extension of the same principle.

In going from holor to tensor, we take two independent steps, both of which are arbitrary:

(I) Choose a group of transformations of the coordinates x^i. For certain purposes, one may find it advisable to limit himself to linear transformations or to projective transformations. In the present chapter, however, we shall employ the general functional transformation of coordinates, Equation (4.11).

(II) Choose a definite transformation of holors, with the idea of selecting holors that have invariant geometric significance.

Suppose that transformation I has been decided and that we are now selecting transformation II to give invariant geometric objects. A geometric object[7] is defined as a holor that transforms according to a certain pattern, so that the new merates at a fixed point P are determined by the old merates and the coordinate transformation. In other words, the transformed merates ω' of a geometric object at a given point P are a function of the old merates ω at that point and the transformation I, or

$$\omega' = f\left(\omega, \frac{\partial x^{i'}}{\partial x^i}, \frac{\partial^2 x^{i'}}{\partial x^i \partial x^j}, \ldots\right). \tag{4.17}$$

As in transformation I, Equation (4.11), it seems advisable to impose some restrictions on the generality of the function f. In particular, if index balance is to be retained, allowable forms of transformation II are severely limited. Consider, for instance, possible geometric objects that are univalent. One might choose as the simplest transformations that fit Equation (4.17)

$$v^{i'} = v^i \quad \text{or} \quad v^{i'} = kv^i \quad \text{or} \quad v^{i'} = \partial x^{i'}/\partial x^i.$$

These are rejected because they do not give index balance. Evidently, the simplest form of Equation (4.17) that achieves index balance is the product

$$v^{i'} = \frac{\partial x^{i'}}{\partial x^i} v^i. \tag{4.18}$$

Other possibilities are

$$\begin{aligned} v^{i'} &= \sigma \frac{\partial x^{i'}}{\partial x^i} v^i, \\ v^{i'} &= \frac{\partial^2 x^{i'}}{\partial x^i \partial x^j} v^i v^j, \end{aligned} \tag{4.19}$$

$$v^{i'} = \frac{\partial^3 x^{i'}}{\partial x^i \, \partial x^j \, \partial x^k} v^i v^j v^k,$$

$$v^{i'} = \alpha \frac{\partial x^{i'}}{\partial x^i} v^i + \beta \frac{\partial^2 x^{i'}}{\partial x^i \, \partial x^j} v^i v^j, \quad \text{etc.}$$

Equation (4.18) is arbitrarily selected as the definition of a contravariant vector[1] or contravariant univalent tensor.[8] Given a holor v^i, this holor is a tensor if and only if it transforms as

$$v^{i'} = \frac{\partial x^{i'}}{\partial x^i} v^i. \tag{4.18}$$

The tensor v^i is a geometric object represented at a given point by an arrow that is invariant with respect to coordinate transformation.

Note particularly that *two* arbitrary choices must be made:
(I) transformation of coordinates and
(II) transformation of merates.
As stated previously, for choice I we have taken the general coordinate transformation for this chapter. In later chapters, other coordinate transformations are sometimes used, which may affect the division line between tensors and nontensors. Regarding II, the tensor concept is not sufficient for some applications, and more complicated geometric objects must be introduced. Chapter 5, for instance, is devoted to the more general transformation, Equation (4.19), where a factor $\sigma(x^i)$ is introduced into the transformation of merates. Another geometric object, whose transformation includes the second derivative, is the linear connection Γ^i_{jk} (Chapter 7):

$$\Gamma^{i'}_{j'k'} = \frac{\partial x^{i'}}{\partial x^i} \frac{\partial x^j}{\partial x^{j'}} \frac{\partial x^k}{\partial x^{k'}} \Gamma^i_{jk} + \frac{\partial x^{i'}}{\partial x^i} \frac{\partial^2 x^i}{\partial x^{j'} \, \partial x^{k'}}. \tag{4.20}$$

4.06. Univalent tensors

The contravariant univalent tensor v^i has been defined by the transformation

$$v^{i'} = \frac{\partial x^{i'}}{\partial x^i} v^i. \tag{4.18}$$

An important example of such a tensor is the infinitesimal increment dx^i in the coordinates of a point. Let the point P be represented by the coordinates $(x^1, x^2, \ldots, x^n)$. Then, by the chain rule of ordinary calculus,[9]

$$dx^{1'} = \frac{\partial x^{1'}}{\partial x^1} dx^1 + \frac{\partial x^{1'}}{\partial x^2} dx^2 + \cdots + \frac{\partial x^{1'}}{\partial x^n} dx^n,$$

$$dx^{2'} = \frac{\partial x^{2'}}{\partial x^1} dx^1 + \frac{\partial x^{2'}}{\partial x^2} dx^2 + \cdots + \frac{\partial x^{2'}}{\partial x^n} dx^n,$$

$$\vdots$$

$$dx^{n'} = \frac{\partial x^{n'}}{\partial x^1} dx^1 + \frac{\partial x^{n'}}{\partial x^2} dx^2 + \cdots + \frac{\partial x^{n'}}{\partial x^n} dx^n.$$

In index notation, these equations are written

$$dx^{i'} = \frac{\partial x^{i'}}{\partial x^i} dx^i. \tag{4.21}$$

Therefore, an infinitesimal increment in the coordinates is a tensor, according to Equation (4.18).

The statement is often made in elementary textbooks that a point P may be considered as a vector from the origin to P. But the transformation

$$x^{i'} = F^{i'}(x^i) \tag{4.11}$$

shows that, in general, the coordinates x^i do not transform according to Equation (4.18), and therefore x^i is not a vector. Only if we limit ourselves to linear transformations, where the merates $\{\partial x^{i'}/\partial x^i\}$ are independent of position, do we obtain x^i as a vector.

We must also define the covariant vector u_i, which occurs, for example, in the potential gradient

$$u_i = \partial\phi/\partial x^i.$$

The simplest choice of transformation for a geometric object of this kind is

$$u_{i'} = \frac{\partial x^i}{\partial x^{i'}} u_i. \tag{4.22}$$

The choice again is arbitrary. It is dictated largely by simplicity, but experience shows that these definitions of univalent tensors, Equations (4.18) and (4.22), are useful.

4.07. Other tensors

Having defined the univalent tensors, we are in a position to extend our definitions to tensors of any valence. The contracted product of two univalent tensors is

$$u_{i'}v^{i'} = S',$$

which defines a nilvalent holor S'. Substitution of Equations (4.18) and (4.22) gives

$$\frac{\partial x^{i}}{\partial x^{i'}}\frac{\partial x^{i'}}{\partial x^{j}}u_i v^j = S' = \delta^i_j u_i v^j = u_i v^i = S.$$

Thus, the definition of a scalar or nilvalent tensor, consistent with the previous definitions of univalent tensors, is

$$S' = S, \tag{4.23}$$

or a nilvalent tensor at a given point P is a single number that is invariant with respect to coordinate transformation.

Similarly, the product of two contravariant vectors may be written

$$u^{i'}v^{j'} = w^{i'j'}.$$

But since $u^{i'}$ and $v^{j'}$ are tensors, they must transform as

$$u^{i'} = \frac{\partial x^{i'}}{\partial x^{i}}u^i, \qquad v^{j'} = \frac{\partial x^{j'}}{\partial x^{j}}v^j.$$

Substitution gives

$$w^{i'j'} = \frac{\partial x^{i'}}{\partial x^{i}}\frac{\partial x^{j'}}{\partial x^{j}}u^i v^j = \frac{\partial x^{i'}}{\partial x^{i}}\frac{\partial x^{j'}}{\partial x^{j}}w^{ij}, \tag{4.24}$$

which is the transformation that defines the contravariant bivalent tensor. A bivalent holor w^{ij} is a tensor if and only if it transforms according to Equation (4.24).

This procedure can be continued for covariant and mixed tensors of any valence. Transformations are listed in Table 4.01. Further examples can be written immediately by merely remembering to achieve index balance and to use a number of transformation matrices, $\partial x^{i'}/\partial x^{i}$ or $\partial x^{i}/\partial x^{i'}$, equal to the valence.

As indicated in considering u_i, an important example of a covariant tensor is the partial derivative of a nilvalent tensor ϕ. Consider a nilvalent holor ϕ that is a function of position. The partial of ϕ with respect to the coordinates x^i is evidently a univalent holor, say u_i. Thus, in the unprimed system,

$$\begin{aligned} u_i &= \frac{\partial \phi(x^i)}{\partial x^i} \\ &= \frac{\partial \phi(x^i)}{\partial x^{i'}}\frac{\partial x^{i'}}{\partial x^{i}}. \end{aligned} \tag{4.25}$$

Table 4.01. *Tensors*

Valence	Symbol	Name	Transformation
0	ϕ	Nilvalent tensor (scalar)	$\phi'(x^{i'}) = \phi(x^i)$
1	v^i	Contravariant univalent tensor (vector)	$v^{i'} = \dfrac{\partial x^{i'}}{\partial x^i} v^i$
1	v_i	Covariant univalent tensor	$v_{i'} = \dfrac{\partial x^i}{\partial x^{i'}} v_i$
2	W^{ij}	Contravariant bivalent tensor	$W^{i'j'} = \dfrac{\partial x^{i'}}{\partial x^i} \dfrac{\partial x^{j'}}{\partial x^j} W^{ij}$
2	W_{ij}	Covariant bivalent tensor	$W_{i'j'} = \dfrac{\partial x^i}{\partial x^{i'}} \dfrac{\partial x^j}{\partial x^{j'}} W_{ij}$
2	W^i_j	Mixed bilvalent tensor	$W^{i'}_{j'} = \dfrac{\partial x^{i'}}{\partial x^i} \dfrac{\partial x^j}{\partial x^{j'}} W^i_j$
	W^j_i		$W^{j'}_{i'} = \dfrac{\partial x^i}{\partial x^{i'}} \dfrac{\partial x^{j'}}{\partial x^j} W^j_i$
$3, \ldots, n$		Tensors of higher valence	

But if ϕ is an invariant,

$$\phi' = \phi$$

Equation (4.25) becomes

$$\frac{\partial \phi}{\partial x^i} = \frac{\partial x^{i'}}{\partial x^i} \frac{\partial \phi'}{\partial x^{i'}}. \tag{4.26}$$

Since Equation (4.26) is the transformation equation of a covariant univalent tensor, the partial derivative of a nilvalent tensor is a covariant univalent tensor. In fact, Equation (4.25) gives the generalized gradient, which degenerates in certain cases to the familiar grad ϕ of vector analysis.[1]

An example of a bivalent tensor occurs in the specification of the dielectric properties of crystals. For a homogeneous isotropic medium, the relation between the electric field strength **E** (volts per meter) and the electric flux density **D** (coulombs per meters squared) may be written

$$\mathbf{D} = \epsilon \mathbf{E},$$

where ϵ, the permittivity, is a constant for a given dielectric under given conditions.

For an anisotropic medium, such as a crystal, the vectors **D** and **E** need no longer be in the same direction, and the complete specification of the material requires a matrix.[10] In index notation,

$$D^i = \epsilon^{ij} E_j, \tag{4.27}$$

where

$$\epsilon^{ij} = \begin{bmatrix} \epsilon^{11} & \epsilon^{12} & \epsilon^{13} \\ \epsilon^{21} & \epsilon^{22} & \epsilon^{23} \\ \epsilon^{31} & \epsilon^{32} & \epsilon^{33} \end{bmatrix}.$$

The quantities D^i and E_j are vector point functions that represent definite physical quantities at each point in the medium. The physics of the subject requires that these holors be independent of the coordinate system. Thus, they must transform as tensors. By a general principle to be enunciated later, Equation (4.27) then requires that ϵ^{ij} also be a tensor.

This conclusion can be checked very easily. Since D^i and E_j are tensors,

$$D^i = \frac{\partial x^i}{\partial x^{i'}} D^{i'}, \qquad E_j = \frac{\partial x^{j'}}{\partial x^j} E_{j'}. \tag{4.28}$$

Also, as will be shown in Section 4.09, tensor equations have the same form in any coordinate system, so

$$D^{i'} = \epsilon^{i'j'} E_{j'}. \tag{4.27a}$$

Substitution of Equation (4.28) into (4.27) gives

$$\frac{\partial x^i}{\partial x^{i'}} D^{i'} = \frac{\partial x^i}{\partial x^{i'}} \epsilon^{i'j'} E_{j'} = \epsilon^{ij} \frac{\partial x^{j'}}{\partial x^j} E_{j'}$$

or

$$E_{j'} \left(\frac{\partial x^i}{\partial x^{i'}} \epsilon^{i'j'} - \epsilon^{ij} \frac{\partial x^{j'}}{\partial x^j} \right) = 0.$$

Since $E_{j'}$ is arbitrary, the quantity within the parentheses must be zero. Multiply it by $\partial x^{k'}/\partial x^i$ to obtain

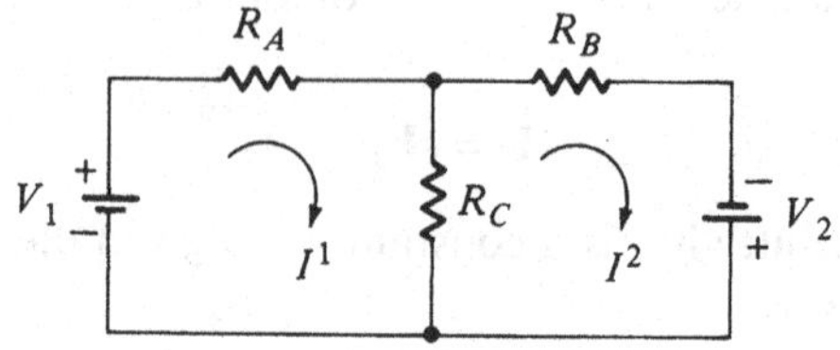

Figure 4.02. Two-loop dc network. Applied voltages are V_1 and V_2; resulting currents are I^1 and I^2. The holors are related by

$$V_i = Z_{ij} I^j.$$

$$\frac{\partial x^{k'}}{\partial x^i}\frac{\partial x^i}{\partial x^{i'}}\epsilon^{i'j'} = \frac{\partial x^{k'}}{\partial x^i}\frac{\partial x^{j'}}{\partial x^j}\epsilon^{ij}.$$

But

$$\frac{\partial x^{k'}}{\partial x^i}\frac{\partial x^i}{\partial x^{i'}} = \delta^{k'}_{i'},$$

so

$$\epsilon^{i'j'} = \frac{\partial x^{i'}}{\partial x^i}\frac{\partial x^{j'}}{\partial x^j}\epsilon^{ij}, \tag{4.29}$$

which proves that the permittivity $\epsilon^{i'j'}$ is a contravariant bivalent tensor.

A quite different application occurs in electric circuit theory.[11] For example, Figure 4.02 shows a two-loop dc network. The loop equations are

$$V_1 = (R_A + R_C)I^1 - R_C I^2,$$

$$V_2 = -R_C I^1 + (R_B + R_C)I^2.$$

Evidently these equations may be written

$$V_i = R_{ij} I^j, \tag{4.30}$$

where the resistance matrix is

$$R_{ij} = \begin{bmatrix} R_A + R_C & -R_C \\ -R_C & R_B + R_C \end{bmatrix}.$$

A network of any number of loops is expressed by Equation (4.30), though, of course, the matrix R_{ij} will be different for each network. In the usual problem of analysis of an l-loop network,

V_i and R_{ij} are given and the currents I^k are to be determined. The result can be written directly from Equation (4.30):

$$I^k = G^{ki} V_i, \tag{4.31}$$

where the conductance matrix G^{ki} is the inverse of the resistance matrix:

$$G^{ki} R_{ij} = \delta^k_j.$$

Evidently the customary problem of network analysis does not involve any question of coordinate transformation, so V_i, I^j, and R_{ij} are merely holors:

$$V_i = (V_1, V_2, \ldots, V_l),$$

$$I^j = (I^1, I^2, \ldots, I^l),$$

$$R_{ij} = \begin{bmatrix} R_{11} & R_{12} & \cdots & R_{1l} \\ \vdots & \vdots & & \vdots \\ R_{l1} & R_{l2} & \cdots & R_{ll} \end{bmatrix}.$$

The extension to the ac case is obvious. In network synthesis, however, the question of transformation may arise.

4.08. Tensor algebra

The algebra of holors was treated in Chapters 2 and 3. Since tensors are merely a subclass of holors, all of this material is directly applicable to tensors. Thus, the reader already knows how to manipulate tensors – how to add, multiply, contract, and invert them. We need study now merely the behavior under coordinate transformations.

First consider equality:

Theorem XII. *Tensors equal to each other in one coordinate system are equal to each other in all coordinate systems.*

This result is a consequence of the way in which tensors transform. For instance, two univalent holors $u^{i'}$ and $v^{i'}$ are equal in a specific coordinate system $(x^{1'}, x^{2'}, \ldots, x^{n'})$:

$$u^{i'} = v^{i'}. \tag{4.32}$$

If these holors are tensors, they transform in any other coordinate system $(x^1, x^2, \ldots, x^n)$ according to the equations

$$u^{i'} = \frac{\partial x^{i'}}{\partial x^i} u^i, \qquad v^{i'} = \frac{\partial x^{i'}}{\partial x^i} v^i. \tag{4.33}$$

Substitution into Equation (4.32) gives

$$\frac{\partial x^{i'}}{\partial x^i}(u^i - v^i) = 0.$$

Since $\partial x^{i'}/\partial x^i$ is arbitrary,

$$u^i = v^i, \tag{4.34}$$

so tensors that are equal in the primed system are also equal in the unprimed system. The same argument can be applied to tensors of any valence and any plethos, contravariant, covariant, or mixed. In all cases, equality established in one coordinate system applies to all coordinate systems.

Consider also the null tensor:

Theorem XIII. *If a tensor vanishes in one coordinate system, it vanishes in all coordinate systems.*

This may be regarded as a corollary to Theorem XII.

We now have various operations on tensors:

(1) addition,
(2) multiplication,
(3) contraction,
(4) symmetrization, and
(5) alternation.

As with the holors of Chapter 2, the addition of tensors is always commutative and associative. Also, the sum of tensors is always a tensor. For example, if $u^{i'}$ and $v^{i'}$ are tensors, index balance allows us to write

$$u^{i'} + v^{i'} = w^{i'}. \tag{4.35}$$

In another coordinate system, index balance requires that

$$u^i + v^i = W^i,$$

where a different base letter is employed because we have not yet proved that $w^{i'}$ transforms as a tensor. Substitution of Equation (4.33) into (4.35) gives

$$\frac{\partial x^{i'}}{\partial x^i}[u^i + v^i] = \frac{\partial x^{i'}}{\partial x^i} W^i = w^{i'}.$$

But this is just the transformation of a contravariant univalent tensor. Thus, $W^i = w^i$ and the sum of tensors is a tensor. Obviously, the tensor character of the sum is not a peculiarity of this particular example but is a result of the tensor transformation and applies to all valences.

Similarly, the product of tensors is always a tensor. For instance, take

$$u^{i'}v^{j'} = w^{i'j'}. \tag{4.36}$$

In another coordinate system, index balance requires

$$u^i v^j = W^{ij}.$$

Substitution of Equation (4.33) into Equation (4.36) gives

$$\frac{\partial x^{i'}}{\partial x^i}\frac{\partial x^{j'}}{\partial x^j}u^i v^j = \frac{\partial x^{i'}}{\partial x^i}\frac{\partial x^{j'}}{\partial x^j}W^{ij} = w^{i'j'},$$

which is the transformation of a bivalent tensor with

$$W^{ij} = w^{ij}.$$

Thus,

$$w^{i'j'} = \frac{\partial x^{i'}}{\partial x^i}\frac{\partial x^{j'}}{\partial x^j}w^{ij},$$

and the product of tensors is a tensor. For simplicity, these examples have dealt with univalent contravariant tensors, but the conclusions apply equally well to covariant or mixed tensors of any valence.

The γ-products of Chapter 3 introduce nothing really new, since they are merely the products of three holors instead of two. For example, for tensors u^i and v^j,

$$u^i \odot v^j = \gamma_{ij} u^i v^j = S,$$

$$u^i \otimes v^j = \gamma_{ij}^{k} u^i v^j = w^k,$$

$$u^i \bigcirc v^j = \gamma_{ij}^{kl} u^i v^j = W^{kl}.$$

Evidently, if the right sides of these equations are to be tensors, the γ's must also be tensors.

From the above examples, one sees that the number of factors is not limited to two but may be any number:

Theorem XIV. *Any sum of products of tensors is a tensor.*

Theorem XIV obviously applies to contracted as well as uncontracted products. By symmetrization of a tensor w^{ij}, we mean

$$w^{(ij)}=\frac{1}{2!}[w^{ij}+w^{ji}]. \tag{4.37}$$

Since w^{ij} and w^{ji} are tensors, their sum is a tensor. Thus, symmetrization of tensors gives tensors.

Similarly, by the operation of alternation of a tensor w^{ij}, we mean

$$w^{[ij]}=\frac{1}{2!}[w^{ij}-w^{ji}]. \tag{4.38}$$

According to Theorem XIV, this operation also gives tensors.

4.09. Tensor equations

Tensors occur in two classes of equations[12]:

(1) transformation equations, which relate two coordinate systems and which therefore contain both primed and unprimed indices;

(2) tensor equations, which relate tensors in a given system, all indices in an equation being either primed or unprimed.

Transformation equations for tensors are always linear in the merates of the tensor and are homogeneous, as indicated in Section 4.07. Take the transformation of a bivalent tensor W^{ij}, for instance:

$$W^{i'j'}=\frac{\partial x^{i'}}{\partial x^{i}}\frac{\partial x^{j'}}{\partial x^{j}}W^{ij}. \tag{4.39}$$

This equation is linear in the W's, though not necessarily in the x's. It contains no squares or higher powers of W, in accordance with the principle of choosing the simplest transformations in defining tensors. The transformation equation is also homogeneous: A transformation such as

$$W^{i'j'}=\frac{\partial x^{i'}}{\partial x^{i}}\frac{\partial x^{j'}}{\partial x^{j}}W^{ij}+C^{i'j'}, \tag{4.40}$$

for instance, does not define a tensor.

The second class of equations – tensor equations – need not be linear or homogeneous.

Definition VII: A tensor equation is an equation having the following properties:

(a) It contains only tensors.

(b) It is expressed in one coordinate system.

(c) It possesses index balance.

What forms can a tensor equation take? Evidently it can be a sum of tensors, or a product, or a sum of products. Examples are

$$u^i + v^i = w^i,$$

$$u^i v^j = w^{ij},$$

$$a_{ij} u^i u^j M^k_l = V^k_l,$$

$$au^i + bA^{ij} U_j = W^i.$$

Transcendental functions are also possible; though the argument of these functions must be a nilvalent tensor because such functions of vectors or of matrices have not been defined. The transcendental function will then be a nilvalent tensor. Some additional tensor equations are

$$u^i \cos\theta + v^i \sin\theta = w^i,$$

$$aY^{ij} + W^{ij} e^{-\alpha z} = X^{ij},$$

$$A\mathcal{J}_0(u_i v^i) + B\mathcal{Y}_0(u_i v^i) = Z,$$

where θ, αz, a, A, B, and Z are nilvalent tensors.

We now consider the important:

Theorem XV. *The form of a tensor equation is invariant with respect to coordinate transformation.*

Take any tensor equation, such as

$$AU_{ij} V^{ij} + Bu_i v^i = C, \tag{4.41}$$

where A, B, and C may include various transcendental functions. Since this is a tensor equation, every holor in it must be a tensor. Thus, the transformations are

$$A = A', \qquad B = B', \qquad C = C',$$

$$U_{ij} = \frac{\partial x^{i'}}{\partial x^i} \frac{\partial x^{j'}}{\partial x^j} U_{i'j'},$$

$$V^{ij} = \frac{\partial x^i}{\partial x^{i'}} \frac{\partial x^j}{\partial x^{j'}} V^{i'j'},$$

$$u_i = \frac{\partial x^{i'}}{\partial x^i} u_{i'},$$

$$v^i = \frac{\partial x^i}{\partial x^{i'}} v^{i'}.$$

Substitution gives

$$A'\frac{\partial x^{i'}}{\partial x^{i}}\frac{\partial x^{j'}}{\partial x^{j}}U_{i'j'}\frac{\partial x^{i}}{\partial x^{k'}}\frac{\partial x^{j}}{\partial x^{l'}}V^{k'l'}+B'\frac{\partial x^{i'}}{\partial x^{i}}\frac{\partial x^{i}}{\partial x^{j'}}u_{i'}v^{j'}=C'$$

or

$$A'U_{i'j'}V^{i'j'}+B'u_{i'}v^{i'}=C'.$$

Evidently the combination of tensor transformation and index balance ensures invariance of any tensor equation under any coordinate transformation, Equation (4.11).

One should remember, however, that the theorem applies only to tensor equations. In particular, x^i is not in general a tensor: It does not transform as

$$x^{i'}=A_i^{i'}x^i$$

but as

$$x^{i'}=F^{i'}(x^i).$$

Thus, unless we limit ourselves to linear transformations, equations containing x^i will not in general keep their form under transformation.

The converse of Theorem XV is not necessarily true: that is, the invariance of an equation need not mean that the equation is a tensor equation. The holors may transform in other ways, still keeping invariance of the equation. As a simple example, take

$$u_i v^i = S \tag{4.42}$$

with

$$u_i = k\frac{\partial x^{i'}}{\partial x^{i}}u_{i'}, \qquad v^i=(k)^{-1}\frac{\partial x^{i}}{\partial x^{i'}}v^{i'}, \qquad S=S'.$$

Evidently the equation is invariant with respect to coordinate transformation, though u_i and v^i are not tensors.

The relation between tensor equations and invariant equations is visualized most easily by means of a Venn diagram[13] (Figure 4.03). All invariant equations are represented by points within the large circle, non-invariant equations being outside this circle. Tensor equations are a special case of invariant equations and are represented by points within the small circle. Thus, Theorem XV may be written

Tensor equations $\subset$ invariant equations,

and

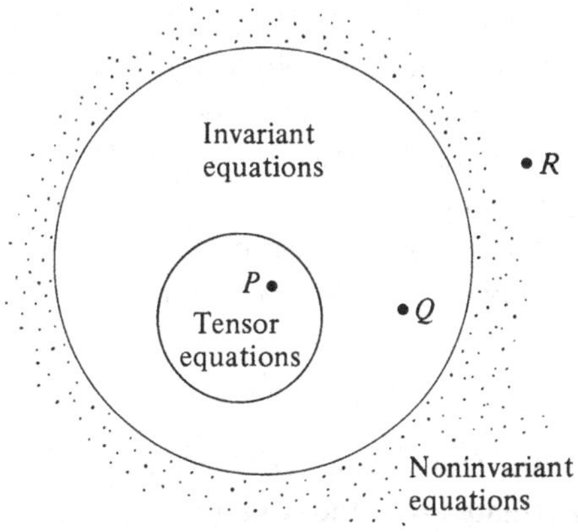

Figure 4.03. Schematic diagram indicating noninvariant and invariant (nontensor and tensor) equations.

$$\text{Equation (4.41)} = \text{tensor equation}$$
$$= \text{invariant equation,} \quad \text{as represented by point } P.$$

But

$$\text{Equation (4.42)} = \text{invariant equation}$$
$$\neq \text{tensor equation,} \quad \text{as represented by point } Q.$$

Most nontensor equations, however, will be in the outer region of Figure 4.03, as at R.

4.10. Tensor character

A decision can often be made as to the tensor character[14] of an unknown holor if the holor occurs in an equation containing tensors. For instance, suppose that we are given the equation

$$M^{ij}u_j + v^i \overset{*}{=} w^i, \tag{4.43}$$

which holds in a known coordinate system.* Let M^{ij}, u_j, and v^i be tensors. Then, according to Theorem XIV, w^i must be a tensor. Obviously this conclusion is equally valid for any arrangement of the equation, such as

$$M^{ij}u_j - w^i \overset{*}{=} -v^i,$$

provided that w^i is a complete term and not merely a factor in a product.

* The asterisk above the equal sign is to emphasize the fact that the equation is not necessarily valid in more than one coordinate system.

Similar results are obtained in many other cases: Usually if an invariant equation relates several holors, and if all these holors with one exception are known to be tensors, then the equation is a tensor equation. But are there exceptions to this rule?

Consider the possibilities:

(a) The unknown holor occurs in a product with an arbitrary tensor, the equation being an invariant equation. For example, take the equation

$$w^i v^j = X^{ij}, \tag{4.44}$$

where X^{ij} is a tensor and v^j is an arbitrary tensor. Is w^i a tensor? Since we have postulated that the equation is invariant,

$$w^{i'} v^{j'} = X^{i'j'}. \tag{4.44a}$$

Also,

$$v^j = \frac{\partial x^j}{\partial x^{j'}} v^{j'}, \qquad X^{ij} = \frac{\partial x^i}{\partial x^{i'}} \frac{\partial x^j}{\partial x^{j'}} X^{i'j'}.$$

Substitution into Equation (4.44) gives

$$w^i \frac{\partial x^j}{\partial x^{j'}} v^{j'} = \frac{\partial x^i}{\partial x^{i'}} \frac{\partial x^j}{\partial x^{j'}} X^{i'j'} = \frac{\partial x^i}{\partial x^{i'}} \frac{\partial x^j}{\partial x^{j'}} w^{i'} v^{j'}$$

or

$$v^{j'} \left[w^i - \frac{\partial x^i}{\partial x^{i'}} w^{i'} \right] = 0. \tag{4.45}$$

Since $v^{j'}$ is arbitrary, the bracket must be zero, and

$$w^i = \frac{\partial x^i}{\partial x^{i'}} w^{i'}. \tag{4.46}$$

Thus, w^i is a tensor. If, on the other hand, some restrictions have been placed on the tensor v^j so that it is not completely arbitrary, then the step from Equation (4.45) to Equation (4.46) cannot be taken and w^i is in general not a tensor.

(b) The unknown holor occurs in a contracted product with a tensor that need not be entirely arbitrary. The equation is an invariant equation. Take, for instance,

$$U_{ij} V^{ij} = S, \tag{4.47}$$

where V^{ij} and S are tensors. Is U_{ij} necessarily a tensor?

Transformations are

$$V^{ij} = \frac{\partial x^i}{\partial x^{i'}} \frac{\partial x^j}{\partial x^{j'}} V^{i'j'}, \qquad S = S'.$$

Since the equation is invariant,

$$U_{i'j'} V^{i'j'} = S'. \tag{4.47a}$$

Substitution into Equation (4.47) gives

$$U_{ij} \frac{\partial x^i}{\partial x^{i'}} \frac{\partial x^j}{\partial x^{j'}} V^{i'j'} = S' = U_{i'j'} V^{i'j'}$$

or

$$V^{i'j'} \left[U_{i'j'} - \frac{\partial x^i}{\partial x^{i'}} \frac{\partial x^j}{\partial x^{j'}} U_{ij} \right] = 0. \tag{4.48}$$

One might be tempted to say that the bracket must be zero, as in the previous example. But such a statement cannot be made in general because $V^{i'j'}$ is not arbitrary. Suppose that it is arbitrary except that it is symmetric. Then expansion of Equation (4.48) in 2-space gives

$$V^{1'1'} \left[U_{1'1'} - \frac{\partial x^i}{\partial x^{1'}} \frac{\partial x^j}{\partial x^{1'}} U_{ij} \right] + V^{1'2'} \left[U_{1'2'} - \frac{\partial x^i}{\partial x^{1'}} \frac{\partial x^j}{\partial x^{2'}} U_{ij} \right]$$

$$+ V^{2'1'} \left[U_{2'1'} - \frac{\partial x^i}{\partial x^{2'}} \frac{\partial x^j}{\partial x^{1'}} U_{ij} \right] + V^{2'2'} \left[U_{2'2'} - \frac{\partial x^i}{\partial x^{2'}} \frac{\partial x^j}{\partial x^{2'}} U_{ij} \right] = 0.$$

Since $V^{1'2'} = V^{2'1'}$, the other V's being arbitrary,

$$V^{1'2'} \left[(U_{1'2'} + U_{2'1'}) - \frac{\partial x^i}{\partial x^{1'}} \frac{\partial x^j}{\partial x^{2'}} (U_{ij} + U_{ji}) \right] = 0, \tag{4.49}$$

and we can write

$$V^{1'2'} \left[U_{(1'2')} - \frac{\partial x^i}{\partial x^{1'}} \frac{\partial x^j}{\partial x^{2'}} U_{(ij)} \right] = 0,$$

$$V^{1'1'} \left[U_{(1'1')} - \frac{\partial x^i}{\partial x^{1'}} \frac{\partial x^j}{\partial x^{1'}} U_{(ij)} \right] = 0,$$

$$V^{2'2'} \left[U_{(2'2')} - \frac{\partial x^i}{\partial x^{2'}} \frac{\partial x^j}{\partial x^{2'}} U_{(ij)} \right] = 0.$$

or

$$U_{(i'j')} = \frac{\partial x^i}{\partial x^{i'}} \frac{\partial x^j}{\partial x^{j'}} U_{(ij)}. \tag{4.50}$$

Thus, U_{ij} of Equation (4.47) is a tensor if it is symmetric, but in general it is not a tensor. We conclude from this example that an invariant equation in which all holors are tensors with one exception does not always allow us to state that the remaining holor is a tensor.

Conclusions are as follows:

Theorem XVI. *If an unknown holor appears as a term in an invariant equation and if every other holor in the equation is a tensor, then the unknown holor is also a tensor.*

Theorem XVII. *If an unknown holor appears in a product with an arbitrary tensor, and if the equation is an invariant equation and all other holors in the equation are tensors, then the unknown is also a tensor and the equation is a tensor equation.*

4.11. Tensors with fixed merates

A special class of tensors occurs when the merates are independent of coordinate transformation.[15] For example, any null tensor remains a null tensor after transformation. If univalent,

$$v^i = (0, 0, \ldots, 0),$$

and after transformation,

$$v^{i'} = (0, 0, \ldots, 0).$$

Similarly for valence 2, a null tensor is

$$w^{ij} = \begin{bmatrix} 0 & 0 & \cdots & 0 \\ 0 & 0 & \cdots & 0 \\ \cdot & \cdot & \cdots & \cdot \\ 0 & 0 & \cdots & 0 \end{bmatrix} = w^{i'j'}.$$

Another tensor with fixed merates is the *Kronecker delta*,

$$\delta^i_j = \begin{bmatrix} 1 & 0 & 0 & \cdots & 0 \\ 0 & 1 & 0 & \cdots & 0 \\ \cdot & \cdot & \cdot & \cdots & \cdot \\ 0 & 0 & 0 & \cdots & 1 \end{bmatrix}.$$

That this holor is a tensor is easily shown. Consider the partial derivative $\partial x^i/\partial x^j$, where x^i and x^j are two coordinates of the same system. If

$i \neq j$, x^i and x^j are independent, so $\partial x^i/\partial x^j = 0$; but if $i = j$, we have $\partial x^i/\partial x^i = 1$. Thus, such a partial derivative may be expressed as an ordinary Kronecker delta:

$$\delta^i_j = \frac{\partial x^i}{\partial x^j},$$
$$\delta^i_j = \frac{\partial x^i}{\partial x^{i'}} \frac{\partial x^{i'}}{\partial x^j}. \tag{4.51}$$

Similarly,

$$\delta^{i'}_{j'} = \frac{\partial x^{i'}}{\partial x^j} \frac{\partial x^j}{\partial x^{j'}}. \tag{4.52}$$

Equation (4.51) is to be changed so that it indicates the transformation between δ^i_j and $\delta^{i'}_{j'}$. This is accomplished by eliminating $\partial x^{i'}/\partial x^j$ between Equations (4.51) and (4.52). Multiply both sides of Equation (4.52) by $\partial x^{j'}/\partial x^k$:

$$\frac{\partial x^{j'}}{\partial x^k} \delta^{i'}_{j'} = \frac{\partial x^{j'}}{\partial x^k} \frac{\partial x^{i'}}{\partial x^j} \frac{\partial x^j}{\partial x^{j'}}$$

or

$$\frac{\partial x^{i'}}{\partial x^k} = \frac{\partial x^{j'}}{\partial x^k} \delta^{i'}_{j'},$$

which can be written

$$\frac{\partial x^{i'}}{\partial x^j} = \frac{\partial x^{j'}}{\partial x^j} \delta^{i'}_{j'}.$$

Substitution into Equation (4.51) gives

$$\delta^i_j = \frac{\partial x^i}{\partial x^{i'}} \frac{\partial x^{j'}}{\partial x^j} \delta^{i'}_{j'}, \tag{4.53}$$

which is the transformation of a bivalent mixed tensor. Therefore, the ordinary Kronecker delta is a tensor and its merates are unchanged by coordinate transformation.

Other tensors with fixed merates can be obtained as products of Kronecker deltas, such as

$$\delta^i_j \delta^k_l \cdots \delta^p_q.$$

Since each factor of the product is a tensor, the product is a tensor (Theorem XIV). For instance, $\delta^i_j \delta^k_l$ is a quadrivalent tensor. As a tensor, its transformation must be

$$\delta^{i'}_{j'}\delta^{k'}_{l'} = \frac{\partial x^{i'}}{\partial x^{i}}\frac{\partial x^{j}}{\partial x^{j'}}\frac{\partial x^{k'}}{\partial x^{k}}\frac{\partial x^{l}}{\partial x^{l'}}\delta^{i}_{j}\delta^{k}_{l},$$

and with $n=2$ and $i \neq j$,

$$\delta^{i}_{j}\delta^{k}_{k} = \delta^{i}_{j} = \begin{bmatrix} 1 & 0 \\ 0 & 1 \end{bmatrix}, \quad \text{for } k=1 \text{ or } k=2,$$

$$\delta^{i}_{j}\delta^{k}_{l} = \begin{bmatrix} 0 & 0 \\ 0 & 0 \end{bmatrix}, \quad \text{for } k \neq l,$$

$$\delta^{i'}_{j'}\delta^{k'}_{k'} = \delta^{i'}_{j'} = \begin{bmatrix} 1 & 0 \\ 0 & 1 \end{bmatrix}, \quad \text{for } k'=1' \text{ or } k'=2',$$

$$\delta^{i'}_{j'}\delta^{k'}_{l'} = \begin{bmatrix} 0 & 0 \\ 0 & 0 \end{bmatrix}, \quad \text{for } k' \neq l'.$$

Now consider the transformation of the general Kronecker delta δ^{ijk}_{lmn}. This holor may be written as a determinant, Equation (1.28):

$$\delta^{ijk}_{lmn} = \begin{vmatrix} \delta^{i}_{l} & \delta^{i}_{m} & \delta^{i}_{n} \\ \delta^{j}_{l} & \delta^{j}_{m} & \delta^{j}_{n} \\ \delta^{k}_{l} & \delta^{k}_{m} & \delta^{k}_{n} \end{vmatrix}$$

$$= \delta^{i}_{l}\delta^{j}_{m}\delta^{k}_{n} + \delta^{i}_{m}\delta^{j}_{n}\delta^{k}_{l} + \delta^{i}_{n}\delta^{j}_{l}\delta^{k}_{m}$$

$$- \delta^{i}_{n}\delta^{j}_{m}\delta^{k}_{l} - \delta^{i}_{m}\delta^{j}_{l}\delta^{k}_{n} - \delta^{i}_{l}\delta^{j}_{n}\delta^{k}_{m}. \tag{4.54}$$

Since the general Kronecker delta is merely a sum of products of tensors, it is itself a tensor and transforms as

$$\delta^{i'j'k'}_{l'm'n'} = \frac{\partial x^{i'}}{\partial x^{i}}\frac{\partial x^{j'}}{\partial x^{j}}\frac{\partial x^{k'}}{\partial x^{k}}\frac{\partial x^{l}}{\partial x^{l'}}\frac{\partial x^{m}}{\partial x^{m'}}\frac{\partial x^{n}}{\partial x^{n'}}\delta^{ijk}_{lmn}. \tag{4.55}$$

Similar equations are written for other valences.

There remain the Kronecker deltas δ^{123}_{lmn} and δ^{ijk}_{123}. Replacement of the indices i', j', k' in Equation (4.55) by $1', 2', 3'$ gives

$$\delta^{1'2'3'}_{l'm'n'} = \frac{\partial x^{1'}}{\partial x^{i}}\frac{\partial x^{2'}}{\partial x^{j}}\frac{\partial x^{3'}}{\partial x^{k}}\frac{\partial x^{l}}{\partial x^{l'}}\frac{\partial x^{m}}{\partial x^{m'}}\frac{\partial x^{n}}{\partial x^{n'}}\delta^{ijk}_{lmm}$$

$$= \frac{\partial x^{l}}{\partial x^{l'}}\frac{\partial x^{m}}{\partial x^{m'}}\frac{\partial x^{n}}{\partial x^{n'}}$$

$$\times \left[\frac{\partial x^{1'}}{\partial x^{1}}\frac{\partial x^{2'}}{\partial x^{2}}\frac{\partial x^{3'}}{\partial x^{3}}\delta^{123}_{lmn} + \frac{\partial x^{1'}}{\partial x^{2}}\frac{\partial x^{2'}}{\partial x^{3}}\frac{\partial x^{3'}}{\partial x^{1}}\delta^{231}_{lmn} \right.$$

$$+\frac{\partial x^{1'}}{\partial x^3}\frac{\partial x^{2'}}{\partial x^1}\frac{\partial x^{3'}}{\partial x^2}\delta^{312}_{lmn}+\frac{\partial x^{1'}}{\partial x^3}\frac{\partial x^{2'}}{\partial x^2}\frac{\partial x^{3'}}{\partial x^1}\delta^{321}_{lmn}$$

$$+\frac{\partial x^{1'}}{\partial x^2}\frac{\partial x^{2'}}{\partial x^1}\frac{\partial x^{3'}}{\partial x^3}\delta^{213}_{lmn}+\frac{\partial x^{1'}}{\partial x^1}\frac{\partial x^{2'}}{\partial x^3}\frac{\partial x^{3'}}{\partial x^2}\delta^{132}_{lmn}\bigg].$$

Making use of the relation

$$\delta^{123}_{lmn}=\delta^{231}_{lmn}=\delta^{312}_{lmn}=-\delta^{321}_{lmn}=-\delta^{213}_{lmn}=-\delta^{132}_{lmn},$$

we obtain

$$\delta^{1'2'3'}_{l'm'n'}=\frac{\partial x^l}{\partial x^{l'}}\frac{\partial x^m}{\partial x^{m'}}\frac{\partial x^n}{\partial x^{n'}}$$

$$\times\bigg[\frac{\partial x^{1'}}{\partial x^1}\frac{\partial x^{2'}}{\partial x^2}\frac{\partial x^{3'}}{\partial x^3}+\frac{\partial x^{1'}}{\partial x^2}\frac{\partial x^{2'}}{\partial x^3}\frac{\partial x^{3'}}{\partial x^1}+\frac{\partial x^{1'}}{\partial x^3}\frac{\partial x^{2'}}{\partial x^1}\frac{\partial x^{3'}}{\partial x^2}$$

$$-\frac{\partial x^{1'}}{\partial x^3}\frac{\partial x^{2'}}{\partial x^2}\frac{\partial x^{3'}}{\partial x^1}-\frac{\partial x^{1'}}{\partial x^2}\frac{\partial x^{2'}}{\partial x^1}\frac{\partial x^{3'}}{\partial x^3}-\frac{\partial x^{1'}}{\partial x^1}\frac{\partial x^{2'}}{\partial x^3}\frac{\partial x^{3'}}{\partial x^2}\bigg]\delta^{123}_{lmn}.$$

(4.56)

Here is the transformation equation for δ^{123}_{lmn}. Evidently this holor is *not* a tensor. Notice, however, that the equation would be the transformation of a trivalent covariant tensor if the bracket were equal to unity. And the bracket is simply the expansion of the Jacobian

$$\left|\frac{\partial x^{i'}}{\partial x^i}\right|=\Delta^{-1}=\begin{vmatrix}\frac{\partial x^{1'}}{\partial x^1} & \frac{\partial x^{1'}}{\partial x^2} & \frac{\partial x^{1'}}{\partial x^3}\\ \frac{\partial x^{2'}}{\partial x^1} & \frac{\partial x^{2'}}{\partial x^2} & \frac{\partial x^{2'}}{\partial x^3}\\ \frac{\partial x^{3'}}{\partial x^1} & \frac{\partial x^{3'}}{\partial x^2} & \frac{\partial x^{3'}}{\partial x^3}\end{vmatrix}.$$

Thus, the transformation of the Kronecker delta δ^{123}_{lmn} may be written

$$\delta^{1'2'3'}_{l'm'n'}=\Delta^{-1}\frac{\partial x^l}{\partial x^{l'}}\frac{\partial x^m}{\partial x^{m'}}\frac{\partial x^n}{\partial x^{n'}}\delta^{123}_{lmn}. \tag{4.57}$$

It is not a tensor but it is a form of *akinetor* (Chapter 5).

4.12. Summary

This chapter deals with the transition from holor to tensor. The distinction is associated with behavior under coordinate transformation. We deal with two arbitrary selections:

(I) *Transformation of coordinates.* One may select the rotation group, the affine group, the projective group, and so on. In this chapter, however, attention is confined to the general transformation

$$x^{i'} = F^{i'}(x^i), \tag{4.11}$$

where the function $\{F^{i'}\}$ is single valued, analytic, and with a unique inverse.

(II) *Transformation of merates.* A univalent tensor is defined as a holor that transforms as

$$v^{i'} = \frac{\partial x^{i'}}{\partial x^i} v^i, \tag{4.18}$$

or

$$u_{i'} = \frac{\partial x^i}{\partial x^{i'}} u_i. \tag{4.22}$$

Tensor algebra follows the rules of holor algebra already stated in Chapter 2. We have also:

Theorem XII. *Tensors equal to each other in one coordinate system are equal to each other in all coordinate systems.*

Theorem XIII. *If a tensor vanishes in one coordinate system, it vanishes in all coordinate systems.*

Theorem XIV. *Any sum of products of tensors is a tensor.*

Tensors occur in two classes of equations:

(1) transformation equations and

(2) tensor equations, which contain only tensors, are expressed in one coordinate system, and have index balance.

Theorem XV. *The form of a tensor equation is invariant with respect to coordinate transformation.*

The converse of this theorem is not necessarily true. A convenient visualization of the relation between tensor equations and invariant equations is given by the Venn diagram, Figure 4.03.

A way of determining if an unknown holor is a tensor is to consider an equation containing the unknown, other holors in the equation being tensors. Usually this fixes the tensor character of the unknown. If the unknown occurs in a product, however, the question may require careful consideration.

Theorem XVI. *If an unknown holor appears as a term in an equation that is valid in at least one coordinate system, and if every other holor in the equation is a tensor, then the unknown holor is also a tensor and the equation is a tensor equation.*

Theorem XVII. *If an unknown holor appears in a product with an arbitrary tensor, and if the equation is an invariant equation and all the other holors in the equation are tensors, then the unknown is also a tensor and the equation is a tensor equation.*

The merates of most tensors change when the coordinate system changes. Tensors exist, however, whose merates are fixed, irrespective of the coordinate system. Examples are the null tensors, the Kronecker delta δ^i_j, and products such as

$$\delta^i_j \delta^k_l \cdots \delta^p_q.$$

Another tensor with fixed merates is the general Kronecker delta, such as δ^{ijk}_{lmn}. But the alternators, δ^{ijk}_{123} and δ^{123}_{lmn}, are not tensors.

Problems

4.01 Consider two arbitrary points $(x^i)_1$ and $(x^j)_2$ in a plane, and take the combination

$$\phi = (x^1)_1 (x^1)_2 + (x^2)_1 (x^2)_2.$$

(a) Determine if ϕ is an invariant under the linear transformation of Equation (4.01).
(b) What conditions must be satisfied so that ϕ is an invariant with respect to Equation (4.01)?

4.02 Repeat Problem 4.01 for

$$\phi = \begin{bmatrix} (x^1)_1 & (x^2)_1 \\ (x^1)_2 & (x^2)_2 \end{bmatrix}.$$

4.03 The Grassmann matrix for two points in a plane may be written

$$\begin{bmatrix} (x^1)_1 & (x^2)_1 & 1 \\ (x^1)_2 & (x^2)_2 & 1 \end{bmatrix}.$$

(a) Consider the three determinants obtained from this matrix, and find transformations for which the three determinants are invariants.
(b) What are the geometric representations of these invariants?

4.04 The Grassmann matrix for two points in 3-space is

$$\begin{bmatrix} (x^1)_1 & (x^2)_1 & (x^3)_1 & 1 \\ (x^1)_2 & (x^2)_2 & (x^3)_2 & 1 \end{bmatrix}.$$

Give geometric interpretation for each determinant obtained from this matrix.

4.05 Repeat Problem 4.04 for the Grassmann matrix involving three arbitrary points in 3-space:

$$\begin{bmatrix} (x^1)_1 & (x^2)_1 & (x^3)_1 & 1 \\ (x^1)_2 & (x^2)_2 & (x^3)_2 & 1 \\ (x^1)_3 & (x^2)_3 & (x^3)_3 & 1 \end{bmatrix}.$$

Consider the 3×3 determinants contained in the matrix.

4.06 Repeat Problem 4.04 for the Grassmann matrix involving four arbitrary points in 3-space:

$$\begin{bmatrix} (x^1)_1 & (x^2)_1 & (x^3)_1 & 1 \\ (x^1)_2 & (x^2)_2 & (x^3)_2 & 1 \\ (x^1)_3 & (x^2)_3 & (x^3)_3 & 1 \\ (x^1)_4 & (x^2)_4 & (x^3)_4 & 1 \end{bmatrix}.$$

Consider the 4×4 determinant contained in the matrix.

4.07 Consider the transformation between rectangular coordinates $x^{i'}$ and spherical coordinates x^i. Obtain expressions for

$$\frac{\partial x^{i'}}{\partial x^i}, \quad \frac{\partial x^i}{\partial x^{i'}}, \quad \Delta, \quad \text{and} \quad \Delta^{-1}.$$

4.08 Elliptic coordinates x^i in 2-space may be written

$$x^{1'} = a \cosh x^1 \cos x^2,$$

$$x^{2'} = a \sinh x^1 \sin x^2,$$

where $x^{i'}$ are rectangular coordinates.

(a) Obtain $\partial x^{i'}/\partial x^i$, $\partial x^i/\partial x^{i'}$, Δ, and Δ^{-1}.

(b) Electrical conduction in an anisotropic medium may be expressed by the tensor equation

$$E_{i'} = \rho_{i'j'} J^{j'},$$

where $E_{i'}$ is electric field strength, $J^{j'}$ is current density, and $\rho_{i'j'}$ is the resistivity matrix. Let $J^{j'} = (1, 2)$ at point P, where $x^{i'} = (1, 1)$. Also, $a = 1$ and

$$\rho_{i'j'} = \begin{bmatrix} 1 & 3 \\ 3 & 2 \end{bmatrix}.$$

Evaluate E_i at P.

(c) Find numerical values of $\partial x^{i'}/\partial x^i$ and $\partial x^i/\partial x^{i'}$ at P.

(d) Obtain J^j, E_i, and ρ_{ij} at P.

4.09 Repeat Problem 4.08 for parabolic coordinates x^i, where

$$x^{1'} = \tfrac{1}{2}[(x^1)^2 - (x^2)^2], \qquad x^{2'} = x^1 x^2.$$

4.10 Repeat Problem 4.08 for polar coordinates x^i, where

$$x^{1'} = x^1 \cos x^2, \qquad x^{2'} = x^1 \sin x^2.$$

4.11 Repeat Problem 4.08 for the transformation

$$x^{1'} = \frac{x^1}{(x^1)^2 + (x^2)^2}, \qquad x^{2'} = \frac{x^2}{(x^1)^2 + (x^2)^2}.$$

4.12 By substitution of the transformation equations, show that X^{jk} is a tensor in

$$u_i v^i K^{jk} + S V^j W^k = X^{jk},$$

if u_i, v^i, K^{jk}, S, V^j, and W^k are tensors.

4.13

(a) Expand $u_{(ij)} v^{(jk)}$ in 2-space.

(b) Expand $V^{i[jk]}$ in 2-space.

4.14 Show that $W_k^{[ij]}$ is a tensor if W_k^{ij} is a tensor.

4.15 Since every tensor equation is an invariant equation, can we say that every nontensor equation is a noninvariant equation? Explain and give an example.

4.16 A treatise on tensors gives the following example: Given the product $U_{ij} V^{ij}$, where U_{ij} is a bivalent covariant tensor but is otherwise arbitrary, is V^{ih} a tensor? Index balance requires that this product be a scalar, so we can write

$$U_{ij} V^{ij} = S.$$

Also, since U_{ij} is a tensor,

$$U_{ij} = \frac{\partial x^{i'}}{\partial x^i} \frac{\partial x^{j'}}{\partial x^j} U_{i'j'}.$$

Thus,

$$\frac{\partial x^{i'}}{\partial x^i}\frac{\partial x^{j'}}{\partial x^j}U_{i'j'}V^{ij}=S=U_{i'j'}V^{i'j'}$$

or

$$U_{i'j'}\left[V^{i'j'}-\frac{\partial x^{i'}}{\partial x^i}\frac{\partial x^{j'}}{\partial x^j}V^{ij}\right]=0.$$

Since $U_{i'j'}$ is arbitrary,

$$V^{i'j'}=\frac{\partial x^{i'}}{\partial x^i}\frac{\partial x^{j'}}{\partial x^j}V^{ij},$$

which is a tensor transformation. Therefore, V^{ij} is a bivalent contravariant tensor.

Criticize this derivation. Is the conclusion correct?

4.17 Criticize the following proof: In rectangular coordinates $x^{i'}$, we have an equation

$$V_{i'j'k'}\xi^{k'}=W_{i'j'},$$

where $\xi^{k'}$ and $W_{i'j'}$ are known to be tensors. We now prove that $V_{i'j'k'}$ is necessarily a tensor.

Transformations are

$$\xi^{k'}=\frac{\partial x^{k'}}{\partial x^k}\xi^k,\qquad W_{i'j'}=\frac{\partial x^i}{\partial x^{i'}}\frac{\partial x^j}{\partial x^{j'}}W_{ij}.$$

We substitute into the first equation to obtain

$$V_{i'j'k'}\frac{\partial x^{k'}}{\partial x^k}\xi^k=\frac{\partial x^i}{\partial x^{i'}}\frac{\partial x^j}{\partial x^{j'}}W_{ij}=\frac{\partial x^i}{\partial x^{i'}}\frac{\partial x^j}{\partial x^{j'}}V_{ijk}\xi^k$$

or

$$\xi^k\left[\frac{\partial x^{k'}}{\partial x^k}V_{i'j'k'}-\frac{\partial x^i}{\partial x^{i'}}\frac{\partial x^j}{\partial x^{j'}}V_{ijk}\right]=0.$$

Therefore,

$$V_{i'j'k'}=\frac{\partial x^i}{\partial x^{i'}}\frac{\partial x^j}{\partial x^{j'}}\frac{\partial x^k}{\partial x^{k'}}V_{ijk}$$

and $V_{i'j'k'}$ is a tensor.

4.18 Criticize the following:

(I) The reason that tensors are so useful in physics is that they are independent of the coordinate system. A tensor is an invariant with respect to coordinate transformation.

(II) A vector v^i may be represented by an arrow that is fixed in magnitude and direction. If we change coordinates, the components of the vector change, but the arrow is invariant. Thus, a vector is a tensor, according to I.

(III) A fixed point x^i is also an invariant. We have the point fixed in space, whereas the coordinates change. Since the point is an invariant with respect to coordinate transformation, it is a tensor, in accordance with I.

4.19 Consider the equation

$$V^i \sin\theta + k^{ij} u_j \cos\theta \overset{*}{=} W^{ijk} X_{jk},$$

where k^{ij}, $u_j W^{ijk}$, X_{jk}, and θ are tensors.

What further assumption is required to prove that V^i is a tensor?

4.20 Given the equation

$$u^i v^j + bU^{ij}_k v^k \overset{*}{=} W^{ij},$$

where u^i, v^j, b, and W^{ij} are tensors. What additional assumptions are required in order to prove that U^{ij}_k is a tensor?

4.21 Given the tensor equation

$$P^{i'} \cosh(u_{j'} v^{j'}) + Q^{i'j'} R_{j'} \overset{*}{=} X^{i'},$$

prove that the equation is invariant with respect to coordinate transformation.

4.22 The current density J^i (amperes per meters squared) in an anisotropic medium may be written

$$J^i = \sigma^{ij} E_j,$$

where E_j is the electric field strength (volts per meter) and σ^{ij} is the conductivity matrix. From physical considerations, we decide that J^i and E_j are arbitrary tensors and that the equation is an invariant equation.

(a) Determine if σ^{ij} is a tensor.

(b) If J^i and E_j are not tensors but transform as

$$J^i = A \frac{\partial x^i}{\partial x^{i'}} J^{i'}, \qquad E_j = B \frac{\partial x^{j'}}{\partial x^j} E_{j'},$$

and the equation is an invariant equation, how does σ^{ij} transform?

(c) If J^i and E_j transform as in (b) and σ^{ij} transforms as a tensor, what is the form of the equation relating $E_{j'}$, $J^{i'}$, and $\sigma^{i'j'}$?

4.23 The magnetic flux density $B^{i'}$ is related to the magnetic field strength $H_{j'}$ in an anisotropic medium and rectangular coordinates by

$$B^{i'} \overset{*}{=} \mu^{i'j'} H_{j'},$$

where $\mu^{i'j'}$ is the permeability matrix.
(a) Determine what restrictions on the holors $B^{i'}$, $H_{j'}$, and $\mu^{i'j'}$ will give an invariant equation.
(b) What restrictions on these holors will give a tensor equation?

4.24 A rectangular coordinate system in 2-space is represented by $x^{i'}$, and a directed line segment between fixed points P and Q is

$$V^{i'} \overset{*}{=} [(x^{1'})_Q - (x^{1'})_P, (x^{2'})_Q - (x^{2'})_P].$$

Transform to polar coordinates and determine if V^i is a vector.

4.25 Repeat Problem 4.24 for 3-space with a transformation from rectangular to spherical coordinates.

4.26 Consider Equation (4.47) with S a tensor and V^{ij} an arbitrary antisymmetric tensor. Under what conditions is U_{ij} a tensor?

4.27 In the invariant equation

$$X^{ij}_{kl} Y^{kl} = Z^{ij},$$

Z^{ij} is an arbitrary tensor and v^{kl} is an antisymmetric, but otherwise arbitrary, tensor. What conditions must be satisfied for the holor X^{ij}_{kl} so that it is a tensor?

4.28 In the invariant equation

$$u_i v^i_j + a b_j = w_j,$$

a, b_j, v^i_j, and w_j are arbitrary tensors.
(a) Prove that u_i is a tensor.
(b) If v^i_j is no longer arbitrary but

$$v^i_j = \delta^i_j,$$

obtain an expression for u_i.
(c) The above equation is given without restrictions. But w_j is not a tensor: It transforms as

$$w_j = k \frac{\partial x^{j'}}{\partial x^j} W_{j'} + c_j.$$

Can the equation be an invariant equation if some or all of the other holors are nontensors?

4.29 Write the matrices

$$\delta^i_k \delta^j_l u^k v^l = w^{ij}$$

and

$$\delta^{ij}_{kl} u^k v^l = W^{ij}$$

for 3-space.

4.30 Evaluate

$$\delta^i_l \delta^j_m \delta^k_n U^l_i U^m_j U^n_k \quad \text{and} \quad \delta^{ijk}_{lmn} U^{lmn}_{ijk}$$

for 3-space.

5
Akinetors

In Chapter 4, a subclass of holors having particularly simple coordinate transformations was considered. These holors are called tensors, a univalent tensor transforming as

$$v^{i'} = \frac{\partial x^{i'}}{\partial x^{i}} v^{i}. \tag{5.01}$$

The present chapter deals with a somewhat more general subclass of holors having a function $\sigma(x^i)$ in the transformation equation, as

$$\tilde{v}^{i'} = \sigma \frac{\partial x^{i'}}{\partial x^{i}} \tilde{v}^{i}. \tag{5.02}$$

For such holors, we coin the name *akinetor*.* To emphasize that an akinetor has a different transformation equation than a tensor, a tilde is sometimes written over the base letter, as $\tilde{v}^i$.

Existing nomenclature is anything but satisfactory. A traditional classification[1] of holors is shown in Figure 5.01, geometric objects being divided into tensors, pseudotensors, Christoffel symbols, and more complicated geometric objects. Such a classification gives the impression that tensors, Equation (5.01), are quite separate from pseudotensors,† Equation (5.02), as indicated in the diagram. But this classification is incorrect: Equations (5.01) and (5.02) show that the so-called pseudotensors form the general case, with tensors a subclass having $\sigma = 1$.

* Greek ἀκίνητος = invariant, fixed. Pronounced *a kiń e tor.* Easily remembered from the English *kinetic* with negative prefix *a*-.

† Greek ψευδης = false; therefore, not a tensor.

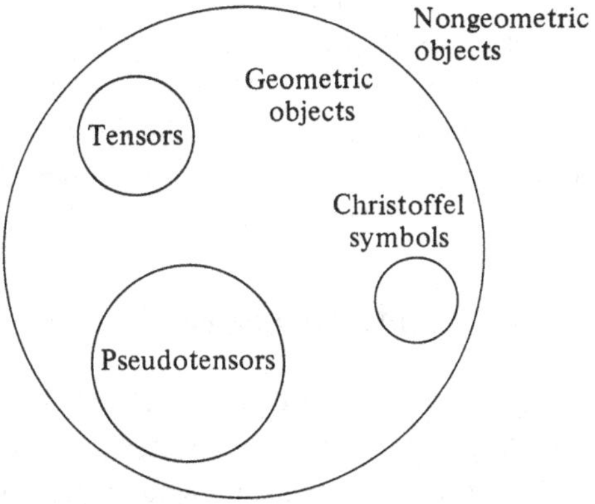

Figure 5.01. A traditional classification of holors.

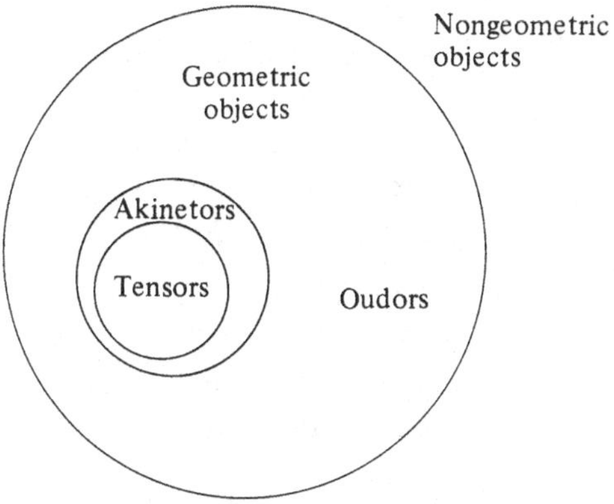

Figure 5.02. The classification of holors used in this book.

Another nomenclature[2] divides tensors into absolute tensors, Equation (5.01), and relative tensors, Equation (5.02). This would make absolute tensors a subclass of relative tensors, which is peculiar, to say the least.

The classification used in this book is shown in Figure 5.02. Geometric objects contain the classes akinetors and oudors.* Tensors are merely the special case of akinetors where $\sigma = 1$. Logically, of

* Greek *οὐ* = not. Pronounced *$\overline{oo}$′ dors.*

course, it would have been better to study akinetors first and to bring in tensors merely as a special case. Pedagogically, however, we believe the present arrangement - tensors in Chapter 4, akinetors in Chapter 5 - to be preferable, since students have enough difficulty in grasping the tensor idea in its simplest form.

5.01. Definitions

In Section 4.11, it was found that the ordinary Kronecker delta δ^i_j transforms as a tensor, as does the generalized Kronecker delta δ^{ijk}_{lmn}. But the alternators δ^{123}_{lmn} and δ^{ijk}_{123} transform in a somewhat more complicated way. Indeed,[3]

$$\delta^{i'j'k'}_{1'2'3'} = \Delta \frac{\partial x^{i'}}{\partial x^i} \frac{\partial x^{j'}}{\partial x^j} \frac{\partial x^{k'}}{\partial x^k} \delta^{ijk}_{123}, \tag{5.03}$$

$$\delta^{1'2'3'}_{l'm'n'} = \Delta^{-1} \frac{\partial x^l}{\partial x^{l'}} \frac{\partial x^m}{\partial x^{m'}} \frac{\partial x^n}{\partial x^{n'}} \delta^{123}_{lmn}, \tag{5.04}$$

where

$$\Delta = \begin{vmatrix} \dfrac{\partial x^1}{\partial x^{1'}} & \dfrac{\partial x^1}{\partial x^{2'}} & \dfrac{\partial x^1}{\partial x^{3'}} \\ \dfrac{\partial x^2}{\partial x^{1'}} & \dfrac{\partial x^2}{\partial x^{2'}} & \dfrac{\partial x^2}{\partial x^{3'}} \\ \dfrac{\partial x^3}{\partial x^{1'}} & \dfrac{\partial x^3}{\partial x^{2'}} & \dfrac{\partial x^3}{\partial x^{3'}} \end{vmatrix}. \tag{5.05}$$

Equations (5.03) and (5.04) are examples of akinetor transformations with $\sigma = \Delta$ or $\sigma = \Delta^{-1}$. They are the same as tensor transformations except for Δ or Δ^{-1}. This idea can be extended to the use of Δ^w, where w is an integer called the weight of the akinetor. To make weight unambiguous, we always take Δ as the Jacobian with the unprimed quantities in the numerator and we write the transformation equation with the unprimed holor and Δ^w on the right side. For instance, Equation (5.03) shows that δ^{ijk}_{123} is an akinetor of weight $+1$, and Equation (5.04) shows that δ^{123}_{lmn} is an akinetor of weight -1.

A nilvalent akinetor of weight w transforms as

$$\tilde{S}'(x^{i'}) = \Delta^w \tilde{S}(x^i). \tag{5.06}$$

A univalent contravariant akinetor of weight w transforms as

$$\tilde{v}^{i'} = \Delta^w \frac{\partial x^{i'}}{\partial x^i} \tilde{v}^i. \tag{5.07}$$

A univalent covariant akinetor of weight w transforms as

$$\tilde{u}_{i'} = \Delta^w \frac{\partial x^i}{\partial x^{i'}} \tilde{u}_i. \tag{5.08}$$

A bivalent contravariant akinetor of weight w transforms as

$$\tilde{V}^{i'j'} = \Delta^w \frac{\partial x^{i'}}{\partial x^i} \frac{\partial x^{j'}}{\partial x^j} \tilde{V}^{ij}. \tag{5.09}$$

The transformations of other akinetors can be written in a similar manner.

5.02. Algebra

Whereas the algebra of tensors is identical with the algebra of holors, the algebra of akinetors is slightly different because of the necessity of considering weight. Two akinetors may differ not only with respect to valence and plethos but also with respect to weight. This fact introduces a few differences in the algebra of Chapter 2 – differences that are obvious but that will be noted in this section.

Definition VIII: Two akinetors are equal if and only if each corresponding pair of merates is equal and if the akinetors are of the same weight.

Since "corresponding pair of merates" is meaningful only if valence is the same and plethos is the same, Definition VIII implies that valence, plethos, and weight must be the same before we can even consider equality.

Similarly, we have:

Definition IX: The sum of two akinetors is an akinetor of the same valence, plethos, and weight, each merate of the sum being obtained by addition of corresponding merates of the original akinetors.

Note that addition of akinetors is defined only for akinetors of the same valence, plethos, and weight. Addition of akinetors is commutative and associative.

Definition X: Contracted and uncontracted products of akinetors are always possible, the weight of the product being the sum of the weights of the factors.

For example, consider the product of the akinetors $\tilde{u}_i$ and $\tilde{V}^{ij}$, where $\tilde{u}_i$ is of weight +2 and $\tilde{V}^{ij}$ is of weight +1. Then

$$\tilde{u}_i \tilde{V}^{ij} = \tilde{W}^j,$$

where $\tilde{W}^j$ is an akinetor of weight +3. According to Section 5.01, transformations are

$$\tilde{u}_{i'} = \Delta^2 \frac{\partial x^i}{\partial x^{i'}} \tilde{u}_i,$$

$$\tilde{V}^{i'j'} = \Delta \frac{\partial x^{i'}}{\partial x^i} \frac{\partial x^{j'}}{\partial x^j} \tilde{V}^{ij},$$

so

$$\tilde{W}^{j'} = \Delta^3 \frac{\partial x^{j'}}{\partial x^j} \tilde{u}_i \tilde{V}^{ij} = \Delta^3 \frac{\partial x^{j'}}{\partial x^j} \tilde{W}^j.$$

An interesting case occurs when the numerical weights of the factors are the same, but the signs of the weights are opposite. Then the product of two akinetors gives a tensor. For instance, suppose that the weight of $\tilde{u}_i$ is +1 and the weight of $\tilde{V}^{ij}$ is −1. Then

$$\tilde{u}_i \tilde{V}^{ij} = W^j,$$

where W^j is a tensor (akinetor of weight zero).

Definition XI: An akinetor equation is defined as an equation having the following properties:
(a) It contains only akinetors, including those of weight zero (tensors).
(b) It is expressed in one coordinate system.
(c) It possesses index balance.
Evidently akinetor equations include tensor equations as a special case.
Index balance applies to akinetors:

Theorem XVIII. *In an akinetor equation, each term must have the same literal indices in the same positions and in the same order, and the weight of each term must be the same.*

As with holors, this theorem is a direct consequence of our definition of addition.

The inverse of a bivalent holor u^{ij} is U_{hi}, where (Section 2.08)

$$U_{hi} u^{ij} = \delta_h^j. \tag{5.10}$$

If $\tilde{u}^{ij}$ is an akinetor of weight w, its inverse must be another akinetor $\tilde{U}_{hi}$ whose weight is $-w$ so that δ_h^j can be a tensor. Suppose that $\tilde{u}^{ij}$ (of weight w) occurs in the equation

$$\tilde{u}^{ij} v_{jk} = \tilde{w}^i_k, \tag{5.11}$$

where v_{jk} is a tensor. Evidently the product must be an akinetor of weight $+w$ to balance the equation with respect to weight. Multiplication of both sides by $\tilde{U}_{hi}$ gives

$$\tilde{U}_{hi} \tilde{u}^{ij} v_{jk} = \tilde{U}_{hi} \tilde{w}^i_k.$$

Thus,

$$\delta^j_h v_{jk} = \tilde{U}_{hi} \tilde{w}^i_k,$$

or

$$v_{hk} = \tilde{U}_{hi} \tilde{w}^i_k. \tag{5.12}$$

Here $\tilde{U}_{hi}$ is of weight $-w$ and $\tilde{w}^i_k$ is of weight $+w$, so their product is a tensor.

5.03. The akinetor as generalization of the tensor

The previous section has stated some of the modifications of holor algebra (Chapter 2) that occur when akinetors are introduced. We now consider how tensor theorems (Chapter 4) are applied to akinetors.

Theorem XIX. *Akinetors equal to each other in one coordinate system are equal to each other in all coordinate systems.*

For example, if

$$\tilde{u}^{i'} = \tilde{v}^{i'}$$

in a specific coordinate system $x^{i'}$, the two akinetors must have the same weight, in accordance with Definition VIII. Thus, the transformations must be

$$\tilde{u}^{i'} = \Delta^w \frac{\partial x^{i'}}{\partial x^i} \tilde{u}^i, \qquad \tilde{v}^{i'} = \Delta^w \frac{\partial x^{i'}}{\partial x^i} \tilde{v}^i.$$

Substitution gives

$$\Delta^w \frac{\partial x^{i'}}{\partial x^i} (\tilde{u}^i - \tilde{v}^i) = 0,$$

and since the transformation is arbitrary,

$$\tilde{u}^i = \tilde{v}^i.$$

Theorem XX. *If an akinetor vanishes in one coordinate system, it vanishes in all coordinate systems.*

Theorem XXI. *Any sum of products of akinetors is an akinetor whose weight is the weight of any term of the sum.*

This, of course, includes the tensor equation as a special case of the akinetor equation. For instance, let

$$a\tilde{u}^i\tilde{v}^j + \tilde{w}^{ij} = \tilde{X}^{ij},$$

where a is a tensor and $\tilde{u}^i$ and $\tilde{v}^j$ are akinetors of weight +1. Then each term of the equation has weight +2, and

$$\tilde{X}^{i'j'} = \Delta^2 \frac{\partial x^{i'}}{\partial x^i} \frac{\partial x^{j'}}{\partial x^j} \tilde{X}^{ij}.$$

On the other hand, if the weight of $\tilde{u}^i$ is $+w$ and the weight of $\tilde{v}^j$ is $-w$, then the weight of each term of the equation is zero. Thus, the equation may be written

$$a\tilde{u}^i\tilde{v}^j + w^{ij} = X^{ij},$$

where w^{ij} and X^{ij} are tensors, and

$$X^{i'j'} = \frac{\partial x^{i'}}{\partial x^i} \frac{\partial x^{j'}}{\partial x^j} X^{ij}.$$

Also, as with tensors, we have:

Theorem XXII. *The form of an akinetor equation is invariant with respect to coordinate transformation.*

For instance, take the equation

$$A'\tilde{U}_{i'j'}\tilde{V}^{i'j'} + B'\tilde{u}_{i'}v^{i'} = \tilde{C}'$$

in a specific coordinate system $x^{i'}$. Here A', B', and $v^{i'}$ are tensors, and the akinetor $\tilde{U}_{i'j'}$ is of weight +1, $\tilde{V}^{i'j'}$ is of weight −2. Since each term must have the same weight, $\tilde{C}'$ and $\tilde{u}_{i'}$ are of weight −1. Therefore, the transformations are

$$A' = A, \qquad B' = B, \qquad \tilde{C}' = \Delta^{-1}C,$$

$$\tilde{U}_{i'j'} = \Delta \frac{\partial x^i}{\partial x^{i'}} \frac{\partial x^j}{\partial x^{j'}} \tilde{U}_{ij},$$

$$\tilde{V}^{i'j'} = \Delta^{-2} \frac{\partial x^{i'}}{\partial x^i} \frac{\partial x^{j'}}{\partial x^j} V^{ij},$$

$$\tilde{u}_{i'} = \Delta^{-1} \frac{\partial x^i}{\partial x^{i'}} \tilde{u}_i,$$

$$v^{i'} = \frac{\partial x^{i'}}{\partial x^i} v^i.$$

Substitution gives

$$A\tilde{U}_{ij}\tilde{V}^{ij}+B\tilde{u}_i v^i=C$$

for any coordinate system x^i.

Finally we have two theorems that are obvious extensions of Theorems XVI and XVII.

Theorem XXIII. *If an unknown holor appears as a term in an invariant equation and if every other holor in the equation is an akinetor, then the unknown holor is also an akinetor.*

Suppose, for instance, that we have the invariant equation

$$\tilde{M}^{i'j'}\tilde{u}_{j'}+v^{i'}=\tilde{w}^{i'},$$

where the holors $\tilde{M}^{i'j'}$, $\tilde{u}_{j'}$, and $\tilde{w}^{i'}$ are known to be akinetors of weights α, β, and $\alpha+\beta$, respectively. Substitution of the transformation equations then shows that the unknown holor $v^{i'}$ must also be an akinetor of weight $\alpha+\beta$.

Theorem XXIV. *If an unknown holor appears in a product with an arbitrary akinetor and if the equation is an invariant equation and all the other holors in the equation are akinetors, then the unknown is also an akinetor and the equation is an akinetor equation.*

As an example, consider the invariant equation

$$\tilde{U}_{i'j'}\tilde{V}^{i'j'}=\tilde{S}', \tag{5.13}$$

where $\tilde{V}^{i'j'}$ is an arbitrary akinetor of weight α, and $\tilde{S}'$ is an akinetor of weight β. Transformations are

$$\begin{aligned}\tilde{V}^{i'j'}&=\Delta^{\alpha}\frac{\partial x^{i'}}{\partial x^{i}}\frac{\partial x^{j'}}{\partial x^{j}}\tilde{V}^{ij},\\ \tilde{S}'&=\Delta^{\beta}\tilde{S}.\end{aligned} \tag{5.14}$$

Since Equation (5.13) is invariant in form with respect to coordinate transformation,

$$\tilde{U}_{ij}\tilde{V}^{ij}=\tilde{S}. \tag{5.15}$$

Substitution of Equation (5.14) into (5.13) and use of (5.15) gives

$$\tilde{V}^{ij}\left[\tilde{U}_{ij}-\Delta^{\alpha-\beta}\frac{\partial x^{i'}}{\partial x^{i}}\frac{\partial x^{j'}}{\partial x^{j}}U_{i'j'}\right]=0.$$

Since $\tilde{V}^{ij}$ is arbitrary, the bracket must be zero and $\tilde{U}_{ij}$ transforms as an akinetor of weight $\beta-\alpha$. If $\tilde{V}^{ij}$ is not arbitrary, however, this conclusion is not necessarily valid.

5.04. Combinations

According to Theorem XXI, any sum of products of akinetors is an akinetor. This principle allows the formation of an unlimited number of new akinetors by combining known akinetors. In 3-space, for instance, nilvalent akinetors can be obtained as products of arbitrary akinetors du^i, dv^j, dw^k:

$$d\tilde{\mathfrak{V}} = \tilde{\gamma}_{ijk}\, du^i\, dv^j\, dw^k, \tag{5.16}$$

where $\tilde{\gamma}_{ijk}$ is an arbitrary akinetor. Similarly, univalent akinetors can be obtained as

$$d\tilde{\mathfrak{A}}_i = dv^j \otimes dw^k = \tilde{\gamma}_{ijk}\, dv^j\, dw^k, \tag{5.17}$$

and bivalent akinetors as

$$d\tilde{W}_{ij} = \tilde{\gamma}_{ijk}\, dw^k. \tag{5.18}$$

Obvious extensions of this method can be made to n-space. For example, in 4-space, Equation (5.16) may be written

$$d\tilde{\mathfrak{V}} = \tilde{\gamma}_{ijkl}\, du^i\, dv^j\, dw^k\, dz^l. \tag{5.16a}$$

According to Theorem XXI, all of these products are akinetors if the holors on the right side of the equations are akinetors.

In talking about tensors and akinetors, we usually consider behavior under the general functional transformation

$$x^{i'} = F^{i'}(x^i). \tag{4.11}$$

With such a general coordinate transformation, familiar concepts such as line, plane, parallelism, orthogonality, distance, area, and volume have no significance. Under these circumstances, it may seem surprising that any combination of akinetors should be meaningful. But akinetors represent invariant geometric objects and are therefore not without geometric significance, even though the exact geometric representation may not be easy to visualize.

First consider Equation (5.17), where $\tilde{\gamma}_{ijk}$ has been replaced by δ^{123}_{ijk} for definiteness:

$$d\tilde{\mathfrak{A}}_i = dv^j \otimes dw^k = \delta^{123}_{ijk}\, dv^j\, dw^k. \tag{5.19}$$

If dv^j and dw^k are tensors (as they will be if taken as increments in coordinates), then $d\tilde{\mathfrak{A}}_i$ is an akinetor of weight -1.

We now show that in the special case of Euclidean 3-space with rectangular coordinates, Equation (5.19) represents area. In this case, the area of a parallelogram may be written as the cross product of vectors **dv** and **dw** along two sides of the parallelogram:

$$d\mathbf{\mathfrak{A}} = d\mathbf{v} \times d\mathbf{w},$$

where $d\mathbf{\mathfrak{A}}$ is a vector that is perpendicular to the plane of $d\mathbf{v}$ and $d\mathbf{w}$ and whose magnitude is equal to the area of the parallelogram. Now return to Equation (5.19). Without loss of generality, the coordinate system may be oriented so that vector dv^j is in the x^2 direction and vector dw^k is in the x^2x^3 plane. Expansion of Equation (5.19) gives

$$d\mathfrak{A}_1 = \delta^{123}_{123}\, dv^2\, dw^3 + \delta^{123}_{132}\, dv^3\, dw^2,$$

$$d\mathfrak{A}_2 = \delta^{123}_{231}\, dv^3\, dw^1 + \delta^{123}_{213}\, dv^1\, dw^3,$$

$$d\mathfrak{A}_3 = \delta^{123}_{312}\, dv^1\, dw^2 + \delta^{123}_{321}\, dv^2\, dw^1.$$

But

$$dv^j = (0, dv, 0), \qquad dw^k = (0, dw \cos\theta, dw \sin\theta),$$

where dv and dw are magnitudes of the vectors. Thus,

$$\begin{aligned} d\tilde{\mathfrak{A}}_1 &= dv^2\, dw^3 - dv^3\, dw^2 = dv\, dw \sin\theta, \\ d\tilde{\mathfrak{A}}_2 &= d\tilde{\mathfrak{A}}_3 = 0, \end{aligned} \tag{5.20}$$

which agrees with the usual definition of area.

We have shown that $d\tilde{\mathfrak{A}}_i$, as defined by Equation (5.19), is an akinetor in general n-space and that it represents an area in the special case of Euclidean 3-space. The general akinetor $d\tilde{\mathfrak{A}}_i$ is called pseudoarea. Ordinarily, of course, it does not represent an area, since the concept of area is meaningless under the transformation of Equation (4.11).

Similarly, a nilvalent akinetor can be obtained from Equation (5.16), and this akinetor may be called pseudovolume[4]:

$$d\tilde{\mathfrak{V}} = \delta^{123}_{ijk}\, du^i\, dv^j\, dw^k. \tag{5.21}$$

If du^i, dv^j, and dw^k are tensors, $d\tilde{\mathfrak{V}}$ is an akinetor of weight -1 because δ^{123}_{ijk} has that weight.

For the special case of orthogonal Cartesian coordinates in Euclidean 3-space, we can take

$$du^i = (du, 0, 0),$$
$$dv^j = (dv\cos(\psi)_1, dv\sin(\psi)_1, 0),$$
$$dw^k = (dw\sin\theta\cos(\psi)_2, dw\sin\theta\sin(\psi)_2, dw\cos\theta).$$

Substitution into Equation (5.21) gives

$$d\tilde{\mathcal{V}} = du\,dv\,dw\cos\theta\sin(\psi)_1, \tag{5.22}$$

which is the true volume of the parallelepiped.

Therefore, Equation (5.21) defines an akinetor in general n-space; and this akinetor represents a volume in the particular case of rectangular coordinates in Euclidean 3-space. In general, of course, pseudovolume is *not* a volume. Evidently other akinetors can be concocted ad libitum to meet the needs of physics.

5.05. Derivatives

Derivatives are also important. Intuition might suggest that the derivative of an akinetor would be an akinetor. With a few notable exceptions, however, such is not the case. Thus, we find it necessary to study the behavior of derivatives under coordinate transformation and to devise methods by which they will, if possible, transform as akinetors. Also, certain combinations of derivatives are found to yield akinetors, even though the individual derivatives are not akinetors.

Let us first consider derivatives of the nilvalent tensor ϕ. Here ϕ is an invariant of coordinate transformation, so

$$\phi = \phi'. \tag{5.23}$$

The derivative with respect to the coordinates is $\partial\phi/\partial x^i$ or

$$\frac{\partial\phi}{\partial x^i} = \frac{\partial\phi}{\partial x^{i'}}\frac{\partial x^{i'}}{\partial x^i}.$$

Substitution of Equation (5.23) gives

$$\frac{\partial\phi}{\partial x^i} = \frac{\partial x^{i'}}{\partial x^i}\frac{\partial\phi'}{\partial x^{i'}}, \tag{5.24}$$

so $\partial\phi/\partial x^i$ transforms as a univalent covariant tensor. It corresponds to gradient in vector analysis and may be written

$$\mathrm{grad}_i\,\phi = \partial\phi/\partial x^i. \tag{5.25}$$

Equation (5.25) represents one of the few derivatives that are akinetors. For instance, there is no general way of writing grad ϕ as a contravariant tensor. Or suppose that $\tilde{\phi}$ is a nilvalent akinetor of weight $-w$. Then Equation (5.23) is replaced by

$$\tilde{\phi} = \Delta^w \tilde{\phi}'. \tag{5.26}$$

Differentiation gives

$$\frac{\partial \tilde{\phi}}{\partial x^i} = \frac{\partial}{\partial x^i} \Delta^w \tilde{\phi}' + \Delta^w \frac{\partial x^{i'}}{\partial x^i} \frac{\partial \tilde{\phi}'}{\partial x^{i'}}. \tag{5.27}$$

The second term on the right is the correct transformation for an akinetor of weight $-w$. But the first term is not zero, and therefore $\partial \tilde{\phi}/\partial x^i$ is *not* an akinetor unless $w = 0$.

Derivatives of univalent akinetors are even more uncooperative. Consider a univalent contravariant tensor v^i:

$$v^{i'} = \frac{\partial x^{i'}}{\partial x^i} v^i. \tag{5.28}$$

Differentiation gives

$$\begin{aligned} \frac{\partial x^{i'}}{\partial x^{j'}} &= \frac{\partial}{\partial x^j}\left[\frac{\partial x^{i'}}{\partial x^i} v^i\right] \frac{\partial x^j}{\partial x^{j'}} \\ &= \frac{\partial x^{i'}}{\partial x^i} \frac{\partial v^j}{\partial x^{j'}} \frac{\partial v^i}{\partial x^j} + \frac{\partial^2 x^{i'}}{\partial x^j \partial x^i} \frac{\partial x^j}{\partial x^{j'}} v^i. \end{aligned} \tag{5.29}$$

Because of the second derivative, $\partial v^i/\partial x^j$ is not an akinetor.

Similarly, with the univalent covariant tensor u_i, the transformation is

$$u_{i'} = \frac{\partial x^i}{\partial x^{i'}} u_i. \tag{5.30}$$

By differentiation,

$$\begin{aligned} \frac{\partial u_{i'}}{\partial x^{j'}} &= \frac{\partial}{\partial x^j}\left[\frac{\partial x^i}{\partial x^{i'}} u_i\right] \frac{\partial x^j}{\partial x^{j'}} \\ &= \frac{\partial x^i}{\partial x^{i'}} \frac{\partial x^j}{\partial x^{j'}} \frac{\partial u_i}{\partial x^j} + \frac{\partial^2 x^i}{\partial x^{j'} \partial x^{i'}} u_i. \end{aligned} \tag{5.31}$$

Again the derivative is not an akinetor.

One more example might be considered. Take the bivalent tensor w^{ij}:

$$w^{i'j'} = \frac{\partial x^{i'}}{\partial x^i}\frac{\partial x^{j'}}{\partial x^j}w^{ij}. \tag{5.32}$$

Then

$$\begin{aligned}\frac{\partial w^{i'j'}}{\partial x^{k'}} &= \frac{\partial}{\partial x^k}\left[\frac{\partial x^{i'}}{\partial x^i}\frac{\partial x^{j'}}{\partial x^j}w^{ij}\right]\frac{\partial x^k}{\partial x^{k'}}\\ &= \frac{\partial x^{i'}}{\partial x^i}\frac{\partial x^{j'}}{\partial x^j}\frac{\partial x^k}{\partial x^{k'}}\frac{\partial w^{ij}}{\partial x^k}\\ &\quad+\left[\frac{\partial^2 x^{i'}}{\partial x^k\partial x^i}\frac{\partial x^{j'}}{\partial x^j}+\frac{\partial^2 x^{j'}}{\partial x^k\partial x^j}\frac{\partial x^{i'}}{\partial x^i}\right]\frac{\partial x^k}{\partial x^{k'}}w^{ij}.\end{aligned} \tag{5.33}$$

The first term, if it existed alone, would make $\partial w^{ij}/\partial x^k$ a tensor; but the second derivatives spoil the picture.

Turning now to akinetors that are not tensors, we find that $\partial\tilde{v}^i/\partial x^i$ is an akinetor if $\tilde{v}^i$ is an akinetor of weight $+1$. The transformation equation for an akinetor $\tilde{v}^i$ of weight w is

$$\tilde{v}^i = \Delta^{-w}\frac{\partial x^i}{\partial x^{i'}}\tilde{v}^{i'}, \tag{5.07}$$

where $\Delta = |\partial x^k/\partial x^{k'}|$ as usual. Differentiation and contraction gives

$$\frac{\partial \tilde{v}^i}{\partial x^i} = \Delta^{-w}\frac{\partial\tilde{v}^{i'}}{\partial x^{i'}} + \Delta^{-w-1}\left[-w\frac{\partial\Delta}{\partial x^{i'}} + \Delta\frac{\partial^2 x^i}{\partial x^{i'}\partial x^{j'}}\frac{\partial x^{j'}}{\partial x^i}\right]\tilde{v}^{i'}.$$

If $\partial\tilde{v}^i/\partial x^i$ is to transform as an akinetor, the bracketed quantity must be zero, or

$$\Delta\frac{\partial^2 x^i}{\partial x^{i'}\partial x^{j'}}\frac{\partial x^{j'}}{\partial x^i} = w\frac{\partial\Delta}{\partial x^{i'}}. \tag{5.34}$$

For $n=2$,

$$\Delta = \left|\frac{\partial x^k}{\partial x^{k'}}\right| = \frac{\partial x^1}{\partial x^{1'}}\frac{\partial x^2}{\partial x^{2'}} - \frac{\partial x^1}{\partial x^{2'}}\frac{\partial x^2}{\partial x^{1'}}$$

and the right side of Equation (5.34) is

$$\begin{aligned}w\frac{\partial\Delta}{\partial x^{i'}} = w\Bigg[&\frac{\partial^2 x^1}{\partial x^{i'}\partial x^{1'}}\frac{\partial x^2}{\partial x^{2'}} + \frac{\partial^2 x^2}{\partial x^{i'}\partial x^{2'}}\frac{\partial x^1}{\partial x^{1'}}\\ &-\frac{\partial^2 x^1}{\partial x^{i'}\partial x^{2'}}\frac{\partial x^2}{\partial x^{1'}} - \frac{\partial^2 x^2}{\partial x^{i'}\partial x^{1'}}\frac{\partial x^1}{\partial x^{2'}}\Bigg].\end{aligned} \tag{5.35}$$

Also,

$$\frac{\partial^2 x^i}{\partial x^{i'}\partial x^{j'}}\frac{\partial x^{j'}}{\partial x^i}=\frac{\partial^2 x^1}{\partial x^{i'}\partial x^{1'}}\frac{\partial x^{1'}}{\partial x^1}+\frac{\partial^2 x^1}{\partial x^{i'}\partial x^{2'}}\frac{\partial x^{2'}}{\partial x^1}$$
$$+\frac{\partial^2 x^2}{\partial x^{i'}\partial x^{1'}}\frac{\partial x^{1'}}{\partial x^2}+\frac{\partial^2 x^2}{\partial x^{i'}\partial x^{2'}}\frac{\partial x^{2'}}{\partial x^{2'}}$$

and the left side of Equation (5.35) becomes

$$\Delta\left[\frac{\partial^2 x^1}{\partial x^{i'}\partial x^{1'}}\frac{\partial x^{1'}}{\partial x^1}+\frac{\partial^2 x^1}{\partial x^{i'}\partial x^{2'}}\frac{\partial x^{2'}}{\partial x^1}+\frac{\partial^2 x^2}{\partial x^{i'}\partial x^{1'}}\frac{\partial x^{1'}}{\partial x^2}+\frac{\partial^2 x^2}{\partial x^{i'}\partial x^{2'}}\frac{\partial x^{2'}}{\partial x^2}\right].$$

But

$$\frac{\partial x^{1'}}{\partial x^1}=\frac{1}{\Delta}\frac{\partial x^2}{\partial x^{2'}},\qquad \frac{\partial x^{1'}}{\partial x^2}=-\frac{1}{\Delta}\frac{\partial x^1}{\partial x^{2'}},$$
$$\frac{\partial x^{2'}}{\partial x^1}=-\frac{1}{\Delta}\frac{\partial x^2}{\partial x^{1'}},\qquad \frac{\partial x^{2'}}{\partial x^2}=\frac{1}{\Delta}\frac{\partial x^1}{\partial x^{1'}}.$$

Therefore,

$$\frac{\partial^2 x^i}{\partial x^{i'}\partial x^{j'}}\frac{\partial x^{j'}}{\partial x^i}=\frac{\partial^2 x^1}{\partial x^{i'}\partial x^{1'}}\frac{\partial x^2}{\partial x^{2'}}-\frac{\partial^2 x^1}{\partial x^{i'}\partial x^{2'}}\frac{\partial x^2}{\partial x^{1'}}$$
$$-\frac{\partial^2 x^2}{\partial x^{i'}\partial x^{1'}}\frac{\partial x^1}{\partial x^{2'}}+\frac{\partial^2 x^2}{\partial x^{i'}\partial x^{2'}}\frac{\partial x^1}{\partial x^{1'}}. \tag{5.36}$$

Comparison with Equation (5.35) shows that

$$\tilde{S}=\partial\tilde{v}^i/\partial x^i \tag{5.37}$$

is an akinetor if and only if $w=+1$. For simplicity, the above proof has been given for 2-space; but it is equally true for n-space.

5.06. Combinations of derivatives

Section 5.05 has shown that most derivatives of akinetors do not transform as akinetors because of the presence of second derivatives in the transformation equations. The possibility of eliminating these second derivatives by subtraction immediately presents itself. For example, take the derivative of a univalent covariant tensor u_i. According to Equation (5.31),

$$\frac{\partial u_{i'}}{\partial x^{j'}}=\frac{\partial^2 x^i}{\partial x^{i'}\partial x^{j'}}u_i+\frac{\partial x^i}{\partial x^{i'}}\frac{\partial x^j}{\partial x^{j'}}\frac{\partial u_i}{\partial x^j}.$$

Interchange of i' and j' gives

$$\frac{\partial u_{j'}}{\partial x^{i'}}=\frac{\partial^2 x^i}{\partial x^{j'}\partial x^{i'}}u_i+\frac{\partial x^i}{\partial x^{j'}}\frac{\partial x^j}{\partial x^{i'}}\frac{\partial u_i}{\partial x^j}.$$

Subtraction of these two expressions eliminates the second derivative, leaving

$$\left[\frac{\partial u_{j'}}{\partial x^{i'}}-\frac{\partial u_{i'}}{\partial x^{j'}}\right]=\frac{\partial x^i}{\partial x^{j'}}\frac{\partial x^j}{\partial x^{i'}}\frac{\partial u_i}{\partial x^j}-\frac{\partial x^i}{\partial x^{i'}}\frac{\partial x^j}{\partial x^{j'}}\frac{\partial u_i}{\partial x^j}.$$

Interchange of the dummy indices i and j in the first term then gives

$$\left[\frac{\partial u_{j'}}{\partial x^{i'}}-\frac{\partial u_{i'}}{\partial x^{j'}}\right]=\frac{\partial x^i}{\partial x^{i'}}\frac{\partial x^j}{\partial x^{j'}}\left[\frac{\partial u_j}{\partial x^i}-\frac{\partial u_i}{\partial x^j}\right],$$

and the bracketed quantity transforms as a tensor. Thus, we have obtained a new tensor, which may be defined as

$$U_{ij}=\left[\frac{\partial u_j}{\partial x^i}-\frac{\partial u_i}{\partial x^j}\right], \tag{5.38}$$

a combination of derivatives that are not akinetors.

This scheme, however, turns out to be of limited applicability. The reader may be interested in trying the same procedure, for instance, on derivatives of the contravariant tensor v^i. Even this slight change from u_i to v^i makes it impossible to eliminate the second derivatives. Likewise, there seems to be no combination of derivatives of bivalent tensors that will yield akinetors.

As a bivalent akinetor, let us take $\tilde{F}^{ij}$ and write

$$\tilde{v}^i=\tilde{F}^{ij}u_j, \tag{5.39}$$

where $\tilde{v}^i$ is an akinetor of weight $+1$ and u_j is a tensor. Then, according to Equation (5.39), $\tilde{F}^{ij}$ is an akinetor of weight $+1$. We now investigate the derivative of $\tilde{F}^{ij}$, making use of the previous results for $\partial\tilde{v}^i/\partial x^i$.

Differentiation of Equation (5.39) gives

$$\frac{\partial\tilde{v}^i}{\partial x^i}=\frac{\partial\tilde{F}^{ij}}{\partial x^i}u_j+\tilde{F}^{ij}\frac{\partial u_j}{\partial x^i}. \tag{5.40}$$

To prove that $\partial\tilde{F}^{ij}/\partial x^i$ is an akinetor, we must either investigate its behavior under coordinate transformation (as was done with $\partial\tilde{v}^i/\partial x^i$) or prove that the other terms in Equation (5.40) are akinetors. That $\partial u_j/\partial x^i$ is not an akinetor was shown in Equation (5.31). However, according to Equation (5.38), the difference of derivatives,

$$U_{ij}=\left[\frac{\partial u_j}{\partial x^i}-\frac{\partial u_i}{\partial x^j}\right], \tag{5.38}$$

is a tensor. Thus, we need merely obtain such a difference in the final term of Equation (5.40).

Interchange of dummy indices in the final term gives

$$\frac{\partial \tilde{v}^i}{\partial x^i} = \frac{\partial \tilde{F}^{ij}}{\partial x^i} u_j + \tilde{F}^{ji} \frac{\partial u_i}{\partial x^j}. \tag{5.41}$$

Adding Equations (5.40) and (5.41),

$$2\frac{\partial \tilde{v}^i}{\partial x^i} = 2\frac{\partial \tilde{F}^{ij}}{\partial x^i} u_j + \tilde{F}^{ij} \frac{\partial u_j}{\partial x^i} + \tilde{F}^{ji} \frac{\partial u_i}{\partial x^j}.$$

If now $\tilde{F}^{ij}$ is antisymmetric so that $\{\tilde{F}^{ji}\} = -\{\tilde{F}^{ij}\}$, then

$$\frac{\partial \tilde{v}^i}{\partial x^i} = \frac{\partial \tilde{F}^{ij}}{\partial x^i} u_j + \frac{\tilde{F}^{ij}}{2} \left[\frac{\partial u_j}{\partial x^i} - \frac{\partial u_i}{\partial x^j} \right]$$

or

$$\frac{\partial \tilde{v}^i}{\partial x^i} = \frac{\partial \tilde{F}^{ij}}{\partial x^i} u_j + \frac{\tilde{F}^{ij}}{2} U_{ij}, \tag{5.41a}$$

where U_{ij} is the tensor given by Equation (5.38). Thus, Equation (5.41a) is an akinetor equation, and

$$\tilde{X}^j = \partial \tilde{F}^{ij} / \partial x^i \tag{5.42}$$

is an antisymmetric akinetor of weight +1.

5.07. Pseudodivergence

In vector analysis, an important combination of derivatives is the *divergence* of a vector. This is a nilvalent holor, which in Euclidean 3-space and rectangular coordinates may be written

$$\operatorname{div} \mathbf{v} = \frac{\partial v^i}{\partial x^i} = \frac{\partial v^1}{\partial x^1} + \frac{\partial v^2}{\partial x^2} + \frac{\partial v^3}{\partial x^3}.$$

An analogous invariant under the general coordinate transformation, Equation (4.11), was found in Section 5.05. It is the derivative $\partial \tilde{v}^i / \partial x^i$ of a contravariant univalent akinetor $\tilde{v}^i$ of weight +1, and it may be called *pseudodivergence.*[5]

An important integral theorem of vector analysis is Gauss's theorem, or the divergence theorem, which equates the volume integral of the divergence to the surface integral of the vector itself. A generalization of this theorem expressed in terms of the pseudodivergence is possible.

Consider the nilvalent holor dF defined by the equation

$$dF = \tilde{\mathfrak{v}}^i \, d\tilde{\mathfrak{a}}_i, \tag{5.43}$$

where $\tilde{\mathfrak{v}}^i$ is a contravariant akinetor of weight $+1$ and $d\tilde{\mathfrak{a}}_i$ is a pseudoarea, Equation (5.19), of weight -1. Evidently the product of these akinetors is of weight zero, so dF is a nilvalent tensor.

Let us visualize a small pseudovolume $\Delta\tilde{\mathfrak{V}}$ in 3-space and obtain the total outward flux ΔF of $\tilde{\mathfrak{v}}^i$, much as we do in considering ordinary divergence.[6] Without loss of generality, we can introduce a coordinate system x^1, x^2, x^3 along the sides of $\Delta\tilde{\mathfrak{V}}$. For the pseudoarea $\Delta\tilde{\mathfrak{a}}_1$, let

$$\Delta u^i = (\Delta x^1, 0, 0), \qquad \Delta v^j = (0, \Delta x^2, 0), \qquad \Delta w^k = (0, 0, \Delta x^3).$$

An infinitesimal pseudoarea $d\tilde{\mathfrak{a}}_1$ is then

$$d\tilde{\mathfrak{a}}_1 = \delta^{123}_{ijk} \, dv^j \, dw^k = dx^2 \, dx^3$$

according to Equation (5.19); and an infinitesimal pseudovolume is

$$d\tilde{\mathfrak{V}} = \delta^{123}_{ijk} \, du^i \, dv^j \, dw^k = dx^1 \, dx^2 \, dx^3$$

according to Equation (5.21).

The outward flux through the back face $\Delta\tilde{\mathfrak{a}}_1$ is

$$\begin{aligned}\Delta F = \int_{\Delta\tilde{\mathfrak{a}}_1} dF &= -\tilde{\mathfrak{v}}^1 \Delta\tilde{\mathfrak{a}}_1 - \epsilon \, \Delta\tilde{\mathfrak{a}}_1 \\ &= -\tilde{\mathfrak{v}}^1 \Delta x^2 \Delta x^3 - \epsilon \, \Delta x^2 \Delta x^3,\end{aligned}$$

where the second term is introduced to take care of any discrepancy between the actual flux and the approximation $\tilde{\mathfrak{v}}^1 \Delta\tilde{\mathfrak{a}}_1$. The outward flux through the corresponding front face at $x^1 = \Delta x^1$ is

$$\Delta F = \left(\tilde{\mathfrak{v}}^1 + \frac{\partial \tilde{\mathfrak{v}}^1}{\partial x^1} \Delta x^1 \right) \Delta x^2 \Delta x^3 + \zeta \, \Delta x^2 \Delta x^3.$$

The total outward flux in the x^1 direction is therefore

$$\begin{aligned}(\Delta F)_1 &= \left(\tilde{\mathfrak{v}}^1 + \frac{\partial \tilde{\mathfrak{v}}^1}{\partial x^1} \Delta x^1 - \tilde{\mathfrak{v}}^1 + \zeta - \epsilon \right) \Delta x^2 \Delta x^3 \\ &= \frac{\partial \tilde{\mathfrak{v}}^1}{\partial x^1} \Delta\tilde{\mathfrak{V}} + (\zeta - \epsilon)_1 \, \Delta x^2 \Delta x^3.\end{aligned}$$

Similarly, for the other pairs of faces,

$$(\Delta F)_2 = \frac{\partial \tilde{\mathfrak{v}}^2}{\partial x^2} \Delta\tilde{\mathfrak{V}} + (\zeta - \epsilon)_2 \, \Delta x^1 \Delta x^3,$$

$$(\Delta F)_3 = \frac{\partial \tilde{v}^3}{\partial x^3} \Delta \tilde{\mathfrak{V}} + (\zeta - \epsilon)_3 \, \Delta x^1 \, \Delta x^2.$$

Thus, the total outward flux is

$$\Delta F = \int_{\Delta \tilde{\mathfrak{a}}} dF = \frac{\partial \tilde{v}^i}{\partial x^i} \Delta \tilde{\mathfrak{V}} + \bar{\epsilon} \, \Delta \tilde{\mathfrak{V}},$$

where $\bar{\epsilon} \, \Delta \tilde{\mathfrak{V}}$ is the sum of the correction terms. This correction approaches zero in the limit, so

$$\lim_{\Delta \tilde{\mathfrak{V}} \to 0} (\Delta F) = \int_{\Delta \tilde{\mathfrak{a}}} dF = \frac{\partial \tilde{v}^i}{\partial x^i} \Delta \tilde{\mathfrak{V}} = \int_{\Delta \tilde{\mathfrak{V}}} \frac{\partial \tilde{v}^i}{\partial x^i} d\tilde{\mathfrak{V}}. \tag{5.44}$$

Now take a finite pseudovolume $\tilde{\mathfrak{V}}$ with pseudosurface $\tilde{\mathfrak{a}}$. This pseudovolume can be considered as made up of elements $\Delta \tilde{\mathfrak{V}}$, and the total volume integral is

$$\int_{\tilde{\mathfrak{V}}} \frac{\partial v^i}{\partial x^i} d\tilde{\mathfrak{V}}.$$

The surface integrals cancel[6] at all boundaries between $\Delta \tilde{\mathfrak{V}}$'s except at the outer pseudosurface $\tilde{\mathfrak{a}}$, so

$$F = \int_{\tilde{\mathfrak{a}}} dF = \int_{\tilde{\mathfrak{a}}} \tilde{v}^i \, d\tilde{\mathfrak{a}}_i$$

and Equation (5.44) becomes:

Theorem XXV. The generalized divergence theorem. *For any contravariant univalent akinetor of weight* +1,

$$\int_{\tilde{\mathfrak{a}}} \tilde{v}^i \, d\tilde{\mathfrak{a}}_i = \int_{\tilde{\mathfrak{V}}} \frac{\partial \tilde{v}^i}{\partial x^i} d\tilde{\mathfrak{V}} \tag{5.45}$$

where $\tilde{\mathfrak{a}}$ is the pseudoarea that bounds the pseudovolume $\tilde{\mathfrak{V}}$.

5.08. Pseudocurl

Three important combinations of derivatives in vector analysis are gradient of a scalar, divergence of a vector, and curl of a vector.[7] The generalization of gradient, applicable for all coordinate transformations, Equation (4.11), is a covariant univalent tensor

$$v_i = \partial \phi / \partial x^i. \tag{5.29}$$

The generalization of divergence applies to a contravariant univalent akinetor of weight +1, the pseudodivergence being expressed as $\partial\tilde{v}^i/\partial x^i$, as shown in Section 5.05. Now consider the generalization of curl.

In Section 5.05, a combination of derivatives was found that gave a bivalent tensor U_{ij}:

$$U_{ij}=\left[\frac{\partial u_j}{\partial x^i}-\frac{\partial u_i}{\partial x^j}\right]. \tag{5.38}$$

In 3-space, a bivalent tensor has in general nine merates. But U_{ij} is antisymmetric, so it has only three independent merates:

$$U_{ij}=\begin{bmatrix} 0 & U_{12} & U_{13} \\ -U_{12} & 0 & U_{23} \\ -U_{13} & -U_{23} & 0 \end{bmatrix}.$$

This fact suggests the possibility of expressing U_{ij} as a univalent holor

$$W^i=(W^1, W^2, W^3).$$

One should realize, of course, that such a possibility is a peculiarity of 3-space. In general, an antisymmetric U_{ij} does not have a number of independent merates equal to the dimensionality of the space.

Let us build a new univalent akinetor expressed as

$$\tilde{W}^i=\delta^{ijk}_{123}\frac{\partial u_k}{\partial x^j}, \tag{5.46}$$

where u_k is a tensor. Expansion gives the general expression

$$\begin{aligned} \tilde{W}^1 &= \delta^{123}_{123}\frac{\partial u_3}{\partial x^2}+\delta^{132}_{123}\frac{\partial u_2}{\partial x^3}=\frac{\partial u_3}{\partial x^2}-\frac{\partial u_2}{\partial x^3}, \\ \tilde{W}^2 &= \frac{\partial u_1}{\partial x^3}-\frac{\partial u_3}{\partial x^1}, \\ \tilde{W}^3 &= \frac{\partial u_2}{\partial x^1}-\frac{\partial u_1}{\partial x^2}. \end{aligned} \tag{5.47}$$

But these are just the merates obtained for curl w in rectangular coordinates.[8] Thus, Equation (5.46) gives a generalization of curl that applies in any coordinate system. One can easily show that this generalization is an akinetor of weight +1.

Now consider an integral theorem involving pseudocurl. Let dI be the integral of u_i about an infinitesimal parallelogram. Evidently, for $\Delta C \to 0$,

$$\oint_{\Delta C} u_i\, dx^i = u_2\, dx^2 + \left(u_3 + \frac{\partial u_3}{\partial x^2}\, dx^2\right) dx^3$$

$$- \left(u_2 + \frac{\partial u_2}{\partial x^3}\, dx^3\right) dx^2 - u_3\, dx^3$$

$$= \left(\frac{\partial u_3}{\partial x^2} - \frac{\partial u_2}{\partial x^3}\right) dx^2\, dx^3.$$

This result is equal to $\tilde{W}^1\, d\tilde{\mathfrak{a}}_1$, according to Equation (5.47):

$$\oint_{\Delta C} u_i\, dx^i = \left(\frac{\partial u_3}{\partial x^2} - \frac{\partial u_2}{\partial x^3}\right) dx^2\, dx^3 = \tilde{W}^1\, d\tilde{\mathfrak{a}}_1.$$

A contour integral in the x^2x^3 plane results in an akinetor $\tilde{W}^1$ in the x^1 direction. For an arbitrarily oriented plane, we add contributions in the directions of the three axes, obtaining

$$\oint_{\Delta C} u_i\, dx^i = \tilde{W}^1\, d\tilde{\mathfrak{a}}_1 + \tilde{W}^2\, d\tilde{\mathfrak{a}}_2 + W^3\, d\tilde{\mathfrak{a}}_3 = \tilde{W}^i\, d\tilde{\mathfrak{a}}_i.$$

Summing over the elementary parallelograms composing a finite pseudoarea $\tilde{\mathfrak{a}}$ and canceling integrals over adjacent boundaries in the usual way,[8] we obtain:

Theorem XXVI. *The generalized Stokes theorem for a covariant vector* u_i,

$$\oint_C u_i\, dx^i = \int_{\tilde{\mathfrak{a}}} \delta_{123}^{ijk} \frac{\partial u_k}{\partial x^j}\, d\tilde{\mathfrak{a}}_i, \qquad (5.48)$$

where C is a contour that bounds the pseudoarea $\tilde{\mathfrak{a}}$.

The foregoing proof can readily be generalized to apply to any holor having at least one covariant index $u_{il\cdots k}^{np\cdots q}$.

Theorem XXVII. *The generalized Stokes theorem for any holor having at least one covariant index,*

$$\oint_C u_{il\cdots m}^{np\cdots q}\, dx^i = \int_{\tilde{\mathfrak{a}}} \delta_{123}^{ijk} \frac{\partial u_{kl\cdots m}^{np\cdots q}}{\partial x^j}\, d\tilde{\mathfrak{a}}_i, \qquad (5.48a)$$

where C is a contour that bounds the pseudoarea $\tilde{\mathfrak{a}}$.

5.09. Other akinetors

As noted previously, any sum of products (contracted or uncontracted) of akinetors is an akinetor. Thus, we can obtain an unlimited supply of akinetors to meet the needs of mathematics or of physical applications. Associated with akinetors of any given valence and weight, we can form new akinetors of the same or different valence, weight, and index position.

As an example of a change in valence, suppose that a univalent akinetor $\tilde{v}^i$ is desired, associated with the given bivalent akinetor $\tilde{V}^{jk}$ of weight β. Let

$$\tilde{v}^i = \tilde{\gamma}^i_{jk} \tilde{V}^{jk}, \tag{5.49}$$

where $\tilde{\gamma}^i_{jk}$ is an arbitrary akinetor of weight α. Evidently $\tilde{v}^i$ is a univalent akinetor of weight $\alpha+\beta$. In particular, if $\alpha=-\beta$, $\tilde{v}^i$ is a tensor. Or if we need a nilvalent akinetor $\tilde{S}$, we may write

$$\tilde{S} = \tilde{\gamma}_{jk} \tilde{V}^{jk}, \tag{5.50}$$

obtaining a nilvalent akinetor of weight $\alpha+\beta$ if $\tilde{\gamma}_{jk}$ is of weight α.

A classical way of obtaining a nilvalent holor from a bivalent holor is to take the determinant associated with the given matrix. For example, suppose that a nilvalent holor $|T^{ij}|$ is to be obtained from a tensor T^{ij} with $n=2$. According to Equation (1.19), the expansion of the determinant may be written

$$|T^{ij}| = \tfrac{1}{2}\delta^{12}_{ij}\delta^{12}_{kl}T^{ik}T^{jl}. \tag{5.51}$$

Since T^{ik} and T^{jl} transform as tensors and each delta is an akinetor of weight -1, $|T^{ij}|$ must be an akinetor of weight -2.

If $\tilde{T}^{ij}$ is an akinetor of weight w, instead of a tensor, it transforms as

$$\tilde{T}^{i'j'} = \Delta^w \frac{\partial x^{i'}}{\partial x^i}\frac{\partial x^{j'}}{\partial x^j}\tilde{T}^{ij}. \tag{5.52}$$

The associated determinant is

$$|\tilde{T}^{i'j'}| = \begin{vmatrix} \Delta^w \dfrac{\partial x^{1'}}{\partial x^i}\dfrac{\partial x^{1'}}{\partial x^j}\tilde{T}^{ij} & \Delta^w \dfrac{\partial x^{1'}}{\partial x^i}\dfrac{\partial x^{2'}}{\partial x^j}\tilde{T}^{ij} \\ \Delta^w \dfrac{\partial x^{2'}}{\partial x^i}\dfrac{\partial x^{1'}}{\partial x^j}\tilde{T}^{ij} & \Delta^w \dfrac{\partial x^{2'}}{\partial x^i}\dfrac{\partial x^{2'}}{\partial x^j}\tilde{T}^{ij} \end{vmatrix}$$

$$= \Delta^{2w} \begin{vmatrix} \dfrac{\partial x^{1'}}{\partial x^i}\dfrac{\partial x^{1'}}{\partial x^j}\tilde{T}^{ij} & \dfrac{\partial x^{1'}}{\partial x^i}\dfrac{\partial x^{2'}}{\partial x^j}\tilde{T}^{ij} \\ \dfrac{\partial x^{2'}}{\partial x^i}\dfrac{\partial x^{1'}}{\partial x^j}\tilde{T}^{ij} & \dfrac{\partial x^{2'}}{\partial x^i}\dfrac{\partial x^{2'}}{\partial x^j}\tilde{T}^{ij} \end{vmatrix}.$$

Thus, the transformation includes Δ^{2w} as well as the weight -2 associated with the tensor, Equation (5.51). Evidently $|T^{ij}|$ is an akinetor of weight $2w-2$.

For an akinetor $\tilde{T}^{ij}$ of plethos n and weight w, Equation (1.19) gives

$$|\tilde{T}^{ij}| = \frac{1}{n!}\delta^{12\cdots n}_{kl\cdots\upsilon}\delta^{12\cdots n}_{\alpha\beta\cdots\omega}\tilde{T}^{k\alpha}\tilde{T}^{l\beta}\cdots\tilde{T}^{\upsilon\omega}. \tag{5.53}$$

An obvious extension of the previous argument shows that $|\tilde{T}^{ij}|$ is now an akinetor with weight $nw-2$. Similarly, if $\tilde{U}_{ij}$ is an akinetor of weight w, its determinant is [Equation (1.18)],

$$|\tilde{U}_{ij}| = \frac{1}{n!}\delta^{kl\cdots\upsilon}_{12\cdots n}\delta^{\alpha\beta\cdots\omega}_{12\cdots n}\tilde{U}_{k\alpha}\tilde{U}_{l\beta}\cdots\tilde{U}_{\upsilon\omega}, \tag{5.54}$$

and $|\tilde{U}_{ij}|$ is an akinetor of weight $nw+2$.

We have considered a number of ways of changing the valence and weight of akinetors. Another possibility is to change between covariance and contravariance.[9] To change an akinetor $\tilde{T}_{pq}$ into $\tilde{T}^{mn}$ in 4-space, for instance, let

$$\tilde{T}^{mn} = \tfrac{1}{2}\delta^{mnpq}_{1234}\tilde{T}_{pq}. \tag{5.55}$$

If $\tilde{T}_{pq}$ is an akinetor of weight w, then the new akinetor $\tilde{T}^{mn}$ is of weight $w+1$, since the Kronecker delta has weight $+1$.

Or we can write

$$\tilde{T}_{pq} = \tfrac{1}{2}\delta^{1234}_{pqrs}\tilde{T}^{rs}. \tag{5.56}$$

Since this delta is of weight -1, the contravariant akinetor has again the higher weight. Akinetors $\tilde{T}^{mn}$ and $\tilde{T}_{pq}$ related in this way are called dual akinetors.[9] Expansion of Equation (5.55) for an arbitrary bivalent covariant akinetor $\tilde{T}_{pq}$ gives

$$\tilde{T}^{mn} = \begin{bmatrix} 0 & \frac{1}{2}(\tilde{T}_{34}-\tilde{T}_{43}) & \frac{1}{2}(\tilde{T}_{42}-\tilde{T}_{24}) & \frac{1}{2}(\tilde{T}_{23}-\tilde{T}_{32}) \\ -\frac{1}{2}(\tilde{T}_{34}-\tilde{T}_{43}) & 0 & \frac{1}{2}(\tilde{T}_{14}-\tilde{T}_{41}) & \frac{1}{2}(\tilde{T}_{31}-\tilde{T}_{13}) \\ -\frac{1}{2}(\tilde{T}_{42}-\tilde{T}_{24}) & -\frac{1}{2}(\tilde{T}_{14}-\tilde{T}_{41}) & 0 & \frac{1}{2}(\tilde{T}_{12}-\tilde{T}_{21}) \\ -\frac{1}{2}(\tilde{T}_{23}-\tilde{T}_{32}) & -\frac{1}{2}(\tilde{T}_{31}-\tilde{T}_{13}) & -\frac{1}{2}(\tilde{T}_{12}-\tilde{T}_{21}) & 0 \end{bmatrix}, \tag{5.55a}$$

which is always an alternating akinetor. Similarly, if $\tilde{T}^{rs}$ is an arbitrary contravariant bivalent akinetor, expansion of Equation (5.56) gives

$$\tilde{T}_{pq} = \begin{bmatrix} 0 & \frac{1}{2}(\tilde{T}^{34}-\tilde{T}^{43}) & \frac{1}{2}(\tilde{T}^{42}-\tilde{T}^{24}) & \frac{1}{2}(\tilde{T}^{23}-\tilde{T}^{32}) \\ -\frac{1}{2}(\tilde{T}^{34}-\tilde{T}^{43}) & 0 & \frac{1}{2}(\tilde{T}^{14}-\tilde{T}^{41}) & \frac{1}{2}(\tilde{T}^{31}-\tilde{T}^{13}) \\ -\frac{1}{2}(\tilde{T}^{42}-\tilde{T}^{24}) & -\frac{1}{2}(\tilde{T}^{14}-\tilde{T}^{41}) & 0 & \frac{1}{2}(\tilde{T}^{12}-\tilde{T}^{21}) \\ -\frac{1}{2}(\tilde{T}^{23}-\tilde{T}^{32}) & -\frac{1}{2}(\tilde{T}^{31}-\tilde{T}^{13}) & -\frac{1}{2}(\tilde{T}^{31}-\tilde{T}^{13}) & 0 \end{bmatrix}, \tag{5.56a}$$

which is also an alternating akinetor.

However, if $\tilde{T}_{pq}$ is an alternating akinetor, Equation (5.55a) becomes

$$\tilde{T}^{mn} = \begin{bmatrix} 0 & \tilde{T}_{34} & \tilde{T}_{42} & \tilde{T}_{23} \\ -\tilde{T}_{34} & 0 & \tilde{T}_{14} & \tilde{T}_{31} \\ -\tilde{T}_{42} & -\tilde{T}_{14} & 0 & \tilde{T}_{12} \\ -\tilde{T}_{23} & -\tilde{T}_{31} & -\tilde{T}_{12} & 0 \end{bmatrix}. \tag{5.55b}$$

Also, if $\tilde{T}^{rs}$ is an alternating tensor,

$$\tilde{T}_{pq} = \begin{bmatrix} 0 & \tilde{T}^{34} & \tilde{T}^{42} & \tilde{T}^{23} \\ -\tilde{T}^{34} & 0 & \tilde{T}^{14} & \tilde{T}^{31} \\ -\tilde{T}^{42} & -\tilde{T}^{14} & 0 & \tilde{T}^{12} \\ -\tilde{T}^{23} & -\tilde{T}^{31} & -\tilde{T}^{12} & 0 \end{bmatrix}. \tag{5.56b}$$

5.10. Summary

This chapter deals with a subclass of holors called akinetors. The univalent contravariant akinetor transforms as

$$\tilde{v}^{i'} = \sigma \frac{\partial x^{i'}}{\partial x^{i}} \tilde{v}^{i}. \tag{5.02}$$

Other akinetors transform in a similar way, reducing to tensors if $\sigma = 1$. The relation is shown graphically in Figure 5.02. With most akinetors, σ is a power of Δ, where Δ is the Jacobian of the transformation:

$$\Delta = \begin{vmatrix} \dfrac{\partial x^1}{\partial x^{1'}} & \dfrac{\partial x^1}{\partial x^{2'}} & \cdots & \dfrac{\partial x^1}{\partial x^{n'}} \\ \dfrac{\partial x^2}{\partial x^{1'}} & \dfrac{\partial x^2}{\partial x^{2'}} & \cdots & \dfrac{\partial x^2}{\partial x^{n'}} \\ \cdot & \cdot & \cdots & \cdot \\ \dfrac{\partial x^n}{\partial x^{1'}} & \dfrac{\partial x^n}{\partial x^{2'}} & \cdots & \dfrac{\partial x^n}{\partial x^{n'}} \end{vmatrix}.$$

For instance, we may have the akinetor transformation

$$\tilde{v}^{i'} = \Delta^{w} \frac{\partial x^{i'}}{\partial x^{i}} \tilde{v}^{i}, \tag{5.07}$$

where w is the weight of the akinetor.

The algebra of akinetors is slightly more complicated than the algebra of tensors because of the necessity of considering weight.

Definition VIII: Two akinetors are equal if and only if each corresponding pair of merates is equal and if the akinetors are of the same weight.

Definition IX: The sum of two akinetors is an akinetor of the same valence, plethos, and weight, each merate of the sum being obtained by addition of corresponding merates of the original akinetors.

Definition X: Contracted and uncontracted products of akinetors are always possible, the weight of the product being the sum of the weights of the factors.

Definition XI: An akinetor equation is defined as an equation having the following properties:
(a) It contains only akinetors.
(b) It is expressed in one coordinate system.
(c) It possesses index balance.

Theorem XVIII. *In an akinetor equation, each term must have the same literal indices in the same positions and in the same order, and the weight of each term must be the same.*

Theorem XIX. *Akinetors equal to each other in one coordinate system are equal to each other in all coordinate systems.*

Theorem XX. *If an akinetor vanishes in one coordinate system, it vanishes in all coordinate systems.*

Theorem XXI. *Any sum of products of akinetors is an akinetor whose weight is the weight of any term of the sum.*

Theorem XXII. *The form of an akinetor equation is invariant with respect to coordinate transformation.*

Theorem XXIII. *If an unknown holor appears as a term in an invariant equation, and if every other holor in the equation is an akinetor, then the unknown holor is also an akinetor.*

Theorem XXIV. *If an unknown holor appears in a product with an arbitrary akinetor, and if the equation is an invariant equation and all the* *other holors in the equation are akinetors, then the unknown is also an akinetor and the equation is an akinetor equation.*

After the development of akinetor algebra, the remainder of the chapter is devoted to obtaining new akinetors (Table 5.01). A great variety of akinetors can be obtained as products of other akinetors. Another method of obtaining a nilvalent akinetor from a bivalent akinetor is to write the determinant.

Derivatives of akinetors are generally not akinetors. There are a few exceptions, however, as indicated in Table 5.01. In this way, we obtain a generalized gradient

$$\text{grad}_i\, \phi = \partial\phi/\partial x^i,$$

a pseudodivergence

$$\partial\tilde{v}^i/\partial x^i,$$

and a pseudocurl

$$\tilde{W}^i = \delta_{123}^{ijk} \frac{\partial u_j}{\partial x^k}.$$

These holors are related by three important theorems:

Theorem XXV. *The generalized divergence theorem. For any contravariant, univalent akinetor of weight +1,*

$$\int_{\tilde{\mathfrak{a}}} \tilde{v}^i\, d\tilde{\mathfrak{a}}_i = \int_{\tilde{\mathfrak{V}}} \frac{\partial \tilde{v}^i}{\partial x^i}\, d\tilde{\mathfrak{V}},$$

where $\tilde{\mathfrak{a}}$ is the pseudoarea that bounds the pseudovolume $\tilde{\mathfrak{V}}$.

Theorem XXVI. *The generalized Stokes theorem for a covariant vector u_i,*

$$\oint_C u_i\, dx^i = \int_{\tilde{\mathfrak{a}}} \delta_{123}^{ijk} \frac{\partial u_k}{\partial x^j}\, d\tilde{\mathfrak{a}}_i,$$

where C is a contour that bounds the pseudoarea $\tilde{\mathfrak{a}}$.

Theorem XXVII. *The generalized Stokes theorem for any holor having at least one covariant index,*

$$\oint_C u_{il\cdots m}^{np\cdots q}\, dx^i = \int_{\tilde{\mathfrak{a}}} \delta_{123}^{ijk} \frac{\partial u_{kl\cdots m}^{np\cdots q}}{\partial x^j}\, d\tilde{\mathfrak{a}}_i,$$

where C is a contour that bounds the pseudoarea $\tilde{\mathfrak{a}}$.

Table 5.01. *Some akinetors*

Name	Equation	Symbol	Valence	Weight
Increment in coordinates	–	dx^i	1	0
Kronecker delta	(4.53)	δ^i_j	2	0
Generalized Kronecker delta	(4.54)	δ^{ijk}_{lmn}	6	0
Alternator	–	δ^{ijk}_{123}	3	+1
Alternator	(4.57)	δ^{123}_{lmn}	3	−1
Pseudoarea	(5.17)	$d\tilde{\mathfrak{A}}_i = \delta^{123}_{ijk}\, dv^j\, dw^k$	1	−1
Pseudovolume	(5.16)	$d\tilde{\mathfrak{V}} = \delta^{123}_{ijk}\, du^i\, dv^j\, dw^k$ (where du^i, dv^j, dw^k are tensors)	0	−1
Gradient	(5.25)	$\mathrm{grad}_i\, \phi = \partial\phi/\partial x^i$ (where ϕ is a nilvalent tensor)	1	0
Pseudodivergence	(5.37)	$\tilde{S} = \partial\tilde{v}^i/\partial x^i$ (where $\tilde{v}^i$ is an akinetor of weight +1)	0	+1
	(5.42)	$\tilde{X}^j = \partial\tilde{F}^{ij}/\partial x^i$ (where $\tilde{F}^{ij}$ is an antisymmetric akinetor of weight +1)	1	+1
	(5.40)	$U_{ij} = \dfrac{\partial u_j}{\partial x^i} - \dfrac{\partial u_i}{\partial x^j}$	2	0
Pseudocurl	(5.46)	$\tilde{W}^i = \delta^{ijk}_{123}\dfrac{\partial u_k}{\partial x^j}$ (where u_k is a tensor)	1	+1

Problems

5.01 Given two akinetors $\tilde{v}^i$ and $\tilde{w}^i$ in 3-space with weights α and β, respectively, show what difficulty is experienced (if any) when these akinetors are added by the familiar rules of holor addition.

5.02 Consider the equation

$$A^i u_i \tilde{V}^{jk} + \tilde{W}^{jk} = \tilde{B}^j X^k,$$

where A^i, X^k, and u_i are tensors and $\tilde{V}^{jk}$ is an akinetor of weight $+2$. Write the transformation equations for all holors in the equation and specify all weights.

5.03 Repeat Problem 5.02 for

$$A^i \tilde{V}^{jk} + \tilde{Y}^{ijk} = \tilde{C}^{ij} X^k.$$

5.04 In the equation

$$\tilde{A} + \tilde{u}_i v^i = \tilde{S},$$

$\tilde{u}_i$ is an akinetor of weight $+w$, v^i is a tensor, and $\tilde{A}$ and $\tilde{S}$ are nilvalent holors. Write the transformation equations for all holors.

5.05 In the akinetor equation

$$\tilde{u}_{ij} \tilde{w}^{jk} = \tilde{U}_i^k,$$

$\tilde{u}_{ij}$ is of weight $+1$ and $\tilde{U}_i^k$ is of weight -2.
(a) Write the transformation equations for the three akinetors.
(b) Find the inverse of $\tilde{u}_{ij}$ and show how it transforms.
(c) Solve the equation for $\tilde{w}^{jk}$.

5.06 In the akinetor equation

$$\tilde{u}_{ij} V^{jkm} = A_i^{km} + W_i^{km},$$

$\tilde{u}_{ij}$ is of weight $+1$ and A_i^{km} is a tensor.
(a) Write the transformation equations for the four akinetors.
(b) Find the inverse of $\tilde{u}_{ij}$ and show how it transforms.
(c) Solve the equation for V^{jkm}.

5.07
(a) A null product is

$$\tilde{U}_{ij} \tilde{V}^{jl} = 0,$$

where $\tilde{U}_{ij}$ is an akinetor of weight $+w$. What is the nature of the holor $\tilde{V}^{kl}$?

(b) Repeat (a) for

$$\tilde{U}_{ij}\tilde{V}^{jk}=0.$$

5.08 A null product

$$\tilde{U}_{ij}\tilde{V}^{jk}=0$$

has $\tilde{U}_{ij}$ an arbitrary antisymmetric akinetor of weight -1. What can be said about $\tilde{V}^{jk}$?

5.09 Repeat Problem 5.08 for $\tilde{U}_{ij}$ symmetric but otherwise arbitrary.

5.10 Suppose that

$$\tilde{U}_{ij}\tilde{V}^{jk}=\tilde{U}_{ij}\tilde{W}^{jk},$$

where $\tilde{U}_{ij}$ is an arbitrary antisymmetric akinetor of weight $+w$. What can be said about the nature and equality of $\tilde{V}^{jk}$ and $\tilde{W}^{jk}$?

5.11 Under what conditions is

$$\tilde{Y}_{jk}=\tilde{Z}_{jk}$$

in the equation

$$X^{ijk}\tilde{Y}_{jk}=X^{ijk}\tilde{Z}_{jk}?$$

5.12 Evaluate $\tilde{S}$ in the equation

$$\tilde{S}=A_{[ij]}\tilde{u}^i\tilde{u}^j,$$

where $A_{[ij]}$ is an antisymmetric tensor and $\tilde{u}^i$ is an akinetor of weight -1. Give the transformation equation for $\tilde{S}$.

5.13 Let $\tilde{u}^i$ be an akinetor of weight w. It is transformed from the coordinate system x^i to the system $x^{i'}$ and then to $x^{i''}$.

(a) If $w=0$, prove that the successive transformations from x^i to $x^{i'}$ and then from $x^{i'}$ to $x^{i''}$ does (or does not) give the same result as direct transformation from x^i to $x^{i''}$.

(b) Repeat (a) for $w\neq 0$.

5.14 If $\tilde{v}^i$ is an akinetor of weight w, determine $\tilde{w}^i_j=\partial\tilde{v}^i/\partial x^j$. Is $\tilde{w}^i_j$ an akinetor? If so, what is its weight? If not, what restrictions must be imposed so that $\tilde{w}^i_j$ transforms as an akinetor?

5.15 If $\tilde{w}_{ij}$ is an akinetor, find the transformation equation for $\partial\tilde{w}_{ij}/\partial x^k$. Under what circumstances does it transform as a tensor?

5.16 Prove that $\partial\tilde{v}^i/\partial x^i$ transforms as an akinetor if $\tilde{v}^i$ is an akinetor of weight $+1$.

5.17 Prove the validity of the generalized Stokes theorem.

5.18 Evaluate the pseudocurl of the gradient of a scalar, or prove that it is meaningless.

5.19 Evaluate the pseudodivergence of the gradient of a scalar, or prove that it is meaningless.

5.20 Evaluate the pseudocurl of the pseudocurl, or prove that it is meaningless.

5.21 Evaluate the pseudodivergence of the pseudocurl of a covariant vector, or prove that it is meaningless.

5.22 Evaluate the gradient of the pseudodivergence of a contravariant akinetor of weight +1, or prove that it is meaningless.

5.23 A bivalent tensor T^{ij} with $n=2$ may be used to form a nilvalent holor $|T^{ij}|$. Write the expansion of this determinant and derive the transformation equation of $|T^{ij}|$.

5.24 Repeat Problem 5.23 for U_{ij}.

5.25 Given a bivalent tensor $V^i_{\ j}$, form the associated holor $|V^i_{\ j}|$. According to Equations (1.15) and (1.16), a determinant can be written either as

$$A=|V^i_{\ j}|=\delta^{kl\cdots v}_{12\cdots n}V^1_{\ k}V^2_{\ l}\cdots V^n_{\ v}$$

or as

$$B=|V^i_{\ j}|=\delta^{12\cdots n}_{kl\cdots v}V^k_{\ 1}V^l_{\ 2}\cdots V^v_{\ n}.$$

Since $\delta^{kl\cdots v}_{12\cdots n}$ has weight +1 and $\delta^{12\cdots n}_{kl\cdots v}$ has weight −1, A is of weight +1 while B is of weight −1.

(a) Explain the discrepancy.

(b) Obtain the correct weight of $|V^i_{\ j}|$.

5.26 An interesting case of valence reduction occurs with bivalent holors in 3-space. The holor has nine merates; but if antisymmetric, only three of the merates are distinct. Thus the bivalent holor $v_{[ij]}$ may be set in 1:1 correspondence with

$$v_i=(v_1,v_2,v_3).$$

(a) Write an equation for β, the number of distinct merates of a bivalent holor $V^{[ij]}$, as a function of n. Determine if there are other cases where $\mu=n$.

(b) Determine if $V^{[ij]k}$ will allow a reduction in valence for some n.

(c) Repeat for the completely antisymmetric $V^{[ijk]}$.

5.27 Prove the validity of Theorem XXVII for a mixed bivalent tensor T_i^n.

5.28 Prove the validity of Theorem XXVII for a covariant akinetor u_i of weight w.

6

Geometric spaces

The word *space* is understood intuitively by all English-speaking people, but a precise definition is not easily obtained.[1] The mathematician appropriates the word and uses it in a special sense, for which the word *manifold* might be less ambiguous:[2]

Definition XII: A space or manifold is a set of points with a structure.

This definition merely replaces the vague word space by the equally vague word *structure.* But at least the definition indicates that a space is somehow associated with a set of points. The meaning of structure will become clearer as we proceed.

In using the word space in the mathematical sense, one must be careful to divest himself of the picture of a Euclidean 3-space. In fact, the purpose of this chapter is to consider various spaces that do not behave at all like the customary Euclidean space. In nonmetric spaces, distance between points is a meaningless concept; and angle, orthogonality, and other familiar geometric properties may have no significance.

Fictitious spaces arise frequently in the practical application of holors. A scientist finds experimentally that a certain phenomenon can be expressed in terms of n independent variables $(x^1, x^2, \ldots, x^n)$:

$$\phi = \phi(x^1, x^2, \ldots, x^n).$$

This gives a holor ϕ and allows the use of holor algebra. In many cases, it is convenient to take a further step and geometrize the problem. Each x^i is then considered as a point in an n-dimensional fictitious space, whose structure is determined by the physical phenomena being represented. Usually the space does not have the properties of Euclidean space.

For example, in the trichromatic specification of color,[3] every possible color is represented by a set of three numbers that specify the amounts of three arbitrary primaries. Geometrically, then, each color is represented by a point x^i in a fictitious color 3-space. Relations among colors are easily visualized in this space, but it is not a Euclidean space.

Another example is circuit space. Here we deal with a current holor I^j (Section 4.07) whose merates represent all the currents in an l-loop electric network.[4] Geometrically, we can take I^j as a point in an artificial l-space, but this space is not Euclidean.

6.01. Arithmetic space

A space has been defined as a set of points with a structure. The simplest possible case occurs where the structure has been reduced to zero. This space, consisting of merely a collection of points, without structure, is called an *arithmetic space.*[5]*

Each point is represented by x^i, where $i = 1, 2, \dots, n$. To visualize these points, we can fix an origin 0 and can draw n axes radiating from 0. Each merate $\{x^i\}_P$ of a given point P is a real number that may be regarded as a "distance" measured along the $\{x^i\}$ axis. Thus, we can formulate an arbitrary scale along each axis. In arithmetic space, however, there is no way of relating the scales along different axes and no way of specifying the angles between axes.

Consider an arithmetic 3-space. We can imagine three straight lines through a common origin 0 and can equip each line with a scale. Such a coordinate system with linear axes and uniform scales is called a *Cartesian coordinate system.* This coordinate system can always be set up if no transformations are allowed. But if a general coordinate transformation

$$x^{i'} = F^{i'}(x^i) \tag{6.01}$$

is introduced, then straight axes will generally transform into curved axes, uniform scales will generally become nonuniform, and orthogonal axes will generally become nonorthogonal. Thus, if one is interested in invariant properties under transformation (6.01), he must abandon the usual geometric properties of Euclidean space R_n. The concepts of distance,

* According to Veblen and Whitehead, "An ordered set of n real numbers $(x^1, \dots, x^n)$ will be called an *arithmetic point.* . . . The set of all arithmetic points, for a given value of n, will be called the *arithmetic space for n dimensions.*" (*Foundations of differential geometry,*[2] p. 1). More elaborate definitions are often given, and a distinction may be made between "arithmetic space" and "geometric space."

angle, straight line, circle, plane, sphere, and so on, are meaningless in arithmetic space X_n.

6.02. Invariants

One might conclude that the generality of arithmetic space would eliminate almost all interesting properties. On the contrary, a surprising number of similarities are found between Euclidean and arithmetic spaces and a large number of invariants exist in the latter. Already in Chapter 4 we found that the elements of vector analysis appear as soon as the definition of a tensor is introduced. This allows a general functional transformation [Equation (6.01)] and thus corresponds to a transformation in arithmetic space. As in Section 4.04, for convenience we limit F^i to single-valued analytic functions with a unique inverse.

What are some geometric invariants of arithmetic space? Evidently, the *point* $(x^1, x^2, \ldots, x^n)$ is a geometric invariant, since Equation (6.01) always transforms a point into a point, never into a line or other object. A collection of points defining a curve will transform into a collection of points; and the analyticity of the transformation ensures that the transformed points also define a curve. Similarly, a surface transforms into a surface. These geometric invariants are listed in Table 6.01.

Perhaps a word of warning should be inserted here. There is a tendency to use the words invariant, geometric object, and tensor interchangeably. Indeed, Veblen introduced the term *geometric object* in 1926 as a synonym for invariant.[6] We have seen, however (Section 4.05), that the modern definition of geometric object is concerned merely with the form of the transformation equation and says nothing about invariance. For instance, a point is a geometric invariant but it is not a geometric object (or a tensor, or an akinetor) in arithmetic space because it does not transform in the prescribed way.

Though x^i is not an akinetor in arithmetic space, dx^i is an akinetor. In fact, it is a tensor, as shown in Chapter 4. Other akinetors in arithmetic space are the gradient $\partial\phi/\partial x^i$, the Kronecker deltas, and the alternators (Table 6.01). Area and volume have no meaning in arithmetic space, but pseudoarea and pseudovolume (Chapter 5) are akinetors in X_n. The divergence of a vector cannot be defined in arithmetic space, but the

Table 6.01. *Arithmetic space*

A. Geometric invariants of arithmetic space

Point x^i, curve, surface

B. Akinetors in arithmetic space

Symbol	Name	Weight
dx^i	Increment in coordinates	0
$\partial\phi/\partial x^i$	Gradient of scalar	0
δ^i_j	Kronecker delta	0
δ^{ijk}_{lmn}	Generalized Kronecker delta	0
δ^{ijk}_{123}	Alternator	+1
δ^{123}_{lmn}	Alternator	−1
$d\tilde{\mathfrak{a}}_i = \delta^{123}_{ijk}\, dv^j\, dw^k$	Pseudoarea	−1
$d\tilde{\mathfrak{V}} = \delta^{123}_{ijk}\, du^i\, dv^j\, dw^k$	Pseudovolume	−1
$\partial\tilde{v}^i/\partial x^i$	Pseudodivergence of contravariant akinetor of weight +1	+1
$\partial\tilde{F}^{ij}/\partial x^i$	Pseudodivergence of a bivalent contravariant alternating akinetor of weight +1	+1
$\tilde{W}^i = \delta^{ijk}_{123} \dfrac{\partial v_k}{\partial x^j}$	Pseudocurl of a covariant vector	+1
$\delta^{ijk}_{123} \dfrac{\partial}{\partial x^j} \dfrac{\partial\phi}{\partial x^k} \equiv 0$	Pseudocurl of the gradient of a scalar	+1
$\dfrac{\partial}{\partial x^i} \delta^{ijk}_{123} \dfrac{\partial v_k}{\partial x^j} \equiv 0$	Pseudodivergence of the pseudo-curl of a covariant vector	+1
$\oint_S \tilde{v}^i\, d\tilde{\mathfrak{a}}_i = \int_{\mathfrak{V}} \dfrac{\partial\tilde{v}^i}{\partial x^i}\, d\tilde{\mathfrak{V}}$	Generalized divergence theorem applied to a contravariant akinetor of weight +1	
$\oint_C v_i\, dx^i = \int_S \tilde{W}^i\, d\tilde{\mathfrak{a}}_i$	Generalized curl theorem applied to a covariant vector	

Note: The generalized Kronecker deltas and alternators are written for X_3 but are equally applicable to X_n.

pseudodivergence $\partial\tilde{v}^i/\partial x^i$ of a contravariant akinetor of weight +1 transforms as a nilvalent akinetor of weight +1. The curl of a contravariant vector cannot be defined, but the pseudocurl $\tilde{W}^i$ of a covariant vector v_k is an akinetor in arithmetic space. The divergence of the gradient cannot be defined in arithmetic space, so the scalar Laplacian is meaningless. Likewise, the vector Laplacian is meaningless in an X_n.

In Euclidean 3-space R_3, we have the vector identities

$$\text{curl grad } \phi \equiv 0, \qquad \text{div curl } \mathbf{A} \equiv 0.$$

The corresponding identities in X_n are

$$\delta^{ijk}_{123} \frac{\partial}{\partial x^j}\left(\frac{\partial \phi}{\partial x^k}\right) \equiv 0$$

and

$$\frac{\partial}{\partial x^i}\left(\delta^{ijk}_{123} \frac{\partial v_k}{\partial x^j}\right) \equiv 0.$$

Finally, Table 6.01 gives the generalized divergence and curl theorems applicable in arithmetic n-space. It is really surprising that so much of vector analysis should be applicable, even though in modified form, to something as general as arithmetic n-space.

Arithmetic space is important in physical applications because it is the starting point for all geometrizations. Suppose that we are investigating a physical phenomenon that depends on n independent variables. A given condition can then be specified by a point x^i in arithmetic n-space. Next, the scientist studies the phenomenon in greater detail to see if further structure cannot be introduced into the space. He may even find that the fictitious space is Euclidean, though such a result would be unusual.

6.03. Transformations

The importance of transformations is evident from Chapters 4 and 5, where the whole tensor idea is based on behavior under coordinate transformation. Transformations also provide a useful way of classifying geometries and spaces. According to Felix Klein,[7] suppose that we are given a manifold and a group of transformations, the set of all invariants of this group is called the *geometry of the group.** Similarly, spaces[8] can be characterized in terms of a group of transformations. One merely

* The original statement is: "Es ist eine Mannigfaltigkeit und in derselben eine Transformationsgruppe gegeben, man entwickle die auf Gruppe bezügliche Invariantentheorie."

picks out objects or relations that are invariant with respect to the group under consideration. Other aspects may occur in a particular case, but they are of minor importance because they are not invariants and therefore do not characterize the space.

We have considered (Section 4.04) the general transformation

$$x^{i'} = F^{i'}(x^{i}), \tag{6.01}$$

where $F^{i'}$ is an arbitrary function. We find it convenient, however, to limit $F^{i'}$ to single-valued analytic functions with a unique inverse. The inverse is written

$$x^{i} = \mathcal{G}^{i}(x^{i'}). \tag{6.02}$$

Evidently, a transformation may be regarded geometrically in two ways:

(a) as a transformation of the coordinate system, the points fixed in space and the coordinate system varied in accordance with Equation (6.01); and

(b) as a transformation of space, the coordinate system being fixed while the points move.

Up to this time, we have employed interpretation (a). The same base letter has been used to indicate that the point remains fixed, and primes have been used on the indices to indicate a change in coordinate system. For interpretation (b), the points are transformed, which may be emphasized by a change in notation[9]:

$$y^{i} = F^{i}(x^{j}), \tag{6.01a}$$

where a new base letter is used to show that y^{i} represents a different point from x^{j}.

In this chapter, the following spaces will be considered:

(i) arithmetic space, based on the general group of transformations [Equation (6.01a)];

(ii) affine space, based on the linear group;

(iii) axonometric space, a special case of (ii);

(iv) projective space, based on the projective group;

(v) perspective space, a special case of (iv); and

(vi) inversive space, based on the inversive group.

The above list is by no means complete. Many other spaces can be studied, but these examples will illustrate how the characteristics of a particular space are determined by invariants of the transformation.

Some of the statements made in Section 6.01 on arithmetic space may now be more obvious. Equation (6.01a) shows that a point goes into a point under the general group of transformations. Does a line go into a line? Obviously not, since F^i is generally not a linear function. Similarly, a plane does not transform into a plane, nor a second-degree surface (such as an ellipsoid) into a second-degree surface. In general, distance is meaningless; and since angle is defined as the ratio of distances, it is also meaningless. In Euclidean space, we are accustomed to thinking of x^i as a point P or as a space vector $0P$ from the origin to P. Evidently, the second alternative is not feasible in arithmetic space since a straight line $0P$ has no invariant meaning and neither does its length.

6.04. Affine space

The most important special case of Equation (6.01a) is the linear or affine transformation. The study of affine space is the study of invariants under affine transformation.[10] A linear transformation may be written

$$y^j = A^j_k x^k + A^j_0, \tag{6.03}$$

where $j, k = 1, 2, \ldots, n$, and $\{A^j_k\}$ are constants. Translation is specified by A^j_0. Since this term introduces no new invariants, however, it is convenient to consider the centro-affine transformation

$$y^j = A^j_k x^k. \tag{6.03a}$$

In this section, the transformation determinant

$$\Delta = |A^j_k|$$

will be assumed to be different from zero. The case of $\Delta = 0$, however, is not without interest and will be considered in Section 6.08.

Equation (6.03) shows that a point x^k always transforms into a point y^j. Thus, an invariant of affine space is the point: Points always transform into points in affine space. Moreover, since all the coefficients $\{A^j_k\}$ are finite constants, every finite point x^k transforms into a finite point, and a point at infinity transforms into a point at infinity. This is another characteristic of affine space: Finite points remain finite and infinite points remain infinite under affine transformation.

In the remainder of this section, we shall confine our attention to affine 3-space. This restriction is merely for convenience in visualization: All the affine properties apply equally to n-space. The equation of a plane in 3-space is written

$$u_j y^j + u_0 = 0. \tag{6.04}$$

This equation imposes a linear constraint on the coordinates of space and thus determines a 2-space. The coordinates u_j determine the direction of the plane, while u_0 distinguishes a particular plane from the family of planes having the same direction. The transformed equation is

$$(u_j A^j_k) x^k + u_0 = 0. \tag{6.05}$$

Equation (6.05) has the same form as Equation (6.04) and also represents a plane. Therefore, planes transform into planes under affine transformation; or planes are an invariant of affine space.

A pair of parallel planes may be designated as

$$u_j y^j + u_0 = 0, \qquad u_j y^j + U_0 = 0. \tag{6.06}$$

If these planes are subject to an affine point transformation,

$$u_j A^j_k x^k + u_0 = 0, \qquad u_j A^j_k x^k + U_0 = 0, \tag{6.07}$$

which again represent parallel planes. Thus, parallelism of planes is an affine invariant.

A line in 3-space can be defined as the intersection of two planes that are not parallel. Since nonparallelism of planes is preserved under affine transformation, lines are always transformed into lines. Parallel lines may be defined as the intersections of two parallel planes by a third plane. Since parallelism of planes is preserved under affine transformation, parallelism of lines is an affine invariant.

Now consider orthogonality of two lines in a plane, say the plane $y^3 =$ const. The lines A and B are

$$(y^2)_A = m(y^1)_A + b, \qquad (y^2)_B = -(y^1)_B/m + c, \tag{6.08}$$

where m is the slope of the first line and $-1/m$ is the slope of the second. Evidently the two are orthogonal. From Equation (6.03a),

$$y^1 = A^1_1 x^1 + A^1_2 x^2, \qquad y^2 = A^2_1 x^1 + A^2_2 x^2,$$

and substitution into Equation (6.08) gives

$$\begin{aligned} (x^2)_A &= \left[\frac{mA^1_1 - A^2_1}{A^2_2 - mA^1_2}\right](x^1)_A + \frac{b}{A^2_2 - mA^1_2}, \\ (x^2)_B &= -\left[\frac{A^2_1 + A^1_1/m}{A^2_2 + A^1_2/m}\right](x^1)_B + \frac{c}{A^2_2 + A^1_2/m}. \end{aligned} \tag{6.09}$$

Since, in general, the coefficients of x^1 in the two equations are not negative reciprocals of one another, the transformed lines are no longer orthogonal. The conclusion is that orthogonality is not an affine invariant and hence has no meaning in affine space.

The foregoing paragraphs have treated the affine point transformation. The corresponding coordinate transformation should also be considered:

$$x^{i'} = A_i^{i'} x^i + A_0^{i'}. \tag{6.10}$$

Differentiation gives

$$\frac{\partial x^{i'}}{\partial x^i} = A_i^{i'},$$

so Equation (6.10) becomes

$$x^{i'} = \frac{\partial x^{i'}}{\partial x^i} x^i + A_0^{i'}, \tag{6.10a}$$

which is not a tensor transformation. If attention is restricted to the centro-affine transformation, however,

$$x^{i'} = \frac{\partial x^{i'}}{\partial x^i} x^i. \tag{6.10b}$$

Thus, under centro-affine coordinate transformations, points transform as contravariant univalent tensors (contravariant vectors).

Now consider the equation of a plane under centro-affine coordinate transformation. The plane may be written

$$u_i x^i + u_0 = 0, \tag{6.11}$$

and the coordinate transformation is given by Equation (6.10b). Substitution gives

$$u_i\left(\frac{\partial x^i}{\partial x^{i'}} x^{i'}\right) + u_0 = 0.$$

If we define $u^{i'}$ as

$$u_{i'} = \frac{\partial x^i}{\partial x^{i'}} u_i,$$

the equation of the plane in primed coordinates is

$$u_{i'} x^{i'} + u_0 = 0. \tag{6.12}$$

The form of equation is invariant under centro-affine transformation, and the holor u_i transforms as a univalent covariant tensor.

In the preceding chapters, the univalent contravariant tensor x^i was interpreted as a point in n-space, but nothing was said about the geometric

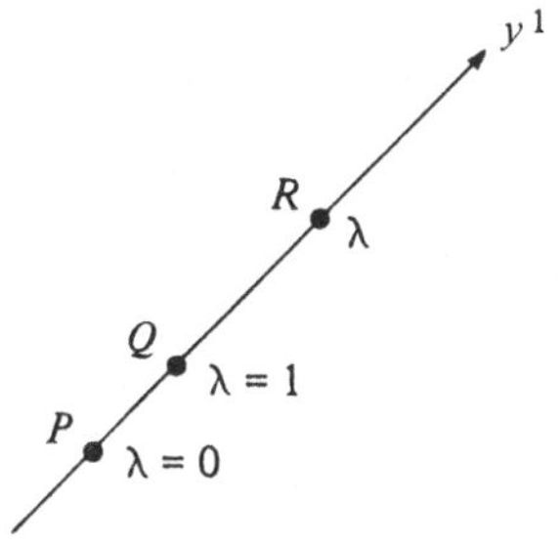

Figure 6.01. Parting λ is an invariant for points on a line:
$(y^j)_R = (1-\lambda)(y^j)_P + \lambda(y^j)_Q$.

meaning of the covariant tensor u_i. We now see that the latter has a simple geometric interpretation. In centro-affine 3-space, μ_i represents a plane. This interpretation is valid, however, only under linear transformation. With a general functional transformation, as in arithmetic space, a plane transforms into some kind of warped surface and therefore has no invariant significance.

6.05. Parting

Consider a straight line *PQR* (Figure 6.01) in affine n-space. A uniform scale y^1 can be established on this line. Another line could be considered as another 1-space, which could be subdivided uniformly with another (generally different) scale. In an n-space, we can have infinitely many lines; but in familiar Euclidean space, all these lines have the same scale. If the space is affine, on the other hand, each line has a different scale and thus there is no unique way of specifying distance in affine space: The concept of distance is not an affine invariant. Evidently, therefore, area, volume, and angle are likewise not affine invariants.

Though distance is not a valid concept in affine space, we can form a related concept that is affine invariant. For the line *PQR* (Figure 6.01), the same scale applies to each segment, and therefore the segment *RP* may be designated as $(y^1)_R - (y^1)_P$ and segment *QP* as $(y^1)_Q - (y^1)_P$. We can now introduce a parameter λ defined as

$$\frac{(y^1)_R - (y^1)_P}{(y^1)_Q - (y^1)_P} = \lambda. \qquad (6.13)$$

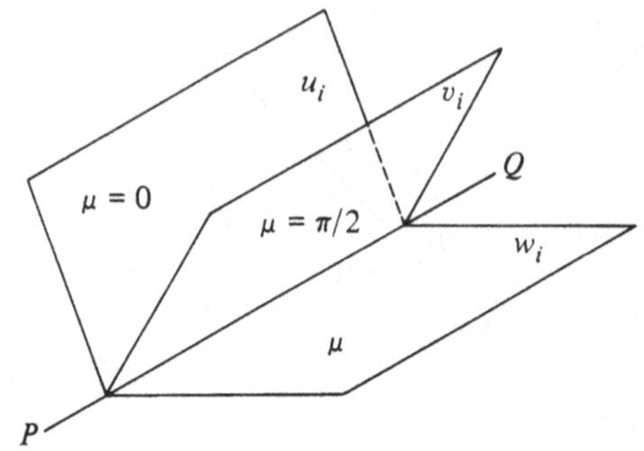

Figure 6.02. Parting can also be applied to planes.

This gives a new linear scale with $-\infty < \lambda < +\infty$. Suppose that P and Q are fixed and R moves. When R is at Q, λ becomes unity, according to Equation (6.13); and when R is at P, $\lambda = 0$. Equation (6.13) may be written in general as

$$(y^j)_R = (1-\lambda)(y^j)_P + \lambda(y^j)_Q. \tag{6.13a}$$

An affine transformation,

$$y^j = A^j_k x^k, \tag{6.14}$$

gives

$$(x^k)_R = (1-\lambda)(x^k)_P + \lambda(x^k)_Q, \tag{6.13b}$$

which has the same form as Equation (6.13a). Thus, we have obtained an affine invariant λ that is analogous to distance. This quantity may be called the *parting* of the points.

Angle can be handled in a similar manner. A line PQ defines a bundle of planes (Figure 6.02). Two of these planes are

$$u_i x^i + u_0 = 0, \qquad v_i x^i + v_0 = 0. \tag{6.15}$$

Any other plane through PQ may be written

$$w_i x^i + w_0 = 0, \tag{6.16}$$

where

$$w_0 = (\cos\mu)u_0 - (\sin\mu)v_0, \qquad w_i = (\cos\mu)u_i - (\sin\mu)v_i.$$

The parameter μ identifies the particular plane of the infinite bundle of planes intersecting in the line PQ and may be called the parting of planes

u_i and w_i. If $\mu = 0$, we specify the plane u_i; if $\mu = \frac{1}{2}\pi$, we specify the plane v_i. All possible planes of the bundle are identified by values of μ in the range $-\frac{1}{2}\pi < \mu \leq \frac{1}{2}\pi$.

It is readily shown that parting of planes is an affine invariant. It corresponds in many ways to the Euclidean concept of angle. But the parameter μ is not the angle between planes. It is merely a parameter that reduces to angle if $A^i_j = \delta^i_j$.

6.06. Ratios of areas and of volumes

Simple equations for area and volume in Euclidean space can be expressed in terms of determinants.[11] In 1-space, the distance between points P and Q is

$$l = \frac{1}{1!}\begin{vmatrix} (y^1)_P & 1 \\ (y^1)_Q & 1 \end{vmatrix}. \tag{6.17}$$

In Euclidean 2-space with orthogonal Cartesian coordinates y^1, y^2, the area of a triangle PQR is

$$(\mathcal{A})_{PQR} = \frac{1}{2!}\begin{vmatrix} (y^1)_P & (y^2)_P & 1 \\ (y^1)_Q & (y^2)_Q & 1 \\ (y^1)_R & (y^2)_R & 1 \end{vmatrix}. \tag{6.18}$$

For 3-space, the volume of a tetrahedron is

$$(\mathcal{V})_{PQRS} = \frac{1}{3!}\begin{vmatrix} (y^1)_P & (y^2)_P & (y^3)_P & 1 \\ (y^1)_Q & (y^2)_Q & (y^3)_Q & 1 \\ (y^1)_R & (y^2)_R & (y^3)_R & 1 \\ (y^1)_S & (y^2)_S & (y^3)_S & 1 \end{vmatrix}. \tag{6.19}$$

These equations are easily extended to n-space. They give not only the magnitude but also a positive or negative sign, depending on the order in which the points are taken.

The foregoing equations apply to Euclidean space. Let us now attempt to extend their use to affine space. The area of a triangle PQR is expressed by Equation (6.18). Now perform an affine transformation to give a new triangle TUV:

$$y^i = A^i_k x^k, \qquad \Delta = |A^i_k| \neq 0.$$

Substitution gives

$$(\mathfrak{A})_{PQR}=\frac{1}{2!}\begin{vmatrix} A_1^1(x^1)_P+A_2^1(x^2)_P & A_1^2(x^1)_P+A_2^2(x^2)_P & 1 \\ A_1^1(x^1)_Q+A_2^1(x^2)_Q & A_1^2(x^1)_Q+A_2^2(x^2)_Q & 1 \\ A_1^1(x^1)_Q+A_2^1(x^2)_R & A_1^2(x^1)_R+A_2^2(x^2)_R & 1 \end{vmatrix}$$

$$=\frac{1}{2!}\begin{vmatrix} (x^1)_P & (x^2)_P & 1 \\ (x^1)_Q & (x^2)_Q & 1 \\ (x^1)_R & (x^2)_R & 1 \end{vmatrix}\begin{vmatrix} A_1^1 & A_1^2 & 0 \\ A_2^1 & A_2^2 & 0 \\ 0 & 0 & 1 \end{vmatrix}.$$

But the second determinant is equal to

$$\begin{vmatrix} A_1^1 & A_1^2 \\ A_2^1 & A_2^2 \end{vmatrix}=\Delta,$$

and the first determinant is equal to the new area $(\mathfrak{A})_{TUV}$. Thus,

$$(\tilde{\mathfrak{A}})_{PQR}=(\tilde{\mathfrak{A}})_{TUV}\Delta$$

or

$$(\tilde{\mathfrak{A}})_{TUV}=\Delta^{-1}(\tilde{\mathfrak{A}})_{PQR}. \tag{6.20}$$

Area, therefore, is not invariant under affine point transformation but involves the transformation determinant Δ. A similar result is obtained with coordinate transformations, showing that area is a nilvalent akinetor of weight -1 under centro-affine coordinate transformation.

For volume, use Equation (6.19). The same procedure is employed as with area, giving

$$(\tilde{\mathfrak{V}})_{TUVW}=\Delta^{-1}(\tilde{\mathfrak{V}})_{PQRS}. \tag{6.21}$$

Therefore, volume is not an affine invariant with respect to space transformation but requires multiplication by the reciprocal of the transformation determinant. Similarly, it is easily shown that volume is not a tensor but transforms as a nilvalent akinetor of weight -1 under centro-affine coordinate transformations (see Section 5.04).

6.07. Second-degree surfaces

Now consider the behavior of second-degree surfaces under affine point transformation. The general equation of a surface of second degree is

$$k_{ij}y^iy^j+k_0=0. \tag{6.22}$$

Under an affine transformation, Equation (6.22) becomes

$$k_{ij}A^i_m A^j_n x^m x^n + k_0 = 0. \tag{6.23}$$

Since this is an equation of the second degree, one can say immediately that second-degree surfaces transform into second-degree surfaces under affine point transformation.

The second-degree surface that has no points at infinity is the ellipsoid, which includes the sphere as a special case. But we have seen that a second-degree surface always transforms into a second-degree surface by affine transformation and also that finite points always transform into finite points. Thus, the ellipsoid is an affine invariant.

Similar results are obtained with other second-degree surfaces. It is easily shown that a paraboloid always transforms into a paraboloid under affine transformation. Also a hyberboloid of one sheet transforms into a hyperboloid of one sheet. These affine invariants are listed in Table 6.02.

Since linear transformations are common in physics, affine space has numerous practical applications. For instance, a fictitious 3-space is used in the trichromatic specification of color,[3] as mentioned at the beginning of the chapter. In terms of a given set of primaries - say a specific red, green, and blue - any color is specified by three real numbers,

$$x^i = (x^1, x^2, x^3),$$

which represent a point P in color space.

According to experiment, however, any set of three colors can be used as primaries (provided that the mixture of two primaries cannot be matched by the third primary). A change in primaries will alter the specification of P, which will become

$$x^{i'} = (x^1, x^2, x^3),$$

where

$$x^{1'} = A^1_1 x^1 + A^1_2 x^2 + A^1_3 x^3,$$

$$x^{2'} = A^2_1 x^1 + A^2_2 x^2 + A^2_3 x^3,$$

$$x^{3'} = A^3_1 x^1 + A^3_2 x^2 + A^3_3 x^3.$$

But this is an affine transformation [Equation (6.03a)]. Therefore, color space is an affine space and we can now speak of lines, planes, and parallelism in this space. Distances and angles in color space, however, are generally meaningless. Thus, the further step to a

Table 6.02. *Affine space*

A. Geometric invariants of affine space

$$y^j = A^j_k x^k, \quad |A^j_k| \neq 0$$

(1) Point
  (a) Finite point
  (b) Infinite point
(2) Plane
(3) Parallelism of planes
(4) Line
(5) Parallelism of lines
(6) Parting of points on a family of parallel lines
(7) Parting of planes intersecting in a family of parallel lines
(8) Second-degree surface
  (a) Ellipsoid
  (b) Paraboloid
  (c) Hyperboloid of one sheet
  (d) Hyperboloid of two sheets

B. Akinetors

$$x^{i'} = A^{i'}_i x^i, \quad |A^{i'}_i| \neq 0$$

(1) Point x^i, a univalent tensor
(2) Plane u_i, a univalent tensor
(3) Parting of points λ, a scalar
(4) Parting of planes μ, a scalar
(5) Area $\mathcal{A}$ of a finite triangle, an akinetor of weight -1
(6) Volume $\mathcal{V}$ of a finite tetrahedron, an akinetor of weight -1

Euclidean color space is not possible unless we can formulate a metric, based perhaps on experimental data regarding minimum perceptible color differences.[12]

6.08. Axonometric space

In Section 6.04, we considered the affine point transformation [Equation (6.14)], where each point x^j of an n-space transforms into a unique point y^i of an n-space. The transformation matrix is

$$A^i_j = \begin{bmatrix} A^1_1 & A^1_2 & \cdots & A^1_n \\ A^2_1 & A^2_2 & \cdots & A^2_n \\ \cdot & \cdot & \cdots & \cdot \\ A^n_1 & A^n_2 & \cdots & A^n_n \end{bmatrix},$$

and the determinant $|A^i_j| = \Delta$ of this matrix must be nonzero to allow a unique inverse.

We now take the special case of an affine transformation where $\Delta = 0$. If we introduce a relationship among the y's, such as

$$k_i y^i = 0, \tag{6.24}$$

the y-space will be reduced to an $(n-1)$-space. Substitution of Equation (6.14) into (6.24) gives

$$k_i A^i_j x^j = 0,$$

and since the x's are arbitrary,

$$k_i A^i_j = 0. \tag{6.25}$$

Equation (6.25) represents a set of homogeneous linear equations. According to algebraic theory,[13] the necessary and sufficient condition that these equations have a solution other than

$$k_1 = k_2 = \cdots = k_n = 0$$

is that the determinant $|A^i_j|$ be zero. This special case of an affine transformation where $\Delta = 0$ is called an axonometric transformation and the resulting $(n-1)$-space is an axonometric space.

For a 3-space, the affine transformation is

$$\begin{aligned} y^1 &= A^1_1 x^1 + A^1_2 x^2 + A^1_3 x^3, \\ y^2 &= A^2_1 x^1 + A^2_2 x^2 + A^2_3 x^3, \\ y^3 &= A^3_1 x^1 + A^2_2 x^2 + A^3_3 x^3. \end{aligned} \tag{6.26}$$

The points x^i are projected onto the picture plane

$$k_1 y^1 + k_2 y^2 + k_3 y^3 = 0, \tag{6.27}$$

and the determinant of the transformation is

$$\Delta = \begin{vmatrix} A^1_1 & A^1_2 & A^1_3 \\ A^2_1 & A^2_2 & A^2_3 \\ A^3_1 & A^3_2 & A^3_3 \end{vmatrix} = 0. \tag{6.28}$$

The relation between the k's and the A's is given by the set of homogeneous equations

$$\begin{aligned} k_1 A^1_1 + k_2 A^2_1 + k_3 A^3_1 &= 0, \\ k_1 A^1_2 + k_2 A^2_2 + k_3 A^3_2 &= 0, \\ k_1 A^1_3 + k_2 A^2_3 + k_3 A^3_3 &= 0. \end{aligned} \tag{6.29}$$

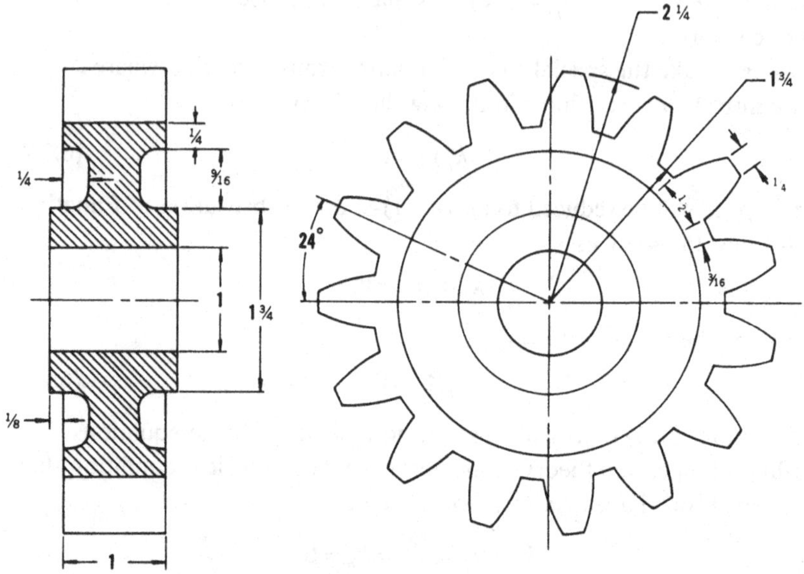

Figure 6.03. Example of axonometric drawing. Orthographic views of a gear wheel. [From T. A. Thomas, *Technical illustration,* third edition (reproduced from McGraw-Hill Book Co., New York, © 1978), Figure 8-2.]

If the rank[13] of the transformation matrix is 2, these equations can be solved uniquely for the ratios $k_1 : k_2 : k_3$. If the rank is 1, the y plane degenerates into a line.

An important practical application of the axonometric transformation occurs in engineering drawing and architectural drawing. In engineering practice, axonometric drawings are generally classified as[15]

Orthographic	(2 or 3 picture planes)
Isometric Dimetric Trimetric Oblique	1 picture plane

An example of the orthographic representation of a machine part is shown in Figure 6.03. A complete working drawing is obtained by parallel projection onto perpendicular planes. For example, let $y^1 = 0$, $y^2 = x^2$, $y^3 = x^3$. Then, according to Equation (6.28),

$$A^i_j = \begin{bmatrix} 0 & 0 & 0 \\ 0 & 1 & 0 \\ 0 & 0 & 1 \end{bmatrix}.$$

Here $\Delta = 0$, but the rank of the matrix is 2 because one of the 2×2 determinants is not zero:

$$\begin{vmatrix} 1 & 0 \\ 0 & 1 \end{vmatrix} \neq 0.$$

Similarly, if $y^1 = x^1$, $y^2 = 0$, and $y^3 = x^3$, we have a mapping by parallel projection onto the y^1y^3 plane. And if $y^1 = x^1$, $y^2 = x^2$, and $y^3 = 0$, we have a mapping onto the y^1y^2 plane. In particular, if axes are taken orthogonal, a solid object is represented by three projections on mutually orthogonal planes. This orthographic projection[14] is widely used in engineering and architectural drawings.

Though orthographic projection is used in most engineering drawings, there is often a demand for pictorial drawings that will give true dimensions in a single view. The isometric, dimetric, trimetric, and oblique drawings satisfy this requirement and provide an approximation to what one sees. The true picture, corresponding to a photograph, is obtained by perspective transformation (Section 6.12); but none of the axonometric transformations give this view. On the other hand, the true perspective transformation does not allow dimensions to be scaled off the drawing; and in this respect, perspective transformation is inferior to axonometric. Figure 6.04 gives an example of perspective transformation, while Figure 6.05 shows the distortion caused by axonometric transformation in Japanese color prints.

Figure 6.06 indicates how a cube is transformed by oblique transformation. On the left, we have plan and elevation of a unit cube. The front elevation is kept the same in y-space as in x-space, but depth in the oblique transformation is represented by lines at the arbitrary angle α. Let us consider points A, B, and C:

Point	x^j	y^i
A	$(1,0,0)$	$(1,0,0)$
B	$(1,1,0)$	$(1,1,0)$
C	$(0,1,1)$	$(\sin\alpha, 1+\cos\alpha, 0)$

The equation of the picture plane is $y^3 = 0$. Substitution into Equation (6.26) gives

Figure 6.04. "Lady and gentleman at the virginals," painting by Johannes Vermeer – an example of perspective. [From L. Goldschneider, *Johannes Vermeer* (Phaidon Press, London, 1967), p. 38.]

$$A:\ \begin{aligned} 1&=A^1_1,\\ 0&=A^2_1,\\ 0&=A^3_1, \end{aligned}\qquad B:\ \begin{aligned} 1&=A^1_1+A^1_2,\\ 1&=A^2_1+A^2_2,\\ 0&=A^3_2, \end{aligned}\qquad C:\ \begin{aligned} \sin\alpha&=A^1_2+A^1_3,\\ 1+\cos\alpha&=A^2_2+A^2_3,\\ 0&=A^3_3. \end{aligned}$$

Thus, the transformation matrix becomes

$$A^i_j=\begin{bmatrix} 1 & 0 & \sin\alpha \\ 0 & 1 & \cos\alpha \\ 0 & 0 & 0 \end{bmatrix},$$

and

$$\begin{aligned} y^1&=x^1+x^3\sin\alpha,\\ y^2&=x^2+x^3\cos\alpha,\\ y^3&=0. \end{aligned}\tag{6.30}$$

Figure 6.05. Example of Japanese color print showing axonometric drawing. [From D. B. Waterhouse, *Harunabu and his age* (Reproduced by Courtesy of the Trustees of the British Museum, London, 1964).]

As another example of axonometric transformation, consider the *isometric* transformation in Figure 6.07. The plan and elevation of a cube are shown on the right, and the transformed cube appears on the left. Parallel projection occurs at the angle β. To obtain the regular hexagonal outline in the y plane,

$$\tan\beta = 1/\sqrt{2}, \quad \text{or} \quad \beta = 35°16'.$$

The transformation matrix can be evaluated by taking three points in Figure 6.07, such as

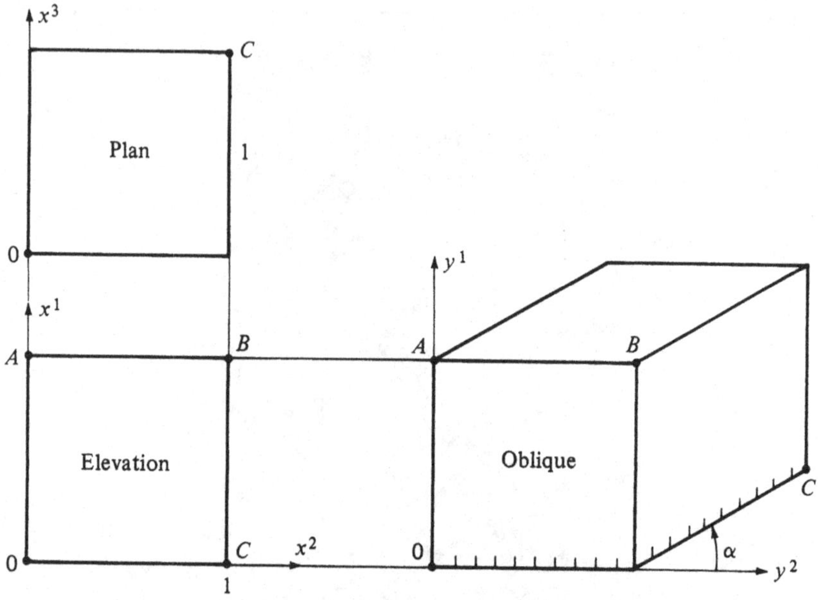

Figure 6.06. Oblique drawing of a cube. The view is distorted, according to our ideas of perspective; but the true dimensions can be scaled from the drawing.

Point	x^i	y^i
A	$(1, \sqrt{2}, 0)$	$(0, a, 0)$
B	$(1, 0, -\sqrt{2})$	$(\frac{1}{2}\sqrt{3}a, \frac{1}{2}a, 0)$
C	$(0, \sqrt{2}, 0)$	$(0, \frac{1}{2}a, 0)$

Substitution into Equation (6.26) then gives

$$A: \quad \begin{aligned} 0 &= A_1^1 + \sqrt{2}A_2^1, \\ a &= A_1^2 + \sqrt{2}A_2^2, \\ 0 &= A_1^3 + \sqrt{2}A_2^3, \end{aligned}$$

$$B: \quad \begin{aligned} \tfrac{1}{2}a\sqrt{3} &= A_1^1 - \sqrt{2}A_3^1, \\ \tfrac{1}{2}a &= A_1^1 - \sqrt{2}A_3^1, \\ 0 &= A_1^3 - \sqrt{2}A_3^3, \end{aligned}$$

$$C: \quad \begin{aligned} 0 &= \sqrt{2}A_2^1, \\ \tfrac{1}{2}a &= \sqrt{2}A_2^2, \\ 0 &= \sqrt{2}A_2^3, \end{aligned}$$

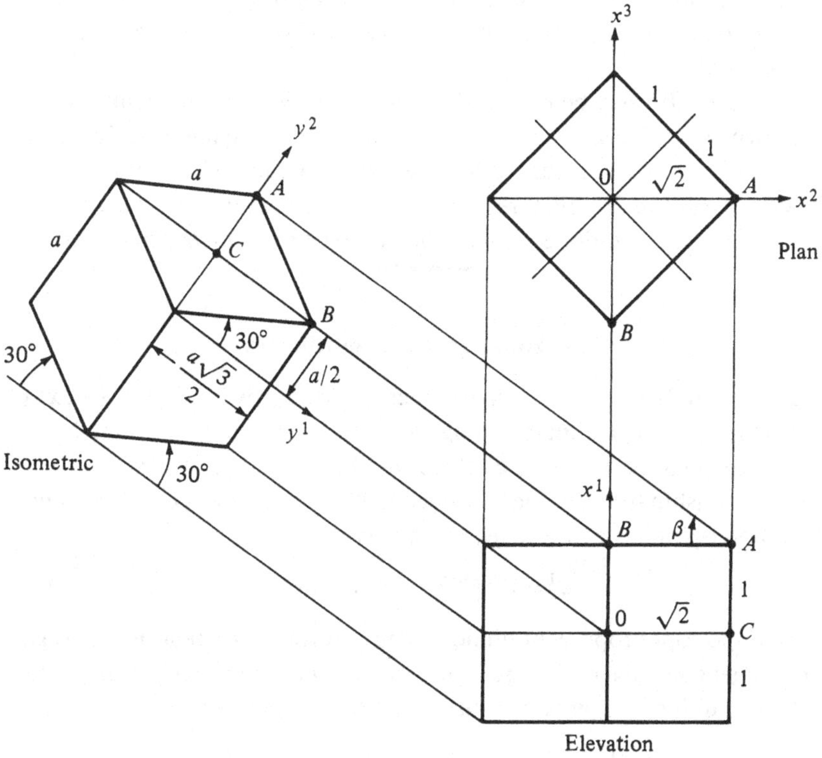

Figure 6.07. Isometric drawing of a cube. Plan and elevation of the cube are shown to the right. Projection at angle β gives the isometric view if $\beta = 35°16'$.

which evaluates the transformation matrix:

$$A^i_j = a \begin{bmatrix} 0 & 0 & -\dfrac{\sqrt{3}}{2\sqrt{2}} \\ \dfrac{1}{2} & \dfrac{1}{2\sqrt{2}} & 0 \\ 0 & 0 & 0 \end{bmatrix}.$$

Thus, the transformation is

$$\begin{aligned} y^1 &= \frac{-\sqrt{3}a}{2\sqrt{2}} x^3, \\ y^2 &= \frac{a}{2}\left(x^1 + \frac{1}{\sqrt{2}} x^2\right), \qquad (6.31) \\ y^3 &= 0. \end{aligned}$$

The constant a may be chosen at will: The important fact is that the scales on all surfaces in the y diagram are the same – the drawing is *isometric.*

Figure 6.07 may be modified to obtain the dimetric and trimetric drawings. If β is not fixed at 35°16′, we obtain a symmetric figure with right and left faces having the same scale but with a different scale on the top face (Figure 6.08b). This is the *dimetric* case. Or symmetry can be abandoned (Figure 6.08c), giving a *trimetric* diagram.

6.09. Homogeneous coordinates

A point P in 2-space is customarily designated by two numbers (x^1, x^2) representing the coordinates along two axes. For some purposes, however, a designation in terms of three numbers $(\tilde{x}^0, \tilde{x}^1, \tilde{x}^2)$ is preferable. The relationship between the homogeneous[16] coordinates $(\tilde{x}^0, \tilde{x}^1, \tilde{x}^2)$ and the inhomogeneous coordinates (x^1, x^2) is

$$x^1 = \tilde{x}^1/\tilde{x}^0, \qquad x^2 = \tilde{x}^2/\tilde{x}^0. \tag{6.32}$$

An important property of homogeneous coordinates is that they change inhomogeneous algebraic equations into homogeneous equations. For instance, an inhomogeneous equation of the second degree is

$$A(x^1)^2 + B(x^2)^2 + Cx^1x^2 + Dx^1 + Ex^2 + F = 0.$$

Substitution of Equation (6.32) gives the homogeneous equation

$$A(\tilde{x}^1)^2 + B(\tilde{x}^2)^2 + C\tilde{x}^1\tilde{x}^2 + D\tilde{x}^0\tilde{x}^1 + E\tilde{x}^0\tilde{x}^2 + F(\tilde{x}^0)^2 = 0.$$

It is easily seen that any algebraic equation of degree p becomes homogeneous of degree p when expressed in homogeneous coordinates.

Homogeneous coordinates can be applied equally well to n-space. Inhomogeneous coordinates of a point in n-space are written

$$x^i = (x^1, x^2, \ldots, x^n), \tag{6.33}$$

and the corresponding homogeneous coordinates are

$$\tilde{x}^\lambda = (\tilde{x}^0, \tilde{x}^1, \ldots, \tilde{x}^n), \tag{6.34}$$

where

$$x^1 = \tilde{x}^1/\tilde{x}^0, \qquad x^2 = \tilde{x}^2/\tilde{x}^0, \ldots, x^n = \tilde{x}^n/\tilde{x}^0.$$

It is customary to employ Roman indices for the range $1, 2, \ldots, n$ and Greek indices for the range $0, 1, 2, \ldots, n$.

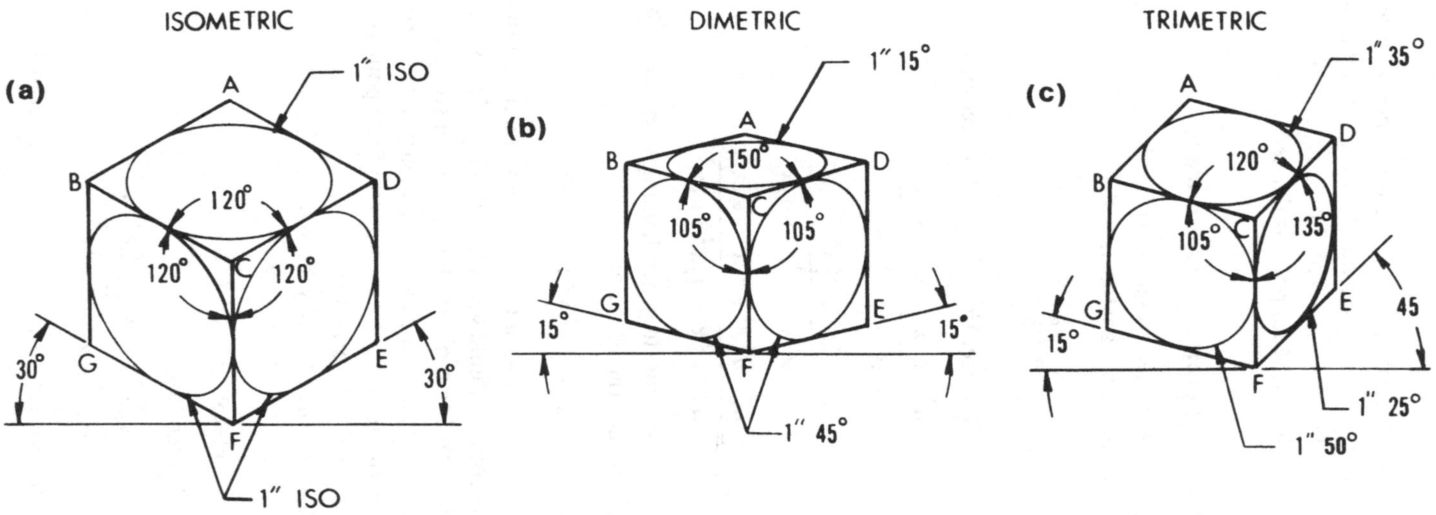

Figure 6.08. (a) Isometric. (b) Dimetric. (c) Trimetric. [From T. A. Thomas, *Technical illustration,* third edition (reproduced from McGraw-Hill Book Co., New York, © 1978), Figures 11–1, 11–2, and 11–3.]

Homogeneous coordinates $\tilde{x}^\lambda$ uniquely define a point in arithmetic n-space, provided that $\tilde{x}^0 \neq 0$. If $x^0 = 0$ but $\tilde{x}^\kappa = 0$, the point at infinity is specified. If all the merates of $\tilde{x}^\lambda$ are zero, however, the ratios of Equation (6.32) are indeterminate and no point is defined.

6.10. The projective transformation

Besides the familiar linear transformation of Section 6.04, we have the linear fractional, or projective, transformation,[17]

$$y^i = \frac{a_0^i + a_k^i x^k}{a_0^0 + a_k^0 x^k}, \tag{6.35}$$

where $i, k = 1, 2, \ldots, n$. In 3-space, Equation (6.35) becomes

$$\begin{aligned} y^1 &= \frac{a_0^1 + a_1^1 x^1 + a_2^1 x^2 + a_3^1 x^3}{a_0^0 + a_1^0 x^1 + a_2^0 x^2 + a_3^0 x^3}, \\ y^2 &= \frac{a_0^2 + a_1^2 x^1 + a_2^2 x^2 + a_3^2 x^3}{a_0^0 + a_1^0 x^1 + a_2^0 x^2 + a_3^0 x^3}, \\ y^3 &= \frac{a_0^3 + a_1^3 x^1 + a_2^3 x^2 + a_3^3 x^3}{a_0^0 + a_1^0 x^1 + a_2^0 x^2 + a_3^0 x^3}. \end{aligned} \tag{6.36}$$

To every point x^i corresponds, in general, a unique point y^i. An exception occurs when x^i satisfies the equation

$$a_0^0 + a_1^0 x^1 + a_2^0 x^2 + a_3^0 x^3 = 0. \tag{6.37}$$

Since the denominator of Equation (6.36) is the same for all three equations, the point y^i becomes infinite when Equation (6.37) is satisfied. But the latter is the equation for a plane. Thus, the transformation moves this plane to infinity; and one of the most characteristic properties of the affine transformation (finite points transform into finite points) no longer holds with the projective transformation.

The projective transformation may be written in terms of homogeneous coordinates $(\tilde{x}^0, \tilde{x}^1, \tilde{x}^2, \tilde{x}^3)$, where

$$x^i = \tilde{x}^i / \tilde{x}^0, \qquad (i = 1, 2, 3). \tag{6.38}$$

Then Equation (6.35) becomes

$$y^i = \frac{a_0^i \tilde{x}^0 + a_1^i \tilde{x}^1 + a_2^i \tilde{x}^2 + a_3^i \tilde{x}^3}{a_0^0 \tilde{x}^0 + a_1^0 \tilde{x}^1 + a_2^0 \tilde{x}^2 + a_3^0 \tilde{x}^3}, \tag{6.39}$$

and if we introduce homogeneous coordinates for the transformed point y^i also,

$$y^i = \tilde{y}^i / \tilde{y}^0, \tag{6.40}$$

with the arbitrary $\tilde{y}^0$ equal to the denominator of Equation (6.39),

$$\tilde{y}^\lambda = a^\lambda_\mu \tilde{x}^\mu. \tag{6.41}$$

Note that Equation (6.41) is an affine transformation in 4-space. Indeed, the introduction of homogeneous coordinates makes a projective transformation in n-space equivalent to an affine transformation in $(n+1)$-space. Since Equation (6.41) is an affine transformation, a unique inverse exists if and only if the determinant of the transformation is not zero:

$$|a^\lambda_\mu| \neq 0.$$

Now consider coordinate transformations in projective space. We know that, in general, a point x^i does not transform as a tensor; but in affine space, a point x^i transforms as a tensor if coordinate transformations are limited to the centro-affine group.

Similarly, a point $\tilde{x}^i$ transforms as an akinetor in projective space if coordinate transformations are limited to

$$x^{i'} = \frac{a^{i'}_0 + a^{i'}_i x^i}{a^{0'}_0 + a^{0'}_i x^i}. \tag{6.42}$$

In terms of homogeneous coordinates, Equation (6.42) becomes

$$\tilde{x}^{\lambda'} = \tau a^{\lambda'}_\lambda \tilde{x}^\lambda, \tag{6.43}$$

where τ is an arbitrary nonvanishing constant. Therefore, for a linear fractional coordinate transformation, the homogeneous coordinates of a point transform as a univalent akinetor.

6.11. Projective space

Just as affine space was characterized by invariant properties under the affine space transformation, so projective space is described by invariant properties under the projective space transformation. The discussion in Section 6.10 has shown that a 1:1 correspondence exists between the original points x^i and the transformed points y^i if $|a^\lambda_\mu| \neq 0$. Thus, points transform into points in projective space. However, finite points do not necessarily remain finite nor do infinite points remain infinite.

That a *plane* transforms into a plane under projective transformation is easily shown. But parallelism is not necessarily preserved.

A line may be defined as passing through two points $\tilde{x}^\kappa$ and $\tilde{y}^\lambda$. The line may be specified by a bivalent alternating akinetor,

$$\tilde{p}^{\kappa\lambda} = \tfrac{1}{2}\tilde{x}^{[\kappa}\tilde{y}^{\lambda]} = \tilde{x}^{\kappa}\tilde{y}^{\lambda} - \tilde{x}^{\lambda}\tilde{y}^{\kappa}. \tag{6.44}$$

The distinct merates are $\tilde{p}^{01}, \tilde{p}^{02}, \tilde{p}^{03}, \tilde{p}^{12}, \tilde{p}^{13}, \tilde{p}^{23}$.

But a line may be specified geometrically by five independent merates (a point and two angles). Thus, we would expect that the six distinct merates of $\tilde{p}^{\kappa\lambda}$ are not linearly independent; a relation must exist among them. The relation was found by Plücker and is called the *Plücker identity*:

$$\tilde{p}^{[\kappa\lambda}\tilde{p}^{\mu\nu]} \equiv 0. \tag{6.45}$$

The square brackets around the four indices are an abbreviation for the sum of the three positive permutations of the indices 0, 1, 2, 3. Thus, Equation (6.45) expands into

$$\tilde{p}^{01}\tilde{p}^{23} + \tilde{p}^{02}\tilde{p}^{31} + \tilde{p}^{03}\tilde{p}^{12} \equiv 0.$$

The validity of the Plücker relation is readily verified by substituting Equation (6.44) into (6.45).

Suppose that a line $\tilde{q}^{\kappa\lambda}$ is defined in terms of two points $\tilde{a}^{\kappa}$ and $\tilde{b}^{\lambda}$ that lie on the line defined by $\tilde{x}^{\kappa}$ and $\tilde{y}^{\lambda}$. Then, since $\tilde{x}^{\kappa}$ and $\tilde{y}^{\lambda}$ are homogeneous Cartesian point coordinates,

$$\tilde{a}^{\kappa} = \mu\tilde{x}^{\kappa} + (1-\mu)\tilde{y}^{\kappa}, \qquad \tilde{b}^{\lambda} = \nu\tilde{x}^{\lambda} + (1-\nu)\tilde{y}^{\lambda}. \tag{6.46}$$

The coordinates of the line passing through $\tilde{a}^{\kappa}$ and $\tilde{b}^{\lambda}$ are

$$\tilde{q}^{\kappa\lambda} = \tfrac{1}{2}\tilde{a}^{[\kappa}\tilde{b}^{\lambda]} = \tilde{a}^{\kappa}\tilde{b}^{\lambda} - \tilde{a}^{\lambda}\tilde{b}^{\kappa}. \tag{6.47}$$

Substitution of Equation (6.46) into (6.47) gives

$$\tilde{q}^{\kappa\lambda} = (\mu-\nu)(\tilde{x}^{\kappa}\tilde{y}^{\lambda} - \tilde{x}^{\lambda}\tilde{y}^{\kappa}) = (\mu-\nu)\tilde{p}^{\kappa\lambda}. \tag{6.48}$$

Since $\tilde{p}^{\kappa\lambda}$ is an akinetor, whose merates can all be multiplied by a constant without changing its meaning, Equation (6.48) indicates that $\tilde{p}^{\kappa\lambda}$ and $\tilde{q}^{\kappa\lambda}$ are one and the same line.

An alternative specification of the line is as the intersection of two planes $\tilde{u}_{\kappa}$ and $\tilde{v}_{\lambda}$. Thus, a line may likewise be defined by the covariant akinetor,

$$\tilde{p}_{\kappa\lambda} = \tfrac{1}{2}\tilde{u}_{[\kappa}\tilde{v}_{\lambda]} = (\tilde{u}_{\kappa}\tilde{v}_{\lambda} - \tilde{u}_{\lambda}\tilde{v}_{\kappa}). \tag{6.49}$$

As with the contravariant representation of the line, the covariant merates of the line satisfy the Plücker identity,

$$\tilde{p}_{[\kappa\lambda}\tilde{p}_{\mu\nu]} = 0. \tag{6.50}$$

Also, the same covariant akinetor is obtained if the line is specified in terms of any other pair of planes that intersect in the same line.

It would appear that we have developed two independent ways of specifying a line: $\tilde{p}^{\kappa\lambda}$ and $\tilde{p}_{\kappa\lambda}$. However, if both refer to the same line, the points $\tilde{x}^\kappa$ and $\tilde{y}^\lambda$ must lie on both the plane $\tilde{u}_\kappa$ and the plane $\tilde{v}_\lambda$:

$$\begin{aligned} \tilde{u}_\kappa \tilde{x}^\kappa &= 0, & \tilde{u}_\kappa \tilde{y}^\kappa &= 0, \\ \tilde{v}_\kappa \tilde{x}^\kappa &= 0, & \tilde{v}_\kappa \tilde{y}^\kappa &= 0. \end{aligned} \tag{6.51}$$

Expansion of these four equations gives

$$\tilde{p}^{01}\tilde{p}_{01} = \tilde{p}^{23}\tilde{p}_{23}, \qquad \tilde{p}^{02}\tilde{p}_{02} = \tilde{p}^{31}\tilde{p}_{31}, \qquad \tilde{p}^{03}\tilde{p}_{03} = \tilde{p}^{12}\tilde{p}_{12}. \tag{6.52}$$

These three equations suggest, but do not prove, a still simpler set of equations:

$$\begin{aligned} \tilde{p}^{01} &= \tilde{p}_{23}, & \tilde{p}^{23} &= \tilde{p}_{01}, \\ \tilde{p}^{02} &= \tilde{p}_{31}, & \tilde{p}^{31} &= \tilde{p}_{02}, \\ \tilde{p}^{03} &= \tilde{p}_{12}, & \tilde{p}^{12} &= \tilde{p}_{03}, \end{aligned}$$

which can be summarized in a single equation,

$$\tilde{p}^{\kappa\lambda} = \tfrac{1}{2}\delta^{\kappa\lambda\mu\nu}_{0123}\tilde{p}_{\mu\nu}, \qquad \tilde{p}_{\kappa\lambda} = \tfrac{1}{2}\delta^{0123}_{\kappa\lambda\mu\nu}\tilde{p}^{\mu\nu}. \tag{6.53}$$

To determine if these simpler equations are indeed valid, we perform a transformation of coordinates so that the line specified by the akinetor $\tilde{p}^{\kappa\lambda}$ coincides with the x^3 axis (Figure 6.09). This line can be specified in terms of any two points on the x^3 axis. Choose the homogeneous coordinates as simply as possible:

$$\tilde{x}^\kappa = (1,0,0,0), \qquad \tilde{y}^\lambda = (1,0,0,1).$$

Then, by Equation (6.44),

$$\tilde{p}^{\kappa\lambda} = \begin{bmatrix} 0 & 0 & 0 & 1 \\ 0 & 0 & 0 & 0 \\ 0 & 0 & 0 & 0 \\ -1 & 0 & 0 & 0 \end{bmatrix}. \tag{6.54}$$

The same line can also be specified in terms of any two planes that intersect in the same line. Suppose we choose $\tilde{u}_\kappa$ and $\tilde{v}_\lambda$ as the two coordinate planes. Then

$$\begin{aligned} \tilde{x}^1 &= 0 \quad \text{or} \quad \tilde{u}_\kappa = (0,1,0,0), \\ \tilde{x}^2 &= 0 \quad \text{or} \quad \tilde{v}_\lambda = (0,0,1,0). \end{aligned}$$

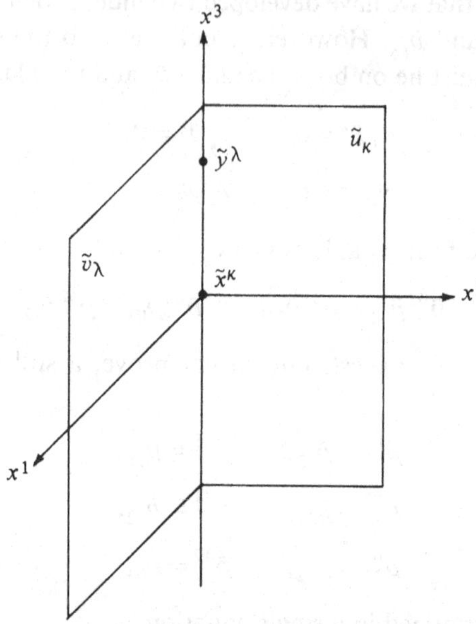

Figure 6.09. Akinetor $\tilde{p}_{\kappa\lambda}$ represented by x^3 axis. Line is specified by the two points

$\tilde{x}^\kappa = (1, 0, 0, 0)$ and $\tilde{y}^\lambda = (1, 0, 0, 1)$

or by the two planes $\tilde{u}_\kappa = (0, 1, 0, 0)$ and $\tilde{v}_\lambda = (0, 0, 1, 0)$.

Thus, by Equation (6.49),

$$\tilde{p}_{\mu\nu} = \begin{bmatrix} 0 & 0 & 0 & 0 \\ 0 & 0 & 1 & 0 \\ 0 & -1 & 0 & 0 \\ 0 & 0 & 0 & 0 \end{bmatrix}. \tag{6.55}$$

Note that for the x^3 axis all merates are zero except for $\tilde{p}^{03}$, $\tilde{p}^{30}$, $\tilde{p}_{12}$, and $\tilde{p}_{21}$. And for this line, not only is Equation (6.52) satisfied but also the more restrictive Equation (6.53). The verification of Equation (6.53) can be completed by writing the contravariant and covariant merates of lines along the other two coordinate axes.

Suppose that the points $\tilde{x}^\kappa$ and $\tilde{y}^\lambda$, in terms of which a line is defined, are transformed into other points $\tilde{X}^\kappa$ and $\tilde{Y}^\lambda$ by a projective transformation. Then

$$\tilde{x}^{\kappa} = a^{\kappa}_{\mu}\tilde{X}^{\mu}, \qquad \tilde{y}^{\lambda} = a^{\lambda}_{\nu}\tilde{Y}^{\nu},$$

and

$$\begin{aligned}\tilde{p}^{\kappa\lambda} &= \tfrac{1}{2}\tilde{x}^{[\kappa}\tilde{y}^{\lambda]} = (\tilde{x}^{\kappa}\tilde{y}^{\lambda} - \tilde{x}^{\lambda}\tilde{y}^{\kappa}) \\ &= a^{\kappa}_{\mu}a^{\lambda}_{\nu}(\tilde{X}^{\mu}\tilde{Y}^{\nu} - \tilde{X}^{\nu}\tilde{Y}^{\mu}) \\ &= a^{\kappa}_{\mu}a^{\lambda}_{\nu}\tilde{X}^{[\mu}\tilde{Y}^{\nu]} = a^{\kappa}_{\mu}a^{\lambda}_{\nu}\tilde{q}^{\mu\nu}. \end{aligned} \tag{6.56}$$

A line $\tilde{p}^{\kappa\lambda}$ always transforms into another line $\tilde{q}^{\mu\nu}$ under projective transformation of the points of space.

On the other hand, if we make a projective transformation of the point coordinates, according to Equation (6.43),

$$\tilde{x}^{\kappa'} = \tau a^{\kappa'}_{\kappa}\tilde{x}^{\kappa}, \qquad \tilde{y}^{\lambda'} = \tau a^{\lambda'}_{\lambda}\tilde{y}^{\lambda}. \tag{6.57}$$

Since by definition,

$$\tilde{p}^{\kappa'\lambda'} = \tilde{x}^{[\kappa'}\tilde{y}^{\lambda']}, \qquad \tilde{p}^{\kappa\lambda} = \tilde{x}^{[\kappa}\tilde{y}^{\lambda]},$$

substitution of Equation (6.57) yields

$$\tilde{p}^{\kappa'\lambda'} = \tau^2 a^{\kappa'}_{\kappa}a^{\lambda'}_{\lambda}\tilde{p}^{\kappa\lambda}. \tag{6.58}$$

Thus, the contravariant representation of a line transforms as a bivalent akinetor.

Likewise, it can be shown that the covariant line coordinates $\tilde{p}_{\kappa\lambda}$ transform as a covariant bivalent akinetor,

$$\tilde{p}_{\kappa\lambda} = \tau^2 a^{\kappa'}_{\kappa}a^{\lambda'}_{\lambda}\tilde{p}_{\kappa'\lambda'}. \tag{6.59}$$

We have seen that in affine space the concept of distance is not valid, though the parting of points on a given line is an affine invariant. In projective space, parting is not invariant.

Consider four points P^i, Q^i, R^i, and S^i (Figure 6.10) that lie on an arbitrary line. According to Equation (6.13a),

$$R^i = (1-\lambda)P^i + \lambda Q^i, \qquad S^i = (1-\mu)P^i + \mu Q^i. \tag{6.60}$$

Then

$$\begin{aligned} R^i - P^i &= \lambda(Q^i - P^i), \qquad & R^i - Q^i &= (\lambda - 1)(Q^i - P^i), \\ S^i - P^i &= \mu(Q^i - P^i), \qquad & S^i - Q^i &= (\mu - 1)(Q^i - P^i). \end{aligned} \tag{6.61}$$

Now transform the line into another line and another set of four points T^i, U^i, V^i, W^i by a projective transformation. Then

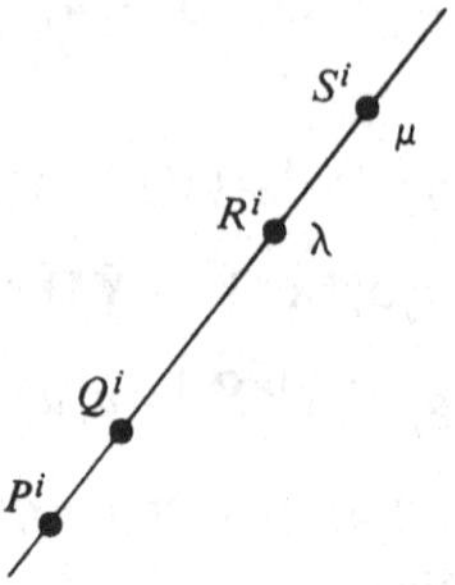

Figure 6.10. Four points on a line with *cross ratio*

$$\frac{\lambda(\mu-1)}{\mu(\lambda-1)}.$$

$$T^i=\frac{a^i_0+a^i_kP^k}{a^0_0+a^0_kP^k},\qquad U^i=\frac{a^i_0+a^i_kQ^k}{a^0_0+a^0_kQ^k},$$
$$V^i=\frac{a^i_0+a^i_kR^k}{a^0_0+a^0_kR^k},\qquad W^i=\frac{a^i_0+a^i_kS^k}{a^0_0+a^0_kS^k}. \tag{6.62}$$

Substitution of Equation (6.60) into (6.62) gives

$$V^i=\frac{a^i_0+a^i_k(1-\lambda)P^k+a^i_k\lambda Q^k}{a^0_0+a^0_k(1-\lambda)P^k+a^0_k\lambda Q^k},$$
$$W^i=\frac{a^i_0+a^i_k(1-\mu)P^k+a^i_k\mu Q^k}{a^0_0+a^0_k(1-\mu)P^k+a^0_k\mu Q^k}. \tag{6.63}$$

From Equations (6.62) and (6.63),

$$V^i-T^i=\frac{\lambda C^i}{D(a^0_0+a^0_kP^k)},\qquad V^i-U^i=\frac{(\lambda-1)C^i}{D(a^0_0+a^0_kQ^k)},$$

$$W^i-T^i=\frac{\mu C^i}{E(a^0_0+a^0_kP^k)},\qquad W^i-U^i=\frac{(\mu-1)C^i}{E(a^0_0+a^0_kQ^k)},$$

where

$$C^i=(a^i_0a^0_k-a^0_0a^i_k)(P^k-Q^k)+(a^i_la^0_k-a^0_la^i_k)P^kQ^l,$$
$$D=a^0_0+(1-\lambda)a^0_kP^k-\lambda a^0_kQ^k,$$
$$E=a^0_0+(1-\mu)a^0_kP^k+\mu a^0_kQ^k.$$

Ratios are

$$\frac{V^i - T^i}{V^i - U^i} = \frac{\lambda}{\lambda - 1} \frac{a_0^0 + a_k^0 Q^k}{a_0^0 + a_k^0 P^k},$$

$$\frac{W^i - T^i}{W^i - U^i} = \frac{\mu}{\mu - 1} \frac{a_0^0 + a_k^0 Q^k}{a_0^0 + a_k^0 P^k},$$

or

$$\frac{\dfrac{V^i - T^i}{V^i - U^i}}{\dfrac{W^i - T^i}{W^i - U^i}} = \frac{\lambda(\mu - 1)}{\mu(\lambda - 1)}. \tag{6.64}$$

Likewise,

$$\frac{R^i - P^i}{R^i - Q^i} = \frac{\lambda}{\lambda - 1}, \qquad \frac{S^i - P^i}{S^i - Q^i} = \frac{\mu}{\mu - 1}.$$

Therefore,

$$\frac{\dfrac{R^i - P^i}{R^i - Q^i}}{\dfrac{S^i - P^i}{S^i - Q^i}} = \frac{\dfrac{V^i - T^i}{V^i - U^i}}{\dfrac{W^i - T^i}{W^i - U^i}} = \frac{\lambda(\mu - 1)}{\mu(\lambda - 1)}. \tag{6.65}$$

Thus, we have found an invariant of projective space. This ratio of ratios is called the *cross ratio* of four points. The invariance of the cross ratio was given by Pappus in his *Mathematical collection* about A.D. 300.[18]

Similarly, the cross ratio of partings of four planes intersecting in a given line is a projective invariant. This is the nearest that we can approach the concept of angle in projective space.

A second-degree surface transforms into a second-degree surface under projective transformation. But since finite points may become infinite and infinite points finite under projective transformations, the classification of second-degree surfaces becomes meaningless.

In terms of homogeneous coordinates, the equation of a quadric surface may be written

$$\tilde{c}_{\kappa\lambda} \tilde{x}^\kappa \tilde{x}^\lambda = 0, \tag{6.66}$$

where

$$\{\tilde{c}_{\kappa\lambda}\} = \{\tilde{c}_{\lambda\kappa}\}$$

is symmetric. Transformation of coordinates shows that

$$\tilde{c}_{\kappa'\lambda'} = \tau^{-2} a^\kappa_{\kappa'} a^\lambda_{\lambda'} \tilde{c}_{\kappa\lambda}, \tag{6.67}$$

Table 6.03. *Projective space*

A. Geometric invariants of projective space

$$\bar{y}^{\lambda} = a^{\lambda}_{\mu}\bar{x}^{\mu}, \quad |a^{\lambda}_{\mu}| \neq 0$$

(1) Point
(2) Plane
(3) Line
(4) Ratio of partings of points on same line
(5) Ratio of partings of planes intersecting in same line
(6) Second-degree surfaces

B. Akinetors

$$\tilde{x}^{\lambda'} = \tau a^{\lambda'}_{\lambda}\tilde{x}^{\lambda}, \quad |a^{\lambda'}_{\lambda}| \neq 0 \qquad \tilde{u}_{\lambda'} = \tau^{-1}a^{\lambda}_{\lambda'}\tilde{u}_{\lambda}$$

(1) Point $\tilde{x}^{\lambda}$, a univalent akinetor
(2) Plane $\tilde{u}_{\lambda}$, a univalent akinetor
(3) Line $\tilde{p}^{\kappa\lambda}$, a bivalent alternating akinetor:

$$\tilde{p}^{\kappa'\lambda'} = \tau^2 a^{\kappa'}_{\kappa} a^{\lambda'}_{\lambda}\tilde{p}^{\kappa\lambda}$$

Line $\tilde{p}_{\kappa\lambda}$, a bivalent alternating akinetor:

$$\tilde{p}_{\kappa'\lambda'} = \tau^{-2} a^{\kappa}_{\kappa'} a^{\lambda}_{\lambda'}\tilde{p}_{\kappa\lambda}$$

$$\tilde{p}^{[\kappa\lambda}\tilde{p}^{\mu\nu]} = 0 \qquad \tilde{p}_{[\kappa\lambda}\tilde{p}_{\mu\nu]} = 0$$

(4) Second-degree surface $\tilde{c}^{\kappa\lambda}$, a bivalent symmetric akinetor:

$$\tilde{c}^{\kappa'\lambda'} = \tau^2 a^{\kappa'}_{\kappa} a^{\lambda'}_{\lambda}\tilde{c}^{\kappa\lambda}$$

Second-degree surface $\tilde{c}_{\kappa\lambda}$, a bivalent symmetric akinetor:

$$\tilde{c}_{\kappa'\lambda'} = \tau^{-2} a^{\kappa}_{\kappa'} a^{\lambda}_{\lambda'}\tilde{c}_{\kappa\lambda}$$

so $\tilde{c}_{\kappa\lambda}$ transforms as a bivalent covariant akinetor under projective transformations of coordinates.

An alternative way of writing the equation of a quadric surface is in terms of the planes $\tilde{u}_{\kappa}$ tangent to the surface:

$$\tilde{c}^{\kappa\lambda}\tilde{u}_{\kappa}\tilde{u}_{\lambda} = 0, \tag{6.68}$$

where

$$\{\tilde{c}^{\kappa\lambda}\} = \{\tilde{c}^{\lambda\kappa}\}$$

and

$$\tilde{c}^{\kappa'\lambda'} = \tau^2 a^{\kappa'}_{\kappa} a^{\lambda'}_{\lambda}\tilde{c}^{\kappa\lambda}. \tag{6.69}$$

The contravariant and covariant representations of the quadric are related by the equation

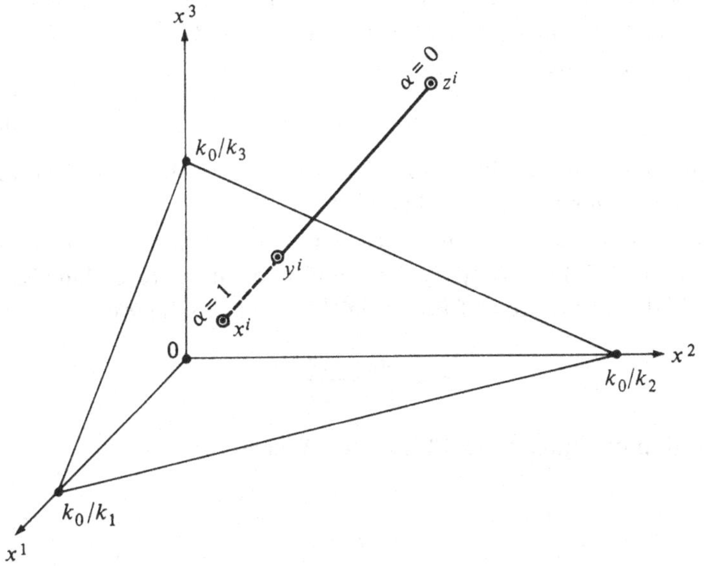

Figure 6.11. Perspective transformation x^i in 3-space to y^i in 2-space (points in the slanting triangle). A fixed point z^i is connected to x^i by a line.

$$\tilde{c}^{\kappa\lambda} = \delta^{\kappa\lambda\mu\nu}_{0123}\tilde{c}_{\mu\nu}. \tag{6.70}$$

Properties of projective space are summarized in Table 6.03.

6.12. Perspective space

Section 6.08 treated the special case of affine transformation in which $\Delta=0$. An analogous degenerate case of projective transformation gives a perspective 2-space that is widely used.

In order to understand exactly how this transformation of space can be developed, consider the arrangement shown in Figure 6.11. A plane is chosen whose equation is

$$k_0 - k_1 y^1 - k_2 y^2 - k_3 y^3 = 0$$

or

$$k_0 - k_i y^i = 0. \tag{6.71}$$

where the intercepts are $k_0/k_1, k_0/k_2, k_0/k_3$. This plane will be called the picture plane. Points x^i of the 3-space are projected onto this plane by connecting x^i with a fixed point z^i called the center of projection. The equation of a line through an arbitrary point x^i and the center of projection z^i can be written

$$y^i = \alpha x^i + (1-\alpha) z^i. \tag{6.72}$$

To each point y^i on the line corresponds a unique value of α. The value $\alpha = 0$ corresponds to z^i, the value $\alpha = 1$ to x^i.

Suppose now that y^i is required to lie in the picture plane. Then the y^i of Equation (6.72) must satisfy the equation of the picture plane [Equation (6.71)]. Substitution of Equation (6.72) into (6.71) gives

$$\alpha = \frac{k_0 - k_i z^i}{k_i (x^i - z^i)}. \tag{6.73}$$

Substitution of Equation (6.73) into (6.72) gives

$$y^i = \frac{k_0 (x^i - z^i) + k_j (x^j z^i - x^i z^j)}{k_i (x^i - z^i)} \tag{6.74}$$

or on expansion

$$\begin{aligned}
y^1 &= \frac{k_0(x^1 - z^1) + k_2(x^2 z^1 - x^1 z^2) + k_3(x^3 z^1 - x^1 z^3)}{k_1(x^1 - z^1) + k_2(x^2 - z^2) + k_3(x^3 - z^3)},\\
y^2 &= \frac{k_0(x^2 - z^2) + k_1(x^1 z^2 - x^2 z^1) + k_3(x^3 z^2 - x^2 z^3)}{k_1(x^1 - z^1) + k_2(x^2 - z^2) + k_3(x^3 - z^3)},\\
y^3 &= \frac{k_0(x^3 - z^3) + k_1(x^1 z^3 - x^3 z^1) + k_2(x^2 z^3 - x^3 z^2)}{k_1(x^1 - z^1) + k_2(x^2 - z^2) + k_3(x^3 - z^3)}.
\end{aligned} \tag{6.74a}$$

Recalling that

$$y^i = \frac{a_0^i + a_j^i x^j}{a_0^0 + a_j^0 x^j}, \tag{6.35}$$

Comparison of Equations (6.35) and (6.74) shows that the perspective transformation for a picture plane k_λ and a center of projection z^i is

$$a_\lambda^\kappa =$$

$$\begin{bmatrix}
-k_1 z^i - k_2 z^2 - k_3 z^3 & k_1 & k_2 & k_3 \\
-k_0 z^1 & k_0 - k_2 z^2 - k_3 z^3 & k_2 z^1 & k_3 z^1 \\
-k_0 z^2 & k_1 z^2 & k_0 - k_1 z^1 - k_3 z^3 & k_3 z^2 \\
-k_0 z^3 & k_1 z^3 & k_2 z^3 & k_0 - k_1 z^1 - k_2 z^2
\end{bmatrix}. \tag{6.75}$$

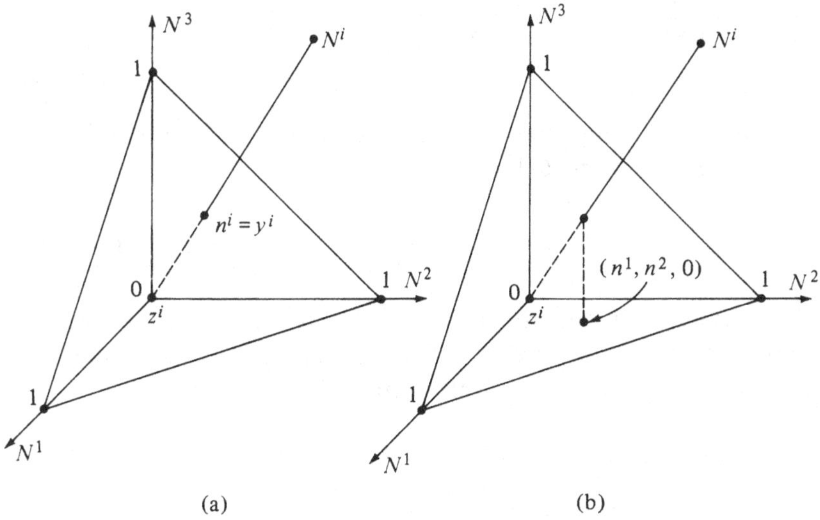

Figure 6.12. An application of perspective transformation to nutrition. The 3-space is *nutrition space* (N^1, N^2, N^3) and the center of projection z^i is arbitrarily selected as the origin. (a) The transformed point $(n^i = y^i)$ is on the unit triangle. (b) The final result is obtained by projecting the previous point parallel to the N^3 axis onto the N^3 plane.

An example used in nutrition[19] is the case in which an arbitrary nutrition vector $N^i = x^i$ is projected onto the unit plane. As shown in Figure 6.12a, $k_0 = 1$, $k_1 = k_2 = k_3 = 1$. The equation of the unit plane, if $n^i = y^i$ is an arbitrary point on the plane, is

$$n^1 + n^2 + n^3 = 1.$$

The center of projection is chosen to be at the origin, $z^i = (0, 0, 0)$. Then substitution into Equation (6.36) gives the matrix

$$a^\kappa_\lambda = \begin{bmatrix} 0 & 1 & 1 & 1 \\ 0 & 1 & 0 & 0 \\ 0 & 0 & 1 & 0 \\ 0 & 0 & 0 & 1 \end{bmatrix}.$$

Note that $|a^\kappa_\lambda| = 0$, but a^κ_λ is a matrix of rank 3. Therefore, the transformation is

$$n^1 = \frac{N^1}{N^1+N^2+N^3}, \qquad n^2 = \frac{N^2}{N^1+N^2+N^3}, \qquad n^3 = \frac{N^3}{N^1+N^2+N^3}. \tag{6.76}$$

However, the transformation most useful in nutrition is not that given in Equation (6.76) but

$$n^1 = \frac{N^1}{N^1+N^2+N^3}, \qquad n^2 = \frac{N^2}{N^1+N^2+N^3}, \qquad n^3 = 0. \tag{6.77}$$

Here all points N^i are projected onto the plane $n^3 = 0$, so the picture plane has the merates $k_0 = k_1 = k_2 = 0$, $k_3 = 1$. Clearly, Equation (6.77) is a projective transformation with matrix

$$a^{\kappa}_{\lambda} = \begin{bmatrix} 0 & 1 & 1 & 1 \\ 0 & 1 & 0 & 0 \\ 0 & 0 & 1 & 0 \\ 0 & 0 & 0 & 0 \end{bmatrix}. \tag{6.78}$$

It would seem reasonable to find the center of projection by substituting the known merates of the picture plane into Equation (6.75),

$$a^{\kappa}_{\lambda} = \begin{bmatrix} -z^3 & 0 & 0 & 1 \\ 0 & -z^3 & 0 & 0 \\ 0 & 0 & -z^3 & z^2 \\ 0 & 0 & 0 & 0 \end{bmatrix}, \tag{6.79}$$

and equating the matrices in Equations (6.78) and (6.79). Inspection of the matrices shows that a solution of the necessary equations is impossible. Thus, for the transformation to the nutrition triangle, no center of projection exists.

How can this be? The difficulty is that the matrix of Equation (6.78) is of rank 2 rather than rank 3. As can be visualized from Figure 6.12b, this transformation can be viewed as a projection of the points of 3-space N^i onto the unit plane with center of projection at the origin, followed by a projection parallel to the N^3 axis onto the N^1N^2 plane. The latter is an affine transformation. A projective transformation followed by an affine transformation yields a degenerate case of a projective transformation, which cannot be replaced by a single perspective transformation from a definite viewpoint.

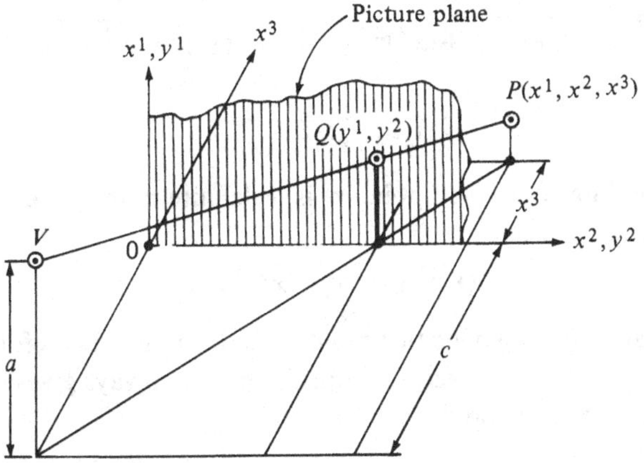

Figure 6.13. Perspective drawing. The eye is at V. The object is at $P(x^1, x^2, x^3)$, while the corresponding point $Q(y^1, y^2)$ is in the vertical picture plane.

Perspective transformations are most widely used to make perspective drawings[20] in mechanical drawing and by every photographer whenever he photographs a scene. Here again it is convenient to choose the picture plane as the coordinate plane $y^3 = 0$, so

$$k_0 = k_1 = k_2 = 0, \qquad k_3 = 1.$$

As shown in Figure 6.13, the center of projection is at $z^i = (a, 0, -c)$. Then, by Equation (6.75), the coefficient matrix is

$$a_\lambda^\kappa = \begin{bmatrix} c & 0 & 0 & 1 \\ 0 & c & 0 & a \\ 0 & 0 & c & 0 \\ 0 & 0 & 0 & 0 \end{bmatrix}$$

and consequently the transformation equations are

$$y^1 = \frac{cx^1 + ax^3}{c + x^3}, \qquad y^2 = \frac{cx^2}{c + x^3}, \qquad y^3 = 0. \tag{6.80}$$

These equations can be used to calculate the locus for the correct perspective drawing y^i of any object whose position x^i in 3-space is known.

6.13. Inversions

The transformations considered in Sections 6.04–6.12 are of the simplest type in that all the variables are of the first degree. A transformation employing second powers is the inversion[21]:

$$y^1 = \frac{x^1}{(x^1)^2+(x^2)^2+(x^3)^2}, \qquad y^2 = \frac{x^2}{(x^1)^2+(x^2)^2+(x^3)^2},$$
$$y^3 = \frac{x^3}{(x^1)^2+(x^2)^2+(x^3)^2}. \tag{6.81}$$

Corresponding to a given point x^i is a definite point y^i, except when x^i satisfies the equation

$$(x^1)^2+(x^2)^2+(x^3)^2=0. \tag{6.82}$$

The only real point satisfying this equation is the origin. Therefore, $y^i \to \infty$ corresponds to $x^i = 0$. In fact, the transformation always gives reciprocal radii. For an arbitrary point x^i,

$$(y^1)^2+(y^2)^2+(y^3)^2=[(x^1)^2+(x^2)^2+(x^3)^2]^{-1},$$

or

$$(r)_y = 1/(r)_x. \tag{6.83}$$

By inversion, spheres transform into spheres. The equation of a sphere may be written

$$A[(y^1)^2+(y^2)^2+(y^3)^2]+By^1+Cy^2+Dy^3+E=0.$$

Transformation by Equation (6.81) gives

$$E[(x^1)^2+(x^2)^2+(x^3)^2]+Bx^1+Cx^2+Dx^3+A=0,$$

which is again the equation for a sphere. The general statement can be made that a sphere always passes into a sphere, considering a plane as the special case of a sphere with infinite radius.

The corresponding theorem for 2-space is that circles always transform into circles (including the circle of infinite radius). This theorem is the basis of the widely used circle diagrams employed in the calculation of transmission lines, induction motors,[22] and so on. The Smith chart is a familiar example.[23] Inversion is also used in obtaining new coordinate systems[24] and in the method of images employed in electrostatics.

It can be shown also that angles are preserved under inversion. This property is valuable in the various circle diagrams and in obtaining new orthogonal coordinate systems by inversion.

The inversion transformation finds practical application in mapping by stereographic projection.[25] Represent the earth as a sphere of radius $\frac{1}{2}$ and place its south pole at the origin and its north pole on the positive x^3 axis. The equation of the sphere is

$$(x^1)^2+(x^2)^2+(x^3-\tfrac{1}{2})^2=\tfrac{1}{4},$$

or

$$(x^1)^2+(x^2)^2+(x^3)^2=x^3. \tag{6.84}$$

Under inversion, this sphere transforms into a plane. The plane is tangent to the sphere at the north pole. A point x^i on the sphere transforms into a point y^i in the plane, where

$$y^1=x^1/x^3, \qquad y^2=x^2/x^3, \qquad y^3=1. \tag{6.85}$$

Since angles are preserved under inversion, the meridians and parallels on the sphere are transformed into an orthogonal map in the plane. Meridians become straight lines in the plane map, and parallels become concentric circles. The south pole becomes the point at infinity; but near the north pole, the map gives a fairly good portrayal of distances.

6.14. Homogeneous coordinates in affine space

We now return to affine space (Section 6.04) but apply homogeneous coordinates.[26] The affine transformation of space may be included as a special case of projective transformation if a^λ_μ is defined as

$$a^\lambda_\mu=\begin{bmatrix} 1 & 0 & 0 & 0 \\ 0 & a^1_1 & a^1_2 & a^1_3 \\ 0 & a^2_1 & a^2_2 & a^2_3 \\ 0 & a^3_1 & a^3_2 & a^3_3 \end{bmatrix}. \tag{6.86}$$

Similarly, the affine coordinate transformation may be included as a special case of the linear fractional coordinate transformation if

$$a^{\lambda'}_\lambda=\begin{bmatrix} 1 & 0 & 0 & 0 \\ 0 & a^{1'}_1 & a^{1'}_2 & a^{1'}_3 \\ 0 & a^{2'}_1 & a^{2'}_2 & a^{2'}_3 \\ 0 & a^{3'}_1 & a^{3'}_2 & a^{3'}_3 \end{bmatrix}. \tag{6.87}$$

Some tensors will now be considered, employing homogeneous coordinates in affine 3-space. For example, one frequently needs a tensor having four merates instead of three; and this tensor can be obtained rather neatly by homogeneous point coordinates. A physical application is the mass point x^λ, where mass is specified by x^0 and position is specified by the other three merates x^i. A similar application is the charged particle x^λ, where charge is specified by the invariant x^0 and position is given by the other merates.

The weighted point x^λ is defined as a holor with homogeneous coordinates (x^0, x^1, x^2, x^3) where $x^0 \neq 0$. Under the centro-affine coordinate transformation,

$$x^\lambda = a^\lambda_{\lambda'} x^{\lambda'} \tag{6.88}$$

or

$$x^0 = x^{0'}, \qquad x^i = a^i_{i'} x^{i'}.$$

The weight x^0 of the point is an invariant. The position x^i of the point transforms as in Equation (6.10b). Note that x^λ is specified by an invariant magnitude x^0, which may be either positive or negative, associated with a position x^i in space. This is quite different from the point x^i studied in Section 6.04. It is only by introducing homogeneous coordinates into affine space that a mathematical representation of weighted points results.

A closely related concept is the free affine vector v^i. Consider the difference of two weighted points x^κ and y^κ whose weights are equal. Let

$$v^\kappa = x^\kappa - y^\kappa.$$

But since v^0 is always zero for pairs of weighted points of equal weight, the entire significant specification of v^κ is given by v^1, v^2, and v^3, or we can define the free affine vector by

$$v^i = x^i - y^i. \tag{6.89}$$

Clearly, the same merates $\{v^i\}$ are obtained for any pair of points of equal parting on any given line or on any parallel lines. Thus, the free affine vector v^i may be thought of as a directed line segment that is free to move parallel to itself.

Corresponding covariant tensors can be interpreted geometrically as planes in affine space. A fixed plane is specified by the covariant tensor u_λ, where

$$u_\lambda x^\lambda = 0. \tag{6.90}$$

Since the point x^λ transforms as a contravariant vector [Equation (6.78)], u_λ transforms as a covariant vector under centro-affine transformations:

$$u_\lambda = a^{\lambda'}_\lambda u_{\lambda'}, \tag{6.91}$$

or

$$u_0 = u_{0'}, \qquad u_i = a^{i'}_i u_{i'}.$$

The invariant merate u_0 is a measure of the separation of the plane from the origin. The merates $\{u_i\}$ define the direction of the plane. Thus, we can say that the geometric representation of u_λ is a fixed plane. If the plane passes through the origin, $u_0=0$. Planes on opposite sides of the origin have values u_0 of opposite signs. If two planes lie on the same side of the origin, one twice as far from the origin as the other, their u_0's will be in the ratio 2:1. Like the weighted point x^κ, the fixed plane u_λ has a geometric position and also an invariant magnitude that may be positive or negative.

Another useful geometric quantity is obtained by the subtraction of two planes whose invariant merates are equal. Then

$$w_\lambda = u_\lambda - v_\lambda.$$

But $u_0 = v_0$, so $w_0 = 0$ and

$$w_i = u_i - v_i. \tag{6.92}$$

The same merates $\{w_i\}$ are obtained for any pair of planes that are parallel to u_i and v_i. A free affine plane, free to move parallel to itself, is specified by w_i. A magnitude and a sign can be associated with w_i, specifying an affine "area" and a sense of rotation. These concepts are illustrated in Table 6.04.

6.15. Bivalent tensors in affine space

Exciting and important results are obtained when bivalent tensors are derived from the univalent tensors studied in Section 6.14. The procedure is to take alternating products of the univalent tensors that have been expressed in terms of homogeneous coordinates.

An alternating bivalent tensor $p^{\kappa\lambda}$ is obtained by taking the alternating product of two weighted points x^κ and y^λ:

$$p^{\kappa\lambda} = 2x^{[\kappa}y^{\lambda]} = x^\kappa y^\lambda - x^\lambda y^\kappa, \tag{6.93}$$

or

$$p^{\kappa\lambda} =$$

$$\begin{bmatrix} 0 & (x^0y^1-x^1y^0) & (x^0y^2-x^2y^0) & (x^0y^3-x^3y^0) \\ -(x^0y^1-x^1y^0) & 0 & (x^1y^2-x^2y^1) & (x^1y^3-x^3y^1) \\ -(x^0y^2-x^2y^0) & -(x^1y^2-x^2y^1) & 0 & (x^2y^3-x^3y^2) \\ -(x^0y^3-x^3y^0) & -(x^1y^3-x^3y^1) & -(x^2y^3-x^3y^2) & 0 \end{bmatrix}. \tag{6.93a}$$

As in Section 6.11, the six distinct merates $p^{01}, p^{02}, p^{03}, p^{12}, p^{13}, p^{23}$ are related by the Plücker identity,

$$p^{[\kappa\lambda}p^{\mu\nu]} \equiv 0.$$

The same merates $\{p^{\kappa\lambda}\}$ are obtained if $p^{\kappa\lambda}$ is defined in terms of any two points that lie on the same line and have the same parting. Thus, the geometric representation of $p^{\kappa\lambda}$ is a directed line segment that is free to slide along a given line. This geometric figure may be called a *fixed rhabdor*[26] or a *sliding vector.*

The merates of the rhabdor assume a particularly simple form if the weights of x^κ and y^λ are unity ($x^0 = y^0 = 1$). Then

$$p^{\kappa\lambda} = \begin{bmatrix} 0 & (x^1 - y^1) & (x^2 - y^2) & (x^3 - y^3) \\ -(x^1 - y^1) & 0 & (x^1y^2 - x^2y^1) & (x^1y^3 - x^3y^1) \\ -(x^2 - y^2) & -(x^1y^2 - x^2y^1) & 0 & (x^2y^3 - x^3y^2) \\ -(x^3 - y^3) & -(x^1y^3 - x^3y^1) & -(x^2y^3 - x^3y^2) & 0 \end{bmatrix}.$$

Here the merates $\{p^{0i}\}$ define a free vector and the merates $\{p^{ij}\}$ restrict the vector to slide along the given line.

Consider the alternating product of two free affine vectors u^k and v^l:

$$q^{kl} = 2u^{[k}v^{l]}. \tag{6.94}$$

This may be regarded as a special case of a proper rhabdor $q^{\kappa\lambda}$ in which $q^{0i} = 0$. It is called a *free rhabdor*. The geometric representation of q^{kl} is a free plane direction defined by the free vectors u^k and v^l. The order of u^k and v^l defines a sense of rotation in the free plane. If two free rhabdors exist with the same free plane direction, their ratio is the same as the ratio of the areas of the parallelograms defined by the free vectors. Note that any free rhabdor automatically satisfies Plücker's relation because $q^{0i} \equiv 0$.

Now suppose that a set of fixed and free rhabdors is added. The result is always a bivalent tensor $P^{\kappa\lambda}$, but it will not generally satisfy Plücker's relation. Such a geometric figure is called a *kineor*.[26]

A kineor $P^{\kappa\lambda}$ can be decomposed into a sum of a fixed rhabdor $p^{\kappa\lambda}$ and a free rhabdor $q^{\kappa\lambda}$ (for which $q^{0i} \equiv 0$) with respect to a given point x^κ:

$$P^{\kappa\lambda} = p^{\kappa\lambda} + q^{\kappa\lambda},$$

where

$$P^{0i} = p^{0i}, \qquad P^{kl} = p^{kl} + q^{kl}.$$

If a point x^κ is given, then the second point y^κ in terms of which $p^{\kappa\lambda}$ is

Table 6.04. *Affine space with homogeneous coordinates*

Geometric invariants of affine space

$$y^\kappa = a^\kappa_\lambda x^\lambda \quad \text{where } a^\kappa_\lambda = \begin{bmatrix} 1 & 0 & 0 & 0 \\ 0 & a^1_1 & a^1_2 & a^1_3 \\ 0 & a^2_1 & a^2_2 & a^2_3 \\ 0 & a^3_1 & a^3_2 & a^3_3 \end{bmatrix} \quad \text{and } |a^\kappa_\lambda| \neq 0$$

Weighted point

$x^\kappa \qquad x^0 = \text{weight} \neq 0$

Free affine vector

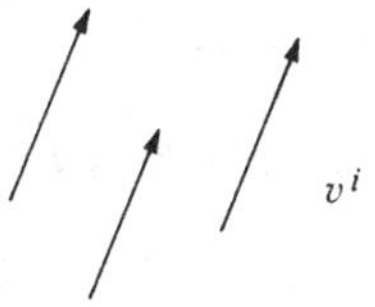

$v^\kappa = x^\kappa - y^\kappa \qquad x^0 = y^0$

$v^0 = 0 \qquad v^i = x^i - y^i$

Fixed plane

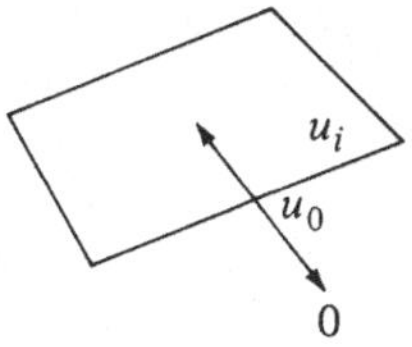

u_λ

Free affine plane

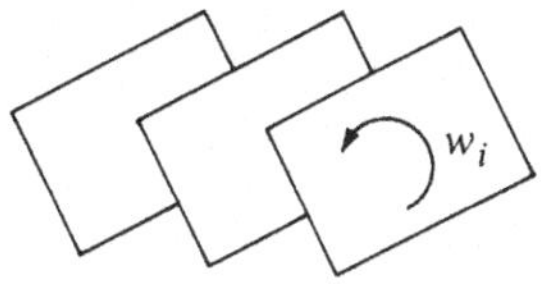

$w_\lambda = u_\lambda - v_\lambda \qquad u_0 = v_0$

$w_0 \equiv 0 \qquad w_i = u_i - v_i$

Fixed rhabdor

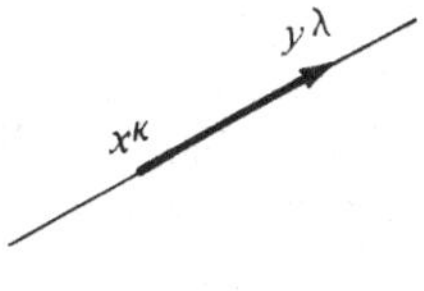

$p^{\kappa\lambda} = 2x^{[\kappa}y^{\lambda]}$

$p^{[\kappa\lambda}p^{\mu\nu]} \equiv 0$

Table 6.04 *(cont.)*

Free rhabdor

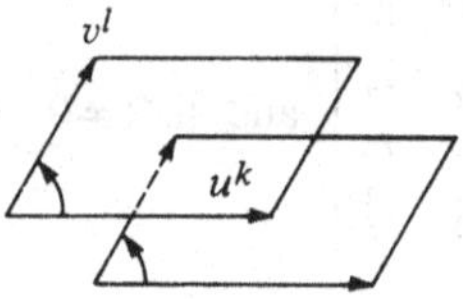

$$q^{kl}=2u^{[k}v^{l]}$$

Kineor

$$P^{0l}=p^{0l}$$
$$P^{kl}=p^{kl}+q^{kl}$$

$$P^{\kappa\lambda}=-P^{\lambda\kappa} \qquad P^{[\kappa\lambda}P^{\mu\nu]}\not\equiv 0$$

Given any x^{κ}, $y^0=x^0$

$$p^{0i}=P^{0i} \qquad q^{0i}=0$$
$$p^{kl}=\frac{x^kP^{0l}-x^lP^{0k}}{x^0}$$
$$q^{kl}=P^{kl}-\frac{x^kP^{0l}-x^lP^{0k}}{x^0}$$

Fixed strophor

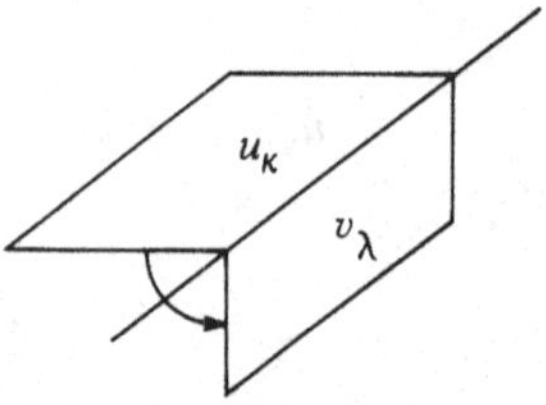

$$p_{\kappa\lambda}=2u_{[\kappa}v_{\lambda]}$$
$$p_{[\kappa\lambda}p_{\mu\nu]}\equiv 0$$

Free strophor

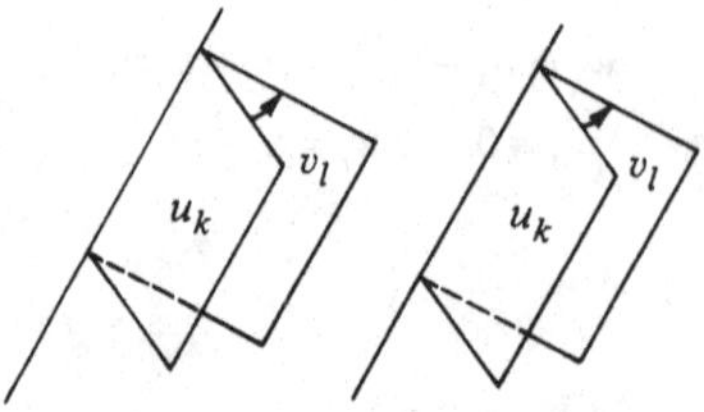

$$q_{kl}=2u_{[k}v_{l]}$$

Helissor

$$P_{0l}=p_{0l}$$
$$P_{kl}=p_{kl}+q_{kl}$$

$$P_{\kappa\lambda}=-P_{\lambda\kappa} \qquad P_{[\kappa\lambda}P_{\mu\nu]}\not\equiv 0$$

Given any u_κ, $u_0=v_0$

Table 6.04 *(cont.)*

	$p_{0i} = P_{0i} \qquad q_{0i} = 0$
	$p_{kl} = \dfrac{u_k P_{0l} - u_l P_{0k}}{u_0}$
	$q_{kl} = P_{kl} - \dfrac{u_k P_{0l} - u_l P_{0k}}{u_0}$

defined can be determined:

$$P^{0i} = p^{0i} = x^0 y^i - y^0 x^i.$$

If the weights of the two points are arbitrarily held equal,

$$y^i = \frac{P^{0i}}{x^0} + x^i$$

and

$$q^{kl} = P^{kl} - p^{kl} = P^{kl} - \left(x^k \frac{P^{0l}}{x^0} - x^l \frac{P^{0k}}{x^0} \right).$$

Thus, the decomposition can be summarized by a single simple equation, given any kineor $P^{\kappa\lambda}$ and any weighted point x^κ:

$$P^{\kappa\lambda} = p^{\kappa\lambda} + q^{\kappa\lambda} \qquad (q^{0l} \equiv 0),$$

where

$$p^{0i} = P^{0i},$$
$$p^{kl} = \frac{x^k p^{0l} - x^l p^{0k}}{x^0}, \qquad (6.95)$$
$$q^{kl} = P^{kl} - \frac{(x^k P^{0l} - x^l P^{0k})}{x^0}.$$

We therefore have a geometric representation for any bivalent alternating tensor $P^{\kappa\lambda}$. If Plücker's relation is satisfied and $P^{0i} \not\equiv 0$, we have a fixed rhabdor; if $P^{0i} \equiv 0$, we have a free rhabdor; if Plücker's relation is not satisfied, the tensor $P^{\kappa\lambda}$ can be expressed as a fixed rhabdor whose axis passes through any given point x^κ, plus a free rhabdor.

Similarly, a covariant bivalent tensor is obtained by taking the alternating product of two fixed planes u_κ and v_λ:

$$p_{\kappa\lambda} = 2u_{[\kappa}v_{\lambda]} = u_\kappa v_\lambda - u_\lambda v_\kappa \tag{6.96}$$

or

$$p_{\kappa\lambda} = \begin{bmatrix} 0 & (u_0v_1-u_1v_0) & (u_0v_2-u_2v_0) & (u_0v_3-u_3v_0) \\ -(u_0v_1-u_1v_0) & 0 & (u_1v_2-u_2v_1) & (u_iv_3-u_3v_1) \\ -(u_0v_2-u_2v_0) & -(u_1v_2-u_2v_1) & 0 & (u_2v_3-u_3v_2) \\ -(u_0v_3-u_3v_0) & -(u_1v_3-u_3v_1) & -(u_2v_3-u_3v_2) & 0 \end{bmatrix}.$$

Thus,

$$\{p_{\kappa\lambda}\} = -\{p_{\kappa\lambda}\}$$

and Plücker's identity is satisfied:

$$p_{[\kappa\lambda}p_{\mu\nu]} \equiv 0.$$

The covariant tensor has a different geometric representation than the contravariant. Again we have a fixed line defined by the intersection of two fixed planes. But instead of a directed line segment measured along the line, we have a sense of direction about the line. The geometric figure defined by $p_{\kappa\lambda}$ may be called a *fixed strophor*.

The alternating product of two free planes u_i and v_j defines a *free strophor* $q_{\kappa\lambda}$.

$$\begin{aligned} q_{0i} &\equiv 0, \\ q_{kl} &= 2u_{[k}v_{l]}. \end{aligned} \tag{6.97}$$

This may be regarded as a special case of the fixed strophor in which $q_{0i} \equiv 0$. The geometric representation is a line direction free to move parallel to itself, a sense of direction and a parting measured about a free line. Note that the free rhabdor and the free strophor are quite different geometrically though very similar algebraically.

The general sum of any set of fixed and free strophors is a bivalent covariant alternating tensor $P_{\kappa\lambda}$ that does not in general satisfy the Plücker identity. This tensor may be called a *helissor*. With respect to any fixed plane u_κ, the helissor $P_{\kappa\lambda}$ may be decomposed into the sum of a fixed strophor $p_{\kappa\lambda}$ and a free strophor $q_{\kappa\lambda}$:

$$P_{\kappa\lambda} = p_{\kappa\lambda} + q_{\kappa\lambda},$$

if all fixed planes have the same first merate ($u_0 = v_0$), where

$$p_{0i} = P_{0i}, \qquad q_{0i} = 0,$$

$$p_{kl} = \frac{u_k P_{0l} - u_l P_{0k}}{u_0} \tag{6.98}$$

$$q_{kl} = P_{kl} - \frac{(u_k P_{0l} - u_l P_{0k})}{u_0}.$$

The properties of the geometric representation of bivalent alternating tensors in affine space are summarized in Table 6.04.

6.16. Summary

In considering a problem in physics, one frequently finds it advantageous to visualize the theory in terms of a fictitious space. But the space is usually not Euclidean. It may be arithmetic, affine, or otherwise, depending on the requirements of the particular physical problem. This chapter covers the details of various geometric spaces.

Problems

6.01 Show that in an arithmetic 2-space, reflection of both axes in the origin is equivalent to a rotation, but that reflection in the x^1 axis is not. Write the transformation determinants for these space transformations.

6.02 Given a right-hand system of orthogonal Cartesian coordinates (x^1, x^2, x^3) (the axes are reflected in the x^1x^2 plane):

(a) Sketch the new axes and indicate if the system is now a right-hand system.

(b) What are the space transformation equations and the transformation determinant?

(c) Repeat (a) and (b) for a reflection in the origin.

6.03 Given an arithmetic n-space:

(a) Prove that a reflection in the origin is equivalent to a rotation of space if n is even but is not equivalent to a rotation if n is odd.

(b) Investigate the same question if the reflection is with respect to a hyperplane ($n-1$ dimensions).

6.04 A set of right-hand, orthogonal, Cartesian axes $x^{i'}$ in 3-space is transformed into a skew set x^i by rotating the $x^{1'}$ axis 30° in a positive direction and the $x^{2'}$ axis 30° in the negative direction. Both rotations are in the $x^{1'}x^{2'}$ plane.

(a) Write the transformation equations and the transformation determinant.
(b) Give the inverse transformation.

6.05 A right-hand orthogonal Cartesian system $x^{i'}$ in 3-space is to be transformed into a left-hand system. The x^3 axis is to be opposite in direction to the $x^{3'}$ axis, and the $x^{1'}$ and $x^{2'}$ axes are to be rotated in the positive direction by 30°.

Write the equations of the transformation. Write the transformation matrix and evaluate the transformation determinant.

6.06 Consider the space transformation

$$y^1 = x^1 + 2x^2, \qquad y^2 = 3x^1 + x^2.$$

(a) Determine the geometric figure into which the unit square with corners at (0,0), (0,1), (1,0), (1,1) transforms.
(b) Does the transformation contain a translation, rotation, reflection, or change of scale?

6.07 A transformation of space consists of a reflection in the x^1x^2 plane and a subsequent rotation of 30° in the positive direction about the x^2 axis. Write the transformation $y^i = A^i_j x^j$ and evaluate $|A^i_j|$.

6.08 Given the rotation

$$x^{1'} = 0.707x^1 + 0.707x^2, \qquad x^{2'} = -0.707x^1 + 0.707x^2,$$

and the transformation

$$x^{1''} = 0.25x^{1'}, \qquad x^{2''} = 3x^{2'}.$$

Express $x^{i''}$ as a function of x^i. Write the matrices and evaluate the determinant.

6.09 Two parallel planes are represented by the equations

$$1.5x^1 + 2.0x^2 + x^3 + 3 = 0,$$

$$3x^1 + 4x^2 + 2x^3 - 4 = 0.$$

An affine transformation is represented by the matrix

$$A^i_j = \begin{bmatrix} 3 & 6 & 9 \\ 4 & 0 & 5 \\ 2 & 0 & 1 \end{bmatrix}.$$

Transform the two planes and show that they remain parallel.

6.10 Given the curve

$$(x^1)^2+(x^2)^2=4,$$

transform by the matrix of Problem 6.09 and identify the transformed curve.

6.11 Repeat Problem 6.09 for the surface

$$(x^1)^2+(x^2)^2+(x^3)^2=a^2.$$

6.12 Given two points P and Q in 3-space, with $(x^i)_p=(0,1,3)$ and $(x^i)_Q=(2,4,8)$:

(a) Find the coordinates y^i of these points when the space is rotated positively about the x^1 axis through an angle of 60° and is stretched by a factor of 2 in the x^3 direction.

(b) Compare the distance between $(x^i)_P$ and $(x^i)_Q$ with the distance between $(y^i)_P$ and $(y^i)_Q$.

(c) Find the coordinates $(x^i)_R$ of all points such that the ratio of the distance $\overline{PQ}$ to the distance $\overline{PR}$ is invariant.

6.13

(a) Prove that the parting λ is an invariant under affine transformations of space.

(b) Prove that λ is an invariant under affine coordinate transformations. Is your conclusion restricted to centro-affine transformations?

6.14 Consider two parallel lines in affine n-space. Show that the ratios of partings of two pairs of points on the two parallel lines is an affine invariant.

6.15

(a) Prove that the parting μ is an invariant under affine transformations of space.

(b) Prove that μ is an invariant under affine coordinate transformations. Is your conclusion restricted to centro-affine transformations?

6.16 Consider two parallel lines. Show that the ratios of partings of two pairs of planes that intersect in the two parallel lines is an affine invariant.

6.17 Consider the triangle whose vertices have coordinates

$$(x^i)_P=(1,2,0), \qquad (x^i)_Q=(2,3,0), \qquad (x^i)_R=(0,0,0).$$

What are the corresponding coordinates when the triangle is altered by the transformation of Problem 6.13?

6.18 Given the plane

$$7x^1+2x^2+4x^3=0,$$

determine the coefficients of an axonometric transformation that will project the points of space into this plane.

6.19 Make a table analogous to Table 6.02 but for axonometric space.

6.20 A special case of axonometric transformation occurs in the pseudo-perspective drawings of oriental artists. Dimensions in all planes parallel to the picture plane are correctly represented. Distances along all lines perpendicular to the picture plane are represented to scale by lines at a fixed angle α with the x^2 axis.

(a) Find the transformation equations

$$y^i = A^i_j x^j.$$

(b) What is the direction of the parallel projection in 3-space?

6.21 Two points have homogeneous coordinates $(\xi^\lambda)_A=(3,2,1,7)$ and $(\xi^\lambda)_B=(1,2,3,4)$.

(a) Locate these points in rectangular coordinates in 3-space.

(b) The point coordinates are then transformed by projective transformation with

$$a^\mu_\lambda = \begin{bmatrix} 0 & 2 & -1 & 4 \\ 3 & 0 & 0 & 0 \\ 4 & 5 & 0 & 0 \\ 0 & 0 & 6 & 7 \end{bmatrix}.$$

Calculate the new coordinates $(\eta^\mu)_A$ and $(\eta^\mu)_B$ of the points and plot the points.

6.22

(a) Prove that a plane $\tilde{v}_\lambda \tilde{y}^\lambda = 0$ transforms into a plane $\tilde{u}_\mu \tilde{x}^\mu = 0$ under projective transformations of the points of space.

(b) Find the relation between $\tilde{u}_\mu$ and $\tilde{v}_\lambda$.

6.23

(a) Prove that under projective coordinate transformations a plane $\tilde{u}_{\lambda'} \tilde{x}^{\lambda'} = 0$ transforms into a plane $\tilde{u}_\lambda \tilde{x}^\lambda = 0$.

(b) Find the transformation equation of $\tilde{u}_\lambda$ under projective transformations of point coordinates.

6.24 Prove that parallel planes $\tilde{u}_i\tilde{y}^i=0$ and $\tilde{u}_i\tilde{y}^i+\tilde{U}_0=0$ do not transform into parallel planes under projective transformation of the points of space.

6.25 Prove the Plücker identity $\tilde{p}^{[\kappa\lambda}\tilde{p}^{\mu\nu]}\equiv 0$.

6.26 Prove the Plücker identity $\tilde{p}_{[\kappa\lambda}\tilde{p}_{\mu\nu]}=0$.

6.27 Prove that if $\tilde{q}_{\kappa\lambda}$ is defined in terms of two planes $\tilde{U}_\kappa$ and $\tilde{V}_\lambda$, which intersect in the line of intersection of the planes $\tilde{u}_\kappa$ and $\tilde{v}_\lambda$, then $\tilde{p}_{\kappa\lambda}$ is proportional to $\tilde{q}_{\kappa\lambda}$.

6.28 Prove Equation (6.52).

6.29 Derive the transformation equation of $\tilde{p}_{\kappa\lambda}$.

6.30 Consider the x^1 axis in 3-space.
(a) Determine $\tilde{p}^{\kappa\lambda}$.
(b) Determine $\tilde{p}_{\mu\nu}$.

6.31 List differences and similarities:
(a) Points x^i and weighted points x^λ.
(b) Points x^i and free affine vectors v^i.
(c) Planes u_λ and free affine planes u_i.

6.32 Prove that the cross ratio of partings of four planes intersecting in a given line is a projective invariant.

6.33 Prove that a second-degree surface

$$k_{ij}y^iy^j+k_0=0$$

transforms into a second-degree surface under projective transformation of the points of space.

6.34 Prove Equation (6.67).

6.35 Prove Equation (6.69).

6.36 Prove that angles are preserved under inversion.

III

Holor calculus

7

The linear connection

Previous chapters have dealt with a few derivatives of akinetors that transform as akinetors in arithmetic space. We have found that there are four such derivatives:

(1) The partial derivative of a scalar φ,

$$\frac{\partial \varphi}{\partial x^i},$$

transforms as a covariant vector.

(2) The contracted partial derivative of a contravariant akinetor $\tilde{D}^i$ of weight $+1$,

$$\frac{\partial \tilde{D}^i}{\partial x^i},$$

transforms as a nilvalent akinetor of weight $+1$.

(3) The difference of two partial derivatives of a covariant vector u_i,

$$\frac{\partial u_i}{\partial x^j} - \frac{\partial u_j}{\partial x^i},$$

transforms as a covariant bivalent tensor.

(4) The contracted partial derivative of a bivalent contravariant alternating tensor T^{ij},

$$\frac{\partial T^{ij}}{\partial x^i},$$

transforms as a contravariant vector.

But except for these four very special derivatives, there are no other known akinetor derivatives of akinetors in arithmetic space.

This chapter will be devoted to the development of a general method of differentiating akinetors that always yields akinetors. The key to this entire subject is the introduction of a holor Γ^i_{jk} that does not transform as a tensor.

In order to understand the reason for introducing Γ^i_{jk}, let us consider what is unsatisfactory about the partial derivative of a contravariant vector v^i. The transformation equation of v^i is the tensor transformation

$$v^{i'} = \frac{\partial x^{i'}}{\partial x^i} v^i. \tag{7.01}$$

A bivalent holor $w^{i'}_{j'}$ can be formed by differentiating Equation (7.01) with respect to $x^{j'}$:

$$\begin{aligned} w^{i'}_{j'} &= \frac{\partial v^{i'}}{\partial x^{j'}} = \frac{\partial}{\partial x^{j'}}\left(\frac{\partial x^{i'}}{\partial x^i} v^i\right) \\ &= \frac{\partial x^{i'}}{\partial x^i}\frac{\partial x^j}{\partial x^{j'}}\frac{\partial v^i}{\partial x^j} + \frac{\partial^2 x^{i'}}{\partial x^j \partial x^i}\frac{\partial x^j}{\partial x^{j'}} v^i. \end{aligned} \tag{7.02}$$

If the second term were absent, $w^{i'}_{j'}$ would transform as a bivalent mixed tensor. The troublesome quantity in Equation (7.02) is the second derivative:

$$\frac{\partial^2 x^{i'}}{\partial x^j \partial x^i}.$$

A similar situation occurs if we take the partial derivative of a covariant vector u_i,

$$u_{i'} = \frac{\partial x^i}{\partial x^{i'}} u_i. \tag{7.03}$$

By taking the partial derivative of $u_{i'}$ with respect to $x^{j'}$, a holor $w_{i'j'}$ is formed:

$$\begin{aligned} w_{i'j'} &= \frac{\partial u_{i'}}{\partial x^{j'}} = \frac{\partial}{\partial x^j}\left(\frac{\partial x^i}{\partial x^{i'}} u_i\right)\frac{\partial x^j}{\partial x^{j'}} \\ &= \frac{\partial x^i}{\partial x^{i'}}\frac{\partial x^j}{\partial x^{j'}}\frac{\partial u_i}{\partial x^j} + \frac{\partial^2 x^i}{\partial x^{i'} \partial x^{j'}} u_i. \end{aligned} \tag{7.04}$$

Once again, the $\partial u_{i'}/\partial x^{j'}$ is prevented from transforming as a tensor by the presence of an additive term containing a second derivative.

For tensors of higher valence, a similar situation prevails, except that there are additional additive terms containing second derivatives. For example, if we take the partial derivative of a bivalent contravariant tensor w^{ij},

$$u^{i'j'} = \frac{\partial x^{i'}}{\partial x^i}\frac{\partial x^{j'}}{\partial x^j}u^{ij}. \tag{7.05}$$

Then the derivative is a trivalent holor $w_{k'}^{i'j'}$,

$$\begin{aligned} w_{k'}^{i'j'} &= \frac{\partial}{\partial x^k}\left(\frac{\partial x^{i'}}{\partial x^i}\frac{\partial x^{j'}}{\partial x^j}u^{ij}\right)\frac{\partial x^k}{\partial x^{k'}} \\ &= \frac{\partial x^{i'}}{\partial x^i}\frac{\partial x^{j'}}{\partial x^j}\frac{\partial x^k}{\partial x^{k'}}\frac{\partial u^{ij}}{\partial x^k} \\ &\quad + \left(\frac{\partial^2 x^{i'}}{\partial x^k \partial x^i}\frac{\partial x^{j'}}{\partial x^j} + \frac{\partial^2 x^{j'}}{\partial x^k \partial x^j}\frac{\partial x^{i'}}{\partial x^i}\right)\frac{\partial x^k}{\partial x^{k'}}u^{ij}. \end{aligned} \tag{7.06}$$

Here there are two extra terms, each containing a second partial derivative of the primed coordinates with respect to the unprimed coordinates.

7.01. The linear connection

A method of differentiating tensors that always results in tensors was devised by Levi-Civita[1] in 1917 and independently by Schouten[2] in 1918. This ingenious procedure was generalized by Weyl,[3] Schouten,[4] and Eddington.[5] An excellent review of the subject is given by Struik.[6]

The basic idea rests on the definition of a holor Γ^i_{jk} called the linear connection. Nothing is defined about the linear connection except its transformation equation, which is

$$\Gamma^i_{jk} = \frac{\partial x^i}{\partial x^{i'}}\frac{\partial x^{j'}}{\partial x^j}\frac{\partial x^{k'}}{\partial x^k}\Gamma^{i'}_{j'k'} + \frac{\partial x^i}{\partial x^{i'}}\frac{\partial^2 x^{i'}}{\partial x^j \partial x^k}. \tag{7.07}$$

This holor Γ^i_{jk} is not a tensor because of the additive term, which contains the second partial derivative. As originally defined by Levi-Civita, the linear connection was, like its ancestor, the Christoffel[7] symbol of the second kind $\{{}_j{}^i{}_k\}$, symmetric in its two subscripts:

$$\{\Gamma^i_{jk}\} = \{\Gamma^i_{kj}\}. \tag{7.08}$$

Note that symmetry in j and k is not essential for the usefulness of the linear connection in establishing tensor calculus. In our development, we will follow Struik[6] in defining the linear connection by Equation (7.07) and by doing our entire development of tensor calculus for the general case in which Γ^i_{jk} is not required to be symmetric.

The torsion S^i_{jk} may be defined as the asymmetric part of the linear connection:

$$S^i_{jk} = \tfrac{1}{2}(\Gamma^i_{jk} - \Gamma^i_{kj}). \tag{7.09}$$

By its definition, S^i_{jk} is antisymmetric in the indices j and k,

$$\{S^i_{jk}\} = -\{S^i_{kj}\}. \tag{7.10}$$

Substitution of Equation (7.07) into (7.09) leads to the transformation equation of S^i_{jk},

$$S^i_{jk} = \frac{1}{2}\left(\frac{\partial x^i}{\partial x^{i'}}\frac{\partial x^{j'}}{\partial x^j}\frac{\partial x^{k'}}{\partial x^k}\Gamma^{i'}_{j'k'} + \frac{\partial x^i}{\partial x^{i'}}\frac{\partial^2 x^{i'}}{\partial x^j \partial x^k}\right.$$
$$\left. - \frac{\partial x^i}{\partial x^{i'}}\frac{\partial x^{j'}}{\partial x^k}\frac{\partial x^{k'}}{\partial x^j}\Gamma^{i'}_{j'k'} - \frac{\partial x^i}{\partial x^{i'}}\frac{\partial^2 x^{i'}}{\partial x^k \partial x^j}\right).$$

Interchange of the dummy indices in the third term and recollection that

$$\frac{\partial^2 x^{i'}}{\partial x^j \partial x^k} = \frac{\partial^2 x^{i'}}{\partial x^k \partial x^j}$$

gives

$$S^i_{jk} = \frac{\partial x^i}{\partial x^{i'}}\frac{\partial x^{j'}}{\partial x^j}\frac{\partial x^{k'}}{\partial x^k}S^{i'}_{j'k'}. \tag{7.11}$$

Thus, the torsion S^i_{jk} transforms as a tensor even though the linear connection does not.

The general linear connection Γ^i_{jk} is not required to be either symmetric or antisymmetric. In the special case in which the torsion tensor vanishes,

$$S^i_{jk} = 0, \qquad \{\Gamma^i_{jk}\} = \{\Gamma^i_{kj}\}. \tag{7.12}$$

The other possible special case in which the linear connection is asymmetric holds no interest in this development. For if Γ^i_{jk} were asymmetric, it would transform as a tensor, and the magic additive term with its second partial derivative would be lost.

7.02. The covariant derivatives of a contravariant vector

As was pointed out in the introduction to this chapter, the transformation equation of a contravariant vector is

$$v^{k'} = \frac{\partial x^{k'}}{\partial x^k} v^k. \tag{7.01}$$

Taking the partial derivative with respect to $x^{j'}$, we have

$$\frac{\partial v^{k'}}{\partial x^{j'}} = \frac{\partial x^{k'}}{\partial x^k}\frac{\partial x^j}{\partial x^{j'}}\frac{\partial v^k}{\partial x^j} + v^k \frac{\partial x^j}{\partial x^{j'}}\frac{\partial^2 x^{k'}}{\partial x^j \partial x^k}.$$

Since

$$\frac{\partial^2 x^{k'}}{\partial x^l \partial x^m} = \frac{\partial^2 x^{k'}}{\partial x^m \partial x^l},$$

the transformation equation of the linear connection [Equation (7.07)] can be written

$$\Gamma^k_{lm} = \Gamma^{k'}_{l'm'}\frac{\partial x^k}{\partial x^{k'}}\frac{\partial x^{l'}}{\partial x^l}\frac{\partial x^{m'}}{\partial x^m} + \frac{\partial x^k}{\partial x^{k'}}\frac{\partial^2 x^{k'}}{\partial x^l \partial x^m} \tag{7.07a}$$

or, interchanging indices l and m,

$$\Gamma^k_{ml} = \Gamma^{k'}_{m'l'}\frac{\partial x^k}{\partial x^{k'}}\frac{\partial x^{m'}}{\partial x^m}\frac{\partial x^{l'}}{\partial x^l} + \frac{\partial x^k}{\partial x^{k'}}\frac{\partial^2 x^{k'}}{\partial x^m \partial x^l}. \tag{7.07b}$$

From Equations (7.07a) and (7.07b), it is possible to solve for $\partial^2 x^{k'}/\partial x^l \partial x^m$. Multiplication by $\partial x^{j'}/\partial x^k$ gives

$$\frac{\partial x^{j'}}{\partial x^k}\Gamma^k_{lm} = \Gamma^{k'}_{l'm'}\frac{\partial x^{j'}}{\partial x^k}\frac{\partial x^k}{\partial x^{k'}}\frac{\partial x^{l'}}{\partial x^l}\frac{\partial x^{m'}}{\partial x^m} + \frac{\partial x^{j'}}{\partial x^k}\frac{\partial x^k}{\partial x^{k'}}\frac{\partial^2 x^{k'}}{\partial x^l \partial x^m} \tag{7.13a}$$

or

$$\frac{\partial x^{j'}}{\partial x^k}\Gamma^k_{ml} = \Gamma^{k'}_{m'l'}\frac{\partial x^{j'}}{\partial x^k}\frac{\partial x^k}{\partial x^{k'}}\frac{\partial x^{m'}}{\partial x^m}\frac{\partial x^{l'}}{\partial x^l} + \frac{\partial x^{j'}}{\partial x^k}\frac{\partial x^k}{\partial x^{k'}}\frac{\partial^2 x^{k'}}{\partial x^l \partial x^m}. \tag{7.13b}$$

Since

$$\frac{\partial x^{j'}}{\partial x^k}\frac{\partial x^k}{\partial x^{k'}} = \delta^{j'}_{k'},$$

Equation (7.13) becomes

$$\frac{\partial x^{j'}}{\partial x^{k}}\Gamma^{k}_{lm}=\Gamma^{k'}_{l'm'}\delta^{j'}_{k'}\frac{\partial x^{l'}}{\partial x^{l}}\frac{\partial x^{m'}}{\partial x^{m}}+\delta^{j'}_{k'}\frac{\partial^2 x^{k'}}{\partial x^{l}\partial x^{m}} \tag{7.14a}$$

or

$$\frac{\partial x^{j'}}{\partial x^{k}}\Gamma^{k}_{ml}=\Gamma^{k'}_{m'l'}\delta^{j'}_{k'}\frac{\partial x^{m'}}{\partial x^{m}}\frac{\partial x^{l'}}{\partial x^{l}}+\delta^{j'}_{k'}\frac{\partial^2 x^{k'}}{\partial x^{l}\partial x^{m}}. \tag{7.14b}$$

Expanding $\delta^{j'}_{k'}$ and solving for the second partial derivatives, we obtain

$$\frac{\partial^2 x^{j'}}{\partial x^{l}\partial x^{m}}=\frac{\partial x^{j'}}{\partial x^{k}}\Gamma^{k}_{lm}-\Gamma^{j'}_{l'm'}\frac{\partial x^{l'}}{\partial x^{l}}\frac{\partial x^{m'}}{\partial x^{m}}, \tag{7.15a}$$

or

$$\frac{\partial^2 x^{j'}}{\partial x^{l}\partial x^{m}}=\frac{\partial x^{j'}}{\partial x^{k}}\Gamma^{k}_{ml}-\Gamma^{j'}_{m'l'}\frac{\partial x^{m'}}{\partial x^{m}}\frac{\partial x^{l'}}{\partial x^{l}}. \tag{7.15b}$$

Before we are ready to substitute these expressions for the second partial derivatives into Equation (7.02), it is necessary to change the names of the indices in Equation (7.15) to match the free indices in the second partial derivatives of Equation (7.02). Thus, Equation (7.15) becomes

$$\frac{\partial^2 x^{k'}}{\partial x^{j}\partial x^{k}}=\frac{\partial x^{k'}}{\partial x^{l}}\Gamma^{l}_{jk}-\Gamma^{k'}_{m'l'}\frac{\partial x^{m'}}{\partial x^{j}}\frac{\partial x^{l'}}{\partial x^{k}} \tag{7.16a}$$

or

$$\frac{\partial^2 x^{k'}}{\partial x^{j}\partial x^{k}}=\frac{\partial x^{k'}}{\partial x^{l}}\Gamma^{l}_{kj}-\Gamma^{k'}_{l'm'}\frac{\partial x^{l'}}{\partial x^{k}}\frac{\partial x^{m'}}{\partial x^{j}}. \tag{7.16b}$$

Now substitute Equation (7.16) into (7.02):

$$\frac{\partial v^{k'}}{\partial x^{j'}}=\frac{\partial x^{k'}}{\partial x^{k}}\frac{\partial x^{j}}{\partial x^{j'}}\frac{\partial v^{k}}{\partial x^{j}}+v^{k}\frac{\partial x^{j}}{\partial x^{j'}}\frac{\partial x^{k'}}{\partial x^{l}}\Gamma^{l}_{jk}$$

$$-v^{k}\frac{\partial x^{j}}{\partial x^{j'}}\Gamma^{k'}_{m'l'}\frac{\partial x^{m'}}{\partial x^{j}}\frac{\partial x^{l'}}{\partial x^{k}} \tag{7.17a}$$

or

$$\frac{\partial v^{k'}}{\partial x^{j'}}=\frac{\partial x^{k'}}{\partial x^{k}}\frac{\partial x^{j}}{\partial x^{j'}}\frac{\partial v^{k}}{\partial x^{j}}+v^{k}\frac{\partial x^{j}}{\partial x^{j'}}\frac{\partial x^{k'}}{\partial x^{l}}\Gamma^{l}_{kj}$$

$$-v^{k}\frac{\partial x^{j}}{\partial x^{j'}}\Gamma^{k'}_{l'm'}\frac{\partial x^{l'}}{\partial x^{k}}\frac{\partial x^{m'}}{\partial x^{j}}. \tag{7.17b}$$

But since

$$\frac{\partial x^j}{\partial x^{j'}}\frac{\partial x^{m'}}{\partial x^j}=\delta^{m'}_{j'} \qquad \text{and} \qquad v^{l'}=v^k\frac{\partial x^{l'}}{\partial x^k},$$

Equation (7.17) becomes

$$\frac{\partial v^{k'}}{\partial x^{j'}}=\frac{\partial x^j}{\partial x^{j'}}\frac{\partial x^{k'}}{\partial x^k}\left(\frac{\partial v^k}{\partial x^j}+v^l\Gamma^k_{jl}\right)-v^{l'}\Gamma^{k'}_{m'l'}\delta^{m'}_{j'} \tag{7.18a}$$

or

$$\frac{\partial v^{k'}}{\partial x^{j'}}=\frac{\partial x^j}{\partial x^{j'}}\frac{\partial x^{k'}}{\partial x^k}\left(\frac{\partial v^k}{\partial x^j}+v^l\Gamma^k_{lj}\right)-v^{l'}\Gamma^{k'}_{l'm'}\delta^{m'}_{j'}. \tag{7.18b}$$

Expanding $\delta^{m'}_{j'}$ and transposing the last term,

$$\left[\frac{\partial v^{k'}}{\partial x^{j'}}+v^{l'}\Gamma^{k'}_{j'l'}\right]=\frac{\partial x^j}{\partial x^{j'}}\frac{\partial x^{k'}}{\partial x^k}\left(\frac{\partial v^k}{\partial x^j}+v^l\Gamma^k_{jl}\right) \tag{7.19a}$$

or

$$\left[\frac{\partial v^{k'}}{\partial x^{j'}}+v^{l'}\Gamma^{k'}_{l'j'}\right]=\frac{\partial x^j}{\partial x^k}\frac{\partial x^{k'}}{\partial x^k}\left(\frac{\partial v^k}{\partial x^j}+v^l\Gamma^k_{lj}\right). \tag{7.19b}$$

Thus, the quantities in the square brackets transform as mixed tensors. Hence, as pointed out by Cavagnero, Roberts, Semagin, and Spencer,[8] there are two distinct derivatives of a contravariant vector that transform as a tensor,

$$\overset{1}{\nabla}_j v^k=\frac{\partial v^k}{\partial x^j}+v^l\Gamma^k_{jl}, \tag{7.20a}$$

and

$$\overset{2}{\nabla}_j v^k=\frac{\partial v^k}{\partial x^j}+v^l\Gamma^k_{lj}. \tag{7.20b}$$

These are called covariant derivatives of the first and second kind.

The second kind of covariant derivative of a contravariant vector can be expressed in terms of the first kind of covariant derivative and the torsion tensor, as shown by Albert, Spencer, and Uma,[9] by the equation

$$\overset{2}{\nabla}_j v^k=\overset{1}{\nabla}_j v^k-2v^l S^k_{jl}. \tag{7.21}$$

The two covariant derivatives become identical if and only if the torsion tensor S^k_{jl} vanishes, that is, if Γ^k_{jl} is symmetric in the two subscripts j and l.

From the definition of the covariant derivative, it is clear that if u^k and v^k are two covariant vectors, the covariant derivative of their sum is

$$\overset{n}{\nabla}_j(u^k+v^k)=\overset{n}{\nabla}_j u^k+\overset{n}{\nabla}_j v^k. \tag{7.22}$$

Thus, the covariant derivatives are distributive with respect to addition.

7.03. The covariant derivatives of a covariant vector

Since the partial derivative of a scalar itself transforms as a tensor, we can call this derivative the covariant derivative of a scalar,

$$\nabla_j \phi=\frac{\partial\phi}{\partial x^j}. \tag{7.23}$$

Note that although there are two covariant derivatives of a contravariant vector, there is a single covariant derivative of a scalar.

Now consider a scalar ϕ formed from the contracted product of a covariant vector u_k and a contravariant vector v^k,

$$\phi=u_k v^k. \tag{7.24}$$

Then, by Equations (7.23) and (7.24),

$$\nabla_j\phi=\frac{\partial\phi}{\partial x^j}=\frac{\partial(u_k v^k)}{\partial x^j}=\frac{\partial u_k}{\partial x^j}v^k+u_k\frac{\partial v^k}{\partial x^j}. \tag{7.25}$$

If we also postulate the ordinary product rule for covariant derivatives,

$$\nabla_j(u_k v^k)=(\overset{n}{\nabla}_j u_k)v^k+u_k(\overset{n}{\nabla}_j v^k). \tag{7.26}$$

Substituting Equation (7.20) into (7.26) and combining with (7.25) gives

$$(\overset{1}{\nabla}_j u_k)v^k+u_k\left(\frac{\partial v^k}{\partial x^j}+v^l\Gamma^k_{jl}\right)=\frac{\partial u_k}{\partial x^j}v^k+u_k\frac{\partial v^k}{\partial x^j}, \tag{7.27a}$$

or

$$(\overset{2}{\nabla}_j u_k)v^k+u_k\left(\frac{\partial v^k}{\partial x^j}+v^l\Gamma^k_{lj}\right)=\frac{\partial u_k}{\partial x^j}v^k+u_k\frac{\partial v^k}{\partial x^j}. \tag{7.27b}$$

Canceling the second term on the left side of Equation (7.27) with the last term on the right side and interchanging dummy indices so that v^k becomes a factor of the remaining terms,

$$v^k\left(\overset{1}{\nabla}_j u_k + u_l \Gamma^l_{jk} - \frac{\partial u_k}{\partial x^j}\right) = 0, \tag{7.28a}$$

or

$$v^k\left(\overset{2}{\nabla}_j u_k + u_l \Gamma^l_{kj} - \frac{\partial u_k}{\partial x^j}\right) = 0. \tag{7.28b}$$

Since v^k is an arbitrary contravariant vector, the quantities contained in the parentheses must vanish. Thus, the two[8] covariant derivatives of a covariant vector may be defined by the equations

$$\overset{1}{\nabla}_j u_k = \frac{\partial u_k}{\partial x^j} - u_l \Gamma^l_{jk} \tag{7.29a}$$

or

$$\overset{2}{\nabla}_j u_k = \frac{\partial u_k}{\partial x^j} - u_l \Gamma^l_{kj}. \tag{7.29b}$$

The second covariant derivative of a covariant vector can be expressed[9] in terms of the first covariant derivative and the torsion tensor by the equation

$$\overset{2}{\nabla}_j u_k = \overset{1}{\nabla}_j u_k + 2u_l S^l_{jk}. \tag{7.30}$$

7.04. The covariant derivatives of bivalent tensors

The idea of the covariant derivative can be extended to tensors of higher valence but with the interesting result that when the valence increases, the number of distinct covariant derivatives that can be defined is greater than the two that are possible for a univalent tensor.

Consider the covariant derivative of a contravariant bivalent tensor T^{ij}. A scalar S can be formed by taking the contracted product of T^{ij} with two arbitrary covariant vectors u_i and v_j:

$$S = T^{ij} u_i v_j. \tag{7.31}$$

Let us postulate that the ordinary product rule for derivatives can be extended to any covariant derivative.

$$\nabla_k S = u_i v_j \overset{n}{\nabla}_k T^{ij} + T^{ij} v_j \overset{p}{\nabla}_k u_i + T^{ij} u_i \overset{q}{\nabla}_k v_j. \tag{7.32}$$

Since, by Equation (7.23),

$$\nabla_k S = \frac{\partial S}{\partial x^k}, \tag{7.23}$$

the left side of Equation (7.32) can be expanded into

$$u_i v_j \frac{\partial T^{ij}}{\partial x^k} + T^{ij} v_j \frac{\partial u_i}{\partial x^k} + T^{ij} u_i \frac{\partial v_j}{\partial x^k}$$
$$= u_i v_j \overset{n}{\nabla}_k T^{ij} + T^{ij} v_j \overset{p}{\nabla}_k u_i + T^{ij} u_i \overset{q}{\nabla}_k v_j. \qquad (7.33)$$

It is necessary to substitute expressions for $\overset{p}{\nabla}_k u_i$ and $\overset{q}{\nabla}_k v_j$ into Equation (7.33). Since there are two possible derivatives of any covariant vector, p can be chosen as 1 or 2. Likewise, q can be chosen as 1 or 2. Thus, there are four possible ways[8] of combining these choices, and therefore there are four distinct covariant derivatives of a bivalent tensor. Which of these four covariant derivatives will be called first, second, third, or fourth is an entirely arbitrary matter. Suppose that we choose

$$n=1 \quad \text{if } p=1 \text{ and } q=1,$$
$$n=2 \quad \text{if } p=1 \text{ and } q=2,$$
$$n=3 \quad \text{if } p=2 \text{ and } q=1,$$
$$n=4 \quad \text{if } p=2 \text{ and } q=2.$$

Then, expansion of Equation (7.33) gives one equation for each of the four choices:

$$u_i v_j \frac{\partial T^{ij}}{\partial x^k} + T^{ij} v_j \frac{\partial u_i}{\partial x^k} + T^{ij} u_i \frac{\partial v_j}{\partial x^k}$$
$$= u_i v_j \overset{1}{\nabla}_k T^{ij} + T^{ij} v_j \frac{\partial u_i}{\partial x^k} - T^{ij} v_j u_l \Gamma^l_{ki} + T^{ij} u_i \frac{\partial v_j}{\partial x^k} - T^{ij} u_i v_l \Gamma^l_{kj}$$
$$= u_i v_j \overset{2}{\nabla}_k T^{ij} + T^{ij} v_j \frac{\partial u_i}{\partial x^k} - T^{ij} v_j u_l \Gamma^l_{ki} + T^{ij} u_i \frac{\partial v_j}{\partial x^k} - T^{ij} u_i v_l \Gamma^l_{jk}$$
$$= u_i v_j \overset{3}{\nabla}_k T^{ij} + T^{ij} v_j \frac{\partial u_i}{\partial x^k} - T^{ij} v_j u_l \Gamma^l_{ik} + T^{ij} u_i \frac{\partial v_j}{\partial x^k} - T^{ij} u_i v_l \Gamma^l_{kj}$$
$$= u_i v_j \overset{4}{\nabla}_k T^{ij} + T^{ij} v_j \frac{\partial u_i}{\partial x^k} - T^{ij} v_j u_l \Gamma^l_{ik} + T^{ij} u_i \frac{\partial v_j}{\partial x^k} - T^{ij} u_i v_l \Gamma^l_{jk}.$$

Canceling the terms containing partial derivatives of u_i and v_j and changing dummy indices so that $u_i v_j$ can be factored, we obtain

$$u_i v_j \left[\frac{\partial T^{ij}}{\partial x^k} - \overset{1}{\nabla}_k T^{ij} + T^{lj} \Gamma^i_{kl} + T^{il} \Gamma^j_{kl} \right] = 0, \qquad (7.34a)$$

$$u_i v_j \left[\frac{\partial T^{ij}}{\partial x^k} - \overset{2}{\nabla}_k T^{ij} + T^{lj} \Gamma^i_{kl} + T^{il} \Gamma^j_{lk} \right] = 0, \qquad (7.34b)$$

$$u_i v_j \left[\frac{\partial T^{ij}}{\partial x^k} - \overset{3}{\nabla}_k T^{ij} + T^{lj}\Gamma^i_{lk} + T^{il}\Gamma^j_{kl} \right] = 0, \tag{7.34c}$$

$$u_i v_j \left[\frac{\partial T^{ij}}{\partial x^k} - \overset{4}{\nabla}_k T^{ij} + T^{lj}\Gamma^i_{lk} + T^{il}\Gamma^j_{lk} \right] = 0. \tag{7.34d}$$

Since u_i and v_j are arbitrary vectors, the quantities in the brackets must be zero. Therefore, there are four covariant derivatives of a bivalent contravariant tensor T^{ij}, all of which transform as tensors:

$$\overset{1}{\nabla}_k T^{ij} = \frac{\partial T^{ij}}{\partial x^k} + T^{lj}\Gamma^i_{kl} + T^{il}\Gamma^j_{kl}, \tag{7.35a}$$

$$\overset{2}{\nabla}_k T^{ij} = \frac{\partial T^{ij}}{\partial x^k} + T^{lj}\Gamma^i_{kl} + T^{il}\Gamma^j_{lk}, \tag{7.35b}$$

$$\overset{3}{\nabla}_k T^{ij} = \frac{\partial T^{ij}}{\partial x^k} + T^{lj}\Gamma^i_{lk} + T^{il}\Gamma^j_{kl}, \tag{7.35c}$$

$$\overset{4}{\nabla}_k T^{ij} = \frac{\partial T^{ij}}{\partial x^k} + T^{lj}\Gamma^i_{lk} + T^{il}\Gamma^j_{lk}. \tag{7.35d}$$

The second, third, and fourth kind of covariant derivatives can all be expressed[9] in terms of the first kind of covariant derivative and the torsion tensor by the equations

$$\overset{2}{\nabla}_k T^{ij} = \overset{1}{\nabla}_k T^{ij} - 2T^{il}S^j_{kl}, \tag{7.36a}$$

$$\overset{3}{\nabla}_k T^{ij} = \overset{1}{\nabla}_k T^{ij} - 2T^{lj}S^j_{kl}, \tag{7.36b}$$

$$\overset{4}{\nabla}_k T^{ij} = \overset{1}{\nabla}_k T^{ij} - 2T^{il}S^j_{kl} - 2T^{lj}S^i_{kl}. \tag{7.36c}$$

Clearly, if the linear connection is symmetric, the torsion tensor vanishes and the four distinct covariant derivatives of T^{ij} collapse into a single covariant derivative that is ordinary called $\nabla_k T^{ij}$.

Similar results can likewise be derived for bivalent covariant tensors T_{ij}:

$$\overset{1}{\nabla}_k T_{ij} = \frac{\partial T_{ij}}{\partial x^k} - T_{lj}\Gamma^l_{ki} - T_{il}\Gamma^l_{kj}, \tag{7.37a}$$

$$\overset{2}{\nabla}_k T_{ij} = \frac{\partial T_{ij}}{\partial x^k} - T_{lj}\Gamma^l_{ki} - T_{il}\Gamma^l_{jk}, \tag{7.37b}$$

$$\overset{3}{\nabla}_k T_{ij} = \frac{\partial T_{ij}}{\partial x^k} - T_{lj}\Gamma^l_{ik} - T_{il}\Gamma^l_{kj}, \tag{7.37c}$$

$$\overset{4}{\nabla}_k T_{ij} = \frac{\partial T_{ij}}{\partial x^k} - T_{lj}\Gamma^l{}_{ik} - T_{il}\Gamma^l{}_{jk}. \tag{7.37d}$$

The equations relating these four covariant derivatives are[9]

$$\overset{2}{\nabla}_k T_{ij} = \overset{1}{\nabla}_k T_{ij} + 2T_{il}S^l_{kj}, \tag{7.38a}$$

$$\overset{3}{\nabla}_k T_{ij} = \overset{1}{\nabla}_k T_{ij} + 2T_{lj}S^l_{ki}, \tag{7.38b}$$

$$\overset{4}{\nabla}_k T_{ij} = \overset{1}{\nabla}_k T_{ij} + 2T_{lj}S^l_{ki} + 2T_{il}S^l_{kj}. \tag{7.38c}$$

For a mixed bivalent tensor T^i_j, a third set of four covariant derivatives transform as tensors:

$$\overset{1}{\nabla}_k T^i_j = \frac{\partial T^i_j}{\partial x^k} + T^l_j\Gamma^i{}_{kl} - T^i_l\Gamma^l{}_{kj}, \tag{7.39a}$$

$$\overset{2}{\nabla}_k T^i_j = \frac{\partial T^i_j}{\partial x^k} + T^l_j\Gamma^i{}_{kl} - T^i_l\Gamma^l{}_{jk}, \tag{7.39b}$$

$$\overset{3}{\nabla}_k T^i_j = \frac{\partial T^i_j}{\partial x^k} + T^l_j\Gamma^i{}_{lk} - T^i_l\Gamma^l{}_{kj}, \tag{7.39c}$$

$$\overset{4}{\nabla}_k T^i_j = \frac{\partial T^i_j}{\partial x^k} + T^l_j\Gamma^i{}_{lk} - T^i_l\Gamma^l{}_{jk}. \tag{7.39d}$$

These four covariant derivatives are related[9] by the equations

$$\overset{2}{\nabla}_k T^i_j = \overset{1}{\nabla}_k T^i_j + 2T^i_l S^l_{kj}, \tag{7.40a}$$

$$\overset{3}{\nabla}_k T^i_j = \overset{1}{\nabla}_k T^i_j - 2T^l_j S^i_{kl}, \tag{7.40b}$$

$$\overset{4}{\nabla}_k T^i_j = \overset{1}{\nabla}_k T^i_j + 2T^i_l S^l_{kj} - 2T^l_j S^i_{kl}. \tag{7.40c}$$

7.05. Covariant derivatives of tensors of valence m

The covariant derivatives of a tensor of arbitrary valence m can be formed in a fashion similar to that already described.[10] To find the covariant derivatives of a tensor

$$T^{r_1 r_2 \cdots r_a}{}_{r_{a+1} r_{a+2} \cdots r_m}$$

with contravariant valence a and covariant valence $m-a$, the first step is to form a scalar by multiplying the tensor by a arbitrary covariant vectors and by $m-a$ arbitrary contravariant vectors,

$$S = T^{r_1 r_2 \cdots r_a}{}_{r_{a+1} r_{a+2} \cdots r_m} \overset{(1)}{u}_{r_1} \overset{(2)}{u}_{r_2} \cdots \overset{(a)}{u}_{r_a} \underset{(a+1)}{v^{r_{a+1}}} \underset{(a+2)}{v^{r_{a+2}}} \cdots \underset{(m)}{v^{r_m}}. \tag{7.41}$$

Since we postulate that the product rule holds for all covariant derivatives,

$$
\begin{aligned}
\nabla_k S = {} & \left(\overset{(1)}{u}_{r_1} \overset{(2)}{u}_{r_2} \cdots \overset{(a)}{u}_{r_a} \underset{(a+1)}{v}^{r_{a+1}} \underset{(a+2)}{v}^{r_{a+2}} \cdots \underset{(m)}{v}^{r_m} \right) \overset{n}{\nabla}_k T^{r_1 r_2 \cdots r_a}_{r_{a+1} r_{a+2} \cdots r_m} \\
& + T^{r_1 r_2 \cdots r_a}_{r_{a+1} r_{a+2} \cdots r_m} \left[\left(\overset{(2)}{u}_{r_2} \cdots \overset{(a)}{u}_{r_a} \underset{(a+1)}{v}^{r_{a+1}} \underset{(a+2)}{v}^{r_{a+2}} \cdots \underset{(m)}{v}^{r_m} \right) \overset{y_1}{\nabla}_k \overset{(1)}{u}_{r_1} \right. \\
& + \left(\overset{(1)}{u}_{r_1} \overset{(3)}{u}_{r_3} \cdots \overset{(a)}{u}_{r_a} \underset{(a+1)}{v}^{r_{a+1}} \underset{(a+2)}{v}^{r_{a+2}} \cdots \underset{(m)}{v}^{r_m} \right) \overset{y_2}{\nabla}_k \overset{(2)}{u}_{r_2} + \cdots \\
& + \left(\overset{(1)}{u}_{r_1} \overset{(2)}{u}_{r_2} \cdots \overset{(a-1)}{u}_{r_{a-1}} \underset{(a+1)}{v}^{r_{a+1}} \underset{(a+2)}{v}^{r_{a+2}} \cdots \underset{(m)}{v}^{r_m} \right) \overset{y_a}{\nabla}_k \overset{(a)}{u}_{r_a} \\
& + \left(\overset{(1)}{u}_{r_1} \overset{(2)}{u}_{r_2} \cdots \overset{(a)}{u}_{r_a} \underset{(a+2)}{v}^{r_{a+2}} \cdots \underset{(m)}{v}^{r_m} \right) \overset{y_{a+1}}{\nabla}_k \underset{(a+1)}{v}^{r_{a+1}} \\
& + \left(\overset{(1)}{u}_{r_1} \overset{(2)}{u}_{r_2} \cdots \overset{(a)}{u}_{r_a} \underset{(a+1)}{v}^{r_{a+1}} \underset{(a+3)}{v}^{r_{a+3}} \cdots \underset{(m)}{v}^{r_m} \right) \overset{y_{a+2}}{\nabla}_k \underset{(a+2)}{v}^{r_{a+2}} + \cdots \\
& \left. + \left(\overset{(1)}{u}_{r_1} \overset{(2)}{u}_{r_2} \cdots \overset{(a)}{u}_{r_a} \underset{(a+1)}{v}^{r_{a+1}} \underset{(a+2)}{v}^{r_{a+2}} \cdots \underset{(m-1)}{v}^{r_{m-1}} \right) \overset{y_m}{\nabla}_k \underset{(m)}{v}^{r_m} \right].
\end{aligned}
\tag{7.42}
$$

But according to Equation (7.23), the covariant derivative of a scalar is merely a partial derivative, so

$$
\begin{aligned}
\nabla_k S = {} & \frac{\partial s}{\partial x^k} \\
= {} & \left(\overset{(1)}{u}_{r_1} \overset{(2)}{u}_{r_2} \cdots \overset{(a)}{u}_{r_a} \underset{(a+1)}{v}^{r_{a+1}} \underset{(a+2)}{v}^{r_{a+2}} \cdots \underset{(m)}{v}^{r_m} \right) \frac{\partial T^{r_1 r_2 \cdots r_a}_{r_{a+1} r_{a+2} \cdots r_m}}{\partial x^k} \\
& + T^{r_1 r_2 \cdots r_a}_{r_{a+1} r_{a+2} \cdots r_m} \left[\left(\overset{(2)}{u}_{r_2} \cdots \overset{(a)}{u}_{r_a} \underset{(a+1)}{v}^{r_{a+1}} \underset{(a+2)}{v}^{r_{a+2}} \cdots \underset{(m)}{v}^{r_m} \right) \frac{\partial \overset{(1)}{u}_{r_1}}{\partial x^k} \right. \\
& + \left(\overset{(1)}{u}_{r_1} \overset{(3)}{u}_{r_3} \cdots \overset{(a)}{u}_{r_a} \underset{(a+1)}{v}^{r_{a+1}} \underset{(a+2)}{v}^{r_{a+2}} \cdots \underset{(m)}{v}^{r_m} \right) \frac{\partial \overset{(2)}{u}_{r_2}}{\partial x^k} + \cdots \\
& + \left(\overset{(1)}{u}_{r_1} \overset{(2)}{u}_{r_2} \cdots \overset{(a-1)}{u}_{r_{a-1}} \underset{(a+1)}{v}^{r_{a+1}} \underset{(a+2)}{v}^{r_{a+2}} \cdots \underset{(m)}{v}^{r_m} \right) \frac{\partial \overset{(a)}{u}_{r_a}}{\partial x^k} \\
& + \left(\overset{(1)}{u}_{r_1} \overset{(2)}{u}_{r_2} \cdots \overset{(a)}{u}_{r_a} \underset{(a+2)}{v}^{r_{a+2}} \cdots \underset{(m)}{v}^{r_m} \right) \frac{\partial \underset{(a+1)}{v}^{r_{a+1}}}{\partial x^k} + \cdots
\end{aligned}
$$

$$+\left(\overset{(1)}{u}_{r_1}\overset{(2)}{u}_{r_2}\cdots\overset{(a)}{u}_{r_a}\underset{(a+1)}{v}^{r_{a+1}}\underset{(a+3)}{v}^{r_{a+3}}\cdots\underset{(m)}{v}^{r_m}\right)\frac{\partial\underset{(a+2)}{v}^{r_{a+2}}}{\partial x^k}+\cdots$$

$$\left.+\left(\overset{(1)}{u}_{r_1}\overset{(2)}{u}_{r_2}\cdots\overset{(a)}{u}_{r_a}\underset{(a+1)}{v}^{r_{a+1}}\underset{(a+2)}{v}^{r_{a+2}}\cdots\underset{(m-1)}{v}^{r_{m-1}}\right)\frac{\partial\underset{(m)}{v}^{r_m}}{\partial x^k}\right]. \tag{7.43}$$

Subtraction of Equations (7.42) and (7.43) gives

$$0=\left(\overset{(1)}{u}_{r_1}\overset{(2)}{u}_{r_2}\cdots\overset{(a)}{u}_{r_a}\underset{(a+1)}{v}^{r_{a+1}}\underset{(a+2)}{v}^{r_{a+2}}\cdots\underset{(m)}{v}^{r_m}\right)\left(\overset{n}{\nabla}_k T^{r_1r_2\cdots r_a}_{r_{a+1}r_{a+2}\cdots r_m}-\frac{\partial T^{r_1r_2\cdots r_a}_{r_{a+1}r_{a+2}\cdots r_m}}{\partial x^k}\right)$$

$$+T^{r_1r_2\cdots r_a}_{r_{a+1}r_{a+2}\cdots r_m}\left[\left(\overset{(2)}{u}_{r_2}\cdots\overset{(a)}{u}_{r_a}\underset{(a+1)}{v}^{r_{a+1}}\underset{(a+2)}{v}^{r_{a+2}}\cdots\underset{(m)}{v}^{r_m}\right)\left(\overset{y_1}{\nabla}_k\overset{(1)}{u}_{r_1}-\frac{\partial\overset{(1)}{u}_{r_1}}{\partial x^k}\right)\right.$$

$$+\left(\overset{(1)}{u}_{r_1}\overset{(3)}{u}_{r_3}\cdots\overset{(a)}{u}_{r_a}\underset{(a+1)}{v}^{r_{a+1}}\underset{(a+2)}{v}^{r_{a+2}}\cdots\underset{(m)}{v}^{r_m}\right)\left(\overset{y_2}{\nabla}_k\overset{(2)}{u}_{r_2}-\frac{\partial\overset{(2)}{u}_{r_2}}{\partial x^k}\right)+\cdots$$

$$+\left(\overset{(1)}{u}_{r_1}\overset{(2)}{u}_{r_2}\cdots\overset{(a-1)}{u}_{r_{a-1}}\underset{(a+1)}{v}^{r_{a+1}}\underset{(a+2)}{v}^{r_{a+2}}\cdots\underset{(m)}{v}^{r_m}\right)\left(\overset{y_a}{\nabla}_k\overset{(a)}{u}_{r_a}-\frac{\partial\overset{(a)}{u}_{r_a}}{\partial x^k}\right)$$

$$+\left(\overset{(1)}{u}_{r_1}\overset{(2)}{u}_{r_2}\cdots\overset{(a)}{u}_{r_a}\underset{(a+2)}{v}^{r_{a+2}}\cdots\underset{(m)}{v}^{r_m}\right)\left(\overset{y_{a+1}}{\nabla}_k\underset{(a+1)}{v}^{r_{a+1}}-\frac{\partial\underset{(a+1)}{v}^{r_{a+1}}}{\partial x^k}\right)$$

$$+\left(\overset{(1)}{u}_{r_1}\overset{(2)}{u}_{r_2}\cdots\overset{(a)}{u}_{r_a}\underset{(a+1)}{v}^{r_{a+1}}\underset{(a+3)}{v}^{r_{a+3}}\cdots\underset{(m)}{v}^{r_m}\right)\left(\overset{y_{a+2}}{\nabla}_k\underset{(a+2)}{v}^{r_{a+2}}-\frac{\partial\underset{(a+2)}{v}^{r_{a+2}}}{\partial x^k}\right)+\cdots$$

$$\left.+\left(\overset{(1)}{u}_{r_1}\overset{(2)}{u}_{r_2}\cdots\overset{(a)}{u}_{r_a}\underset{(a+1)}{v}^{r_{a+1}}\underset{(a+2)}{v}^{r_{a+2}}\cdots\underset{(m-1)}{v}^{r_{m-1}}\right)\left(\overset{y_m}{\nabla}_k\underset{(m)}{v}^{r_m}-\frac{\partial\underset{(m)}{v}^{r_m}}{\partial x^k}\right)\right]. \tag{7.44}$$

According to Equations (7.20) and (7.28), there are two covariant derivatives for each contravariant and for each covariant vector. Each of the covariant derivatives in Equation (7.44) can thus be chosen in two ways. Before we substitute these covariant derivatives into Equation (7.44), it is convenient to introduce the following special notation.[10] Let

$$\begin{aligned}\overset{y_p}{\nabla}_k u_j&=\frac{\partial u_j}{\partial x^k}-u_l\overset{(y_p)}{\Gamma}{}^l_{kj},\\ \overset{y_p}{\nabla}_k v^j&=\frac{\partial v^j}{\partial x^k}+v^l\overset{(y_p)}{\Gamma}{}^j_{kl},\end{aligned} \tag{7.45}$$

where

$$\overset{(y_p)}{\Gamma}{}^{i}_{kl} = \begin{cases} \Gamma^{i}_{kl} & \text{if } y_p = 1, \\ \Gamma^{i}_{kl} & \text{if } y_p = 2. \end{cases}$$

Substitution of Equation (7.45) into Equation (7.44) gives

$$\begin{aligned} 0 = \Big(\overset{(1)}{u}_{r_1}\overset{(2)}{u}_{r_2}\cdots\overset{(a)}{u}_{r_a}\underset{(a+1)}{v}{}^{r_{a+1}}\underset{(a+2)}{v}{}^{r_{a+2}}\cdots\underset{(m)}{v}{}^{r_m}\Big)\Bigg[&\overset{n}{\nabla}_k T^{r_1 r_2\cdots r_a}_{\quad r_{a+1}r_{a+2}\cdots r_m} - \frac{\partial T^{r_1 r_2\cdots r_a}_{\quad r_{a+1}r_{a+2}\cdots r_m}}{\partial x^k} \\ &- T^{l r_2\cdots r_a}_{\quad r_{a+1}r_{a+2}\cdots r_m}\overset{(y_p)}{\Gamma}{}^{r_1}_{kl} - T^{r_1 l\cdots r_a}_{\quad r_{a+1}r_{a+2}\cdots r_m}\overset{(y_p)}{\Gamma}{}^{r_2}_{kl} - \cdots \\ &- T^{r_1 r_2\cdots l}_{\quad r_{a+1}r_{a+2}\cdots r_m}\overset{(y_p)}{\Gamma}{}^{r_2}_{kl} - T^{r_1 r_2\cdots r_a}_{\quad l r_{a+2}\cdots r_m}\overset{(y_p)}{\Gamma}{}^{l}_{kr_{a+1}} \\ &+ T^{r_1 r_2\cdots r_a}_{\quad r_{a+1}l\cdots r_m}\overset{(y_p)}{\Gamma}{}^{l}_{kr_{a+2}} + \cdots + T^{r_1 r_2\cdots r_a}_{\quad r_{a+1}r_{a+2}\cdots l}\overset{(y_p)}{\Gamma}{}^{l}_{kr_m}\Bigg]. \end{aligned} \tag{7.46}$$

Since

$$\overset{(1)}{u}_{r_1}\overset{(2)}{u}_{r_2}\cdots\overset{(a)}{u}_{r_a}\underset{(a+1)}{v}{}^{r_{a+1}}\underset{(a+2)}{v}{}^{r_{a+2}}\cdots\underset{(m)}{v}{}^{r_m}$$

is a product of arbitrary vectors, the quantity inside the square brackets in Equation (7.46) must vanish. Consequently, the covariant derivatives of any tensor can be expressed as

$$\begin{aligned} \overset{n}{\nabla}_k T^{r_1 r_2\cdots r_a}_{\quad r_{a+1}r_{a+2}\cdots r_m} = {} & \frac{\partial T^{r_1 r_2\cdots r_a}_{\quad r_{a+1}r_{a+2}\cdots r_m}}{\partial x^k} + T^{l r_2\cdots r_a}_{\quad r_{a+1}r_{a+2}\cdots r_m}\overset{(y_p)}{\Gamma}{}^{r_1}_{kl} \\ & + T^{r_1 l\cdots r_a}_{\quad r_{a+1}r_{a+2}\cdots r_m}\overset{(y_p)}{\Gamma}{}^{r_2}_{kl} + \cdots + T^{r_1 r_2\cdots l}_{\quad r_{a+1}r_{a+2}\cdots r_m}\overset{(y_p)}{\Gamma}{}^{r_a}_{kl} \\ & - T^{r_1 r_2\cdots r_a}_{\quad r_{a+2}\cdots r_m}\overset{(y_p)}{\Gamma}{}^{l}_{kr_{a+1}} - T^{r_1 r_2\cdots r_a}_{\quad r_{a+1}\cdots r_m}\overset{(y_p)}{\Gamma}{}^{l}_{kr_{a+2}} - \cdots \\ & - T^{r_1 r_2\cdots r_a}_{\quad r_{a+1}r_{a+2}\cdots l}\overset{(y_p)}{\Gamma}{}^{l}_{kr_m}. \end{aligned} \tag{7.47}$$

From inspection of Equation (7.47), it is evident that for each of the covariant derivatives m choices must be made of whether to use Γ^{i}_{jk} or Γ^{i}_{kj}. Consequently, there are 2^m possible covariant derivatives of a tensor of valence m.

In order to define the covariant derivative

$$\overset{n}{\nabla}_k T^{r_1 r_2\cdots r_a}_{\quad r_{a+1}r_{a+2}\cdots r_m},$$

we must find a way of associating n with a set of m choices of whether to use 1 or 2. It has been suggested by Albert and Spencer[10] that the binary number system be employed to set up the desired correspondence:

Step 1: Express the number $n-1$ as an m-digit binary number:

$$\begin{aligned}(n-1)_{10} &= 2^{m-1}(s_1)+\cdots+2(s_{m-1})+1(s_m)\\ &= (s_1 s_2 s_3 \cdots s_m)_2\end{aligned} \tag{7.48}$$

where $s_i = 0$ or 1 and the leading digits are retained.

Step 2: Since y_i must be 1 or 2 rather than 0 or 1, let

$$y_i = s_i + 1. \tag{7.49}$$

For example, if $m=2$ and the tensors are T^{ij}, T^i_j, or T_{ij}, there are $2^2=4$ covariant derivatives of each tensor:

$$\overset{(n)}{\nabla}_k T^{ij} = \frac{\partial T^{ij}}{\partial x^k} + T^{lj}\overset{(y_1)}{\Gamma}{}^{i}_{kl} + T^{il}\overset{(y_2)}{\Gamma}{}^{j}_{kl}, \tag{7.50a}$$

$$\overset{(n)}{\nabla}_k T_{ij} = \frac{\partial T_{ij}}{\partial x^k} + T_{lj}\overset{(y_1)}{\Gamma}{}^{l}_{ki} - T_{il}\overset{(y_2)}{\Gamma}{}^{l}_{ki}, \tag{7.50b}$$

$$\overset{(n)}{\nabla}_k T^i_j = \frac{\partial T^i_j}{\partial x^k} + T^l_j\overset{(y_1)}{\Gamma}{}^{i}_{kl} - T^i_l\overset{(y_2)}{\Gamma}{}^{l}_{ki}. \tag{7.50c}$$

The possible values of n are $1,2,3,4$. Therefore, $n-1$ may take on the values of $0,1,2,3$. According to Equation (7.48),

$$\begin{aligned}(0)_{10} &= 2(0)+1(0) = (00)_2,\\ (1)_{10} &= 2(0)+1(1) = (01)_2,\\ (2)_{10} &= 2(1)+1(0) = (10)_2,\\ (3)_{10} &= 2(1)+1(1) = (11)_2.\end{aligned}$$

Therefore, by Equation (7.49) for

$$\begin{aligned}n=1:&\quad y_1=1,\ y_2=1,\\ n=2:&\quad y_1=1,\ y_2=2,\\ n=3:&\quad y_1=2,\ y_2=1,\\ n=4:&\quad y_1=2,\ y_2=2.\end{aligned}$$

Substitution of these values of y_i into Equation (7.50) gives the covariant derivatives of Equation (7.35), (7.37), and (7.39).

As another example, consider the covariant derivatives of T^i_{jlm}. Since there are four indices, $m=4$, and there must be $2^4=16$ distinct covariant derivatives. The integers $n-1$ take on the values 0, 1, 2, 3, 4, 5, 6, 7, 8, 9, 10, 11, 12, 13, 14, 15. According to step 1, we must write

$$\begin{aligned}
0&=2^3(0)+2^2(0)+2(0)+1(0)=(0000)_2,\\
1&=2^3(0)+2^2(0)+2(0)+1(1)=(0001)_2,\\
2&=2^3(0)+2^2(0)+2(1)+1(0)=(0010)_2,\\
3&=2^3(0)+2^2(0)+2(1)+1(1)=(0011)_2,\\
4&=2^3(0)+2^2(1)+2(0)+1(0)=(0100)_2,\\
5&=2^3(0)+2^2(1)+2(0)+1(1)=(0101)_2,\\
6&=2^3(0)+2^2(1)+2(1)+1(0)=(0110)_2,\\
7&=2^3(0)+2^2(1)+2(1)+1(1)=(0111)_2,\\
8&=2^3(1)+2^2(0)+2(0)+1(0)=(1000)_2,\\
9&=2^3(1)+2^2(0)+2(0)+1(1)=(1001)_2,\\
10&=2^3(1)+2^2(0)+2(1)+1(0)=(1010)_2,\\
11&=2^3(1)+2^2(0)+2(1)+1(1)=(1011)_2,\\
12&=2^3(1)+2^2(1)+2(0)+1(0)=(1100)_2,\\
13&=2^3(1)+2^2(1)+2(0)+1(1)=(1101)_2,\\
14&=2^3(1)+2^2(1)+2(1)+1(0)=(1110)_2,\\
15&=2^3(1)+2^2(1)+2(1)+1(1)=(1111)_2.
\end{aligned}\tag{7.51}$$

And by step 2, the values y^i are

$$\begin{aligned}
n&=1: && y_1=1,\ y_2=1,\ y_3=1,\ y_4=1,\\
n&=2: && y_1=1,\ y_2=1,\ y_3=1,\ y_4=2,\\
n&=3: && y_1=1,\ y_2=1,\ y_3=2,\ y_4=1,\\
n&=4: && y_1=1,\ y_2=1,\ y_3=2,\ y_4=2,\\
n&=5: && y_1=1,\ y_2=2,\ y_3=1,\ y_4=1,\\
n&=6: && y_1=1,\ y_2=2,\ y_3=1,\ y_4=2,\\
n&=7: && y_1=1,\ y_2=2,\ y_3=2,\ y_4=1,\\
n&=8: && y_1=1,\ y_2=2,\ y_3=2,\ y_4=2,\\
n&=9: && y_1=2,\ y_2=1,\ y_3=1,\ y_4=1,\\
n&=10: && y_1=2,\ y_2=1,\ y_3=1,\ y_4=2,\\
n&=11: && y_1=2,\ y_2=1,\ y_3=2,\ y_4=1,\\
n&=12: && y_1=2,\ y_2=1,\ y_3=2,\ y_4=2,\\
n&=13: && y_1=2,\ y_2=2,\ y_3=1,\ y_4=1,\\
n&=14: && y_1=2,\ y_2=2,\ y_3=1,\ y_4=2,\\
n&=15: && y_1=2,\ y_2=2,\ y_3=2,\ y_4=1,\\
n&=16: && y_1=2,\ y_2=2,\ y_3=2,\ y_4=2.
\end{aligned}\tag{7.52}$$

Thus, the 16 covariant derivatives of $T^{i}{}_{jm}$ can be written as

$$\overset{1}{\nabla}_k T^{i}{}_{jlm} = \frac{\partial T^{i}{}_{jlm}}{\partial x^k} + T^{n}{}_{jlm}\Gamma^{i}{}_{kn} - T^{i}{}_{nlm}\Gamma^{n}{}_{kj} - T^{i}{}_{jnm}\Gamma^{n}{}_{kl} - T^{i}{}_{jln}\Gamma^{n}{}_{km}, \tag{7.53a}$$

$$\overset{2}{\nabla}_k T^{i}{}_{jlm} = \frac{\partial T^{i}{}_{jlm}}{\partial x^k} + T^{n}{}_{jlm}\Gamma^{i}{}_{kn} - T^{i}{}_{nlm}\Gamma^{n}{}_{kj} - T^{i}{}_{jnm}\Gamma^{n}{}_{kl} - T^{i}{}_{jln}\Gamma^{n}{}_{mk}, \tag{7.53b}$$

$$\overset{3}{\nabla}_k T^{i}{}_{jlm} = \frac{\partial T^{i}{}_{jlm}}{\partial x^k} + T^{n}{}_{jlm}\Gamma^{i}{}_{kn} - T^{i}{}_{nlm}\Gamma^{n}{}_{kj} - T^{i}{}_{jnm}\Gamma^{n}{}_{lk} - T^{i}{}_{jln}\Gamma^{n}{}_{km}, \tag{7.53c}$$

$$\overset{4}{\nabla}_k T^{i}{}_{jlm} = \frac{\partial T^{i}{}_{jlm}}{\partial x^k} + T^{n}{}_{jlm}\Gamma^{i}{}_{kn} - T^{i}{}_{nlm}\Gamma^{n}{}_{kj} - T^{i}{}_{jnm}\Gamma^{n}{}_{lk} - T^{i}{}_{jln}\Gamma^{n}{}_{mk}, \tag{7.53d}$$

$$\overset{5}{\nabla}_k T^{i}{}_{jlm} = \frac{\partial T^{i}{}_{jlm}}{\partial x^k} + T^{n}{}_{jlm}\Gamma^{i}{}_{kn} - T^{i}{}_{nlm}\Gamma^{n}{}_{jk} - T^{i}{}_{jnm}\Gamma^{n}{}_{kl} - T^{i}{}_{jln}\Gamma^{n}{}_{km}, \tag{7.53e}$$

$$\overset{6}{\nabla}_k T^{i}{}_{jlm} = \frac{\partial T^{i}{}_{jlm}}{\partial x^k} + T^{n}{}_{jlm}\Gamma^{i}{}_{kn} - T^{i}{}_{nlm}\Gamma^{n}{}_{jk} - T^{i}{}_{jnm}\Gamma^{n}{}_{kl} - T^{i}{}_{jln}\Gamma^{n}{}_{mk}, \tag{7.53f}$$

$$\overset{7}{\nabla}_k T^{i}{}_{jlm} = \frac{\partial T^{i}{}_{jlm}}{\partial x^k} + T^{n}{}_{jlm}\Gamma^{i}{}_{kn} - T^{i}{}_{nlm}\Gamma^{n}{}_{jk} - T^{i}{}_{jnm}\Gamma^{n}{}_{lk} - T^{i}{}_{jln}\Gamma^{n}{}_{km}, \tag{7.53g}$$

$$\overset{8}{\nabla}_k T^{i}{}_{jlm} = \frac{\partial T^{i}{}_{jlm}}{\partial x^k} + T^{n}{}_{jlm}\Gamma^{i}{}_{kn} - T^{i}{}_{nlm}\Gamma^{n}{}_{jk} - T^{i}{}_{jnm}\Gamma^{n}{}_{lk} - T^{i}{}_{jln}\Gamma^{n}{}_{mk}, \tag{7.53h}$$

$$\overset{9}{\nabla}_k T^{i}{}_{jlm} = \frac{\partial T^{i}{}_{jlm}}{\partial x^k} + T^{n}{}_{jlm}\Gamma^{i}{}_{nk} - T^{i}{}_{nlm}\Gamma^{n}{}_{kj} - T^{i}{}_{jnm}\Gamma^{n}{}_{kl} - T^{i}{}_{jln}\Gamma^{n}{}_{km}, \tag{7.53i}$$

$$\overset{10}{\nabla}_k T^{i}{}_{jlm} = \frac{\partial T^{i}{}_{jlm}}{\partial x^k} + T^{n}{}_{jlm}\Gamma^{i}{}_{nk} - T^{i}{}_{nlm}\Gamma^{n}{}_{kj} - T^{i}{}_{jnm}\Gamma^{n}{}_{kl} - T^{i}{}_{jln}\Gamma^{n}{}_{mk}, \tag{7.53j}$$

$$\overset{11}{\nabla}_k T^{i}{}_{jlm} = \frac{\partial T^{i}{}_{jlm}}{\partial x^k} + T^{n}{}_{jlm}\Gamma^{i}{}_{nk} - T^{i}{}_{nlm}\Gamma^{n}{}_{kj} - T^{i}{}_{jnm}\Gamma^{n}{}_{lk} - T^{i}{}_{jln}\Gamma^{n}{}_{km}, \tag{7.53k}$$

$$\overset{12}{\nabla}_k T^{i}{}_{jlm} = \frac{\partial T^{i}{}_{jlm}}{\partial x^k} + T^{n}{}_{jlm}\Gamma^{i}{}_{nk} - T^{i}{}_{nlm}\Gamma^{n}{}_{kj} - T^{i}{}_{jnm}\Gamma^{n}{}_{lk} - T^{i}{}_{jln}\Gamma^{n}{}_{mk}, \tag{7.53l}$$

$$\overset{13}{\nabla}_k T^{i}{}_{jlm} = \frac{\partial T^{i}{}_{jlm}}{\partial x^k} + T^{n}{}_{jlm}\Gamma^{i}{}_{nk} - T^{i}{}_{nlm}\Gamma^{n}{}_{jk} - T^{i}{}_{jnm}\Gamma^{n}{}_{kl} - T^{i}{}_{jln}\Gamma^{n}{}_{km}, \tag{7.53m}$$

$$\overset{14}{\nabla}_k T^{i}{}_{jlm} = \frac{\partial T^{i}{}_{jlm}}{\partial x^k} + T^{n}{}_{jlm}\Gamma^{i}{}_{nk} - T^{i}{}_{nlm}\Gamma^{n}{}_{jk} - T^{i}{}_{jnm}\Gamma^{n}{}_{kl} - T^{i}{}_{jln}\Gamma^{n}{}_{mk}, \tag{7.53n}$$

$$\overset{15}{\nabla}_k T^{i}{}_{jlm} = \frac{\partial T^{i}{}_{jlm}}{\partial x^k} + T^{n}{}_{jlm}\Gamma^{i}{}_{nk} - T^{i}{}_{nlm}\Gamma^{n}{}_{jk} - T^{i}{}_{jnm}\Gamma^{n}{}_{lk} - T^{i}{}_{jln}\Gamma^{n}{}_{km}, \tag{7.53o}$$

$$\overset{16}{\nabla}_k T^{i}{}_{jlm} = \frac{\partial T^{i}{}_{jlm}}{\partial x^k} + T^{n}{}_{jlm}\Gamma^{i}{}_{nk} - T^{i}{}_{nlm}\Gamma^{n}{}_{jk} - T^{i}{}_{jnm}\Gamma^{n}{}_{lk} - T^{i}{}_{jln}\Gamma^{n}{}_{mk}. \tag{7.53p}$$

All of these covariant derivatives can be expressed in terms of the first covariant derivative and the torsion tensor,

$$\overset{2}{\nabla}_k T^{i}_{\;jlm} = \overset{1}{\nabla}_k T^{i}_{\;jlm} + 2T^{i}_{\;jln} S^{n}_{km}, \tag{7.54a}$$

$$\overset{3}{\nabla}_k T^{i}_{\;jlm} = \overset{1}{\nabla}_k T^{i}_{\;jlm} + 2T^{i}_{\;jnm} S^{n}_{kl}, \tag{7.54b}$$

$$\overset{4}{\nabla}_k T^{i}_{\;jlm} = \overset{1}{\nabla}_k T^{i}_{\;jlm} + 2T^{i}_{\;jnm} S^{n}_{kl} + 2T^{i}_{\;jln} S^{n}_{km}, \tag{7.54c}$$

$$\overset{5}{\nabla}_k T^{i}_{\;jlm} = \overset{1}{\nabla}_k T^{i}_{\;jlm} + 2T^{i}_{\;nlm} S^{n}_{kj}, \tag{7.54d}$$

$$\overset{6}{\nabla}_k T^{i}_{\;jlm} = \overset{1}{\nabla}_k T^{i}_{\;jlm} + 2T^{i}_{\;nlm} S^{n}_{kj} + 2T^{i}_{\;jln} S^{n}_{km}, \tag{7.54e}$$

$$\overset{7}{\nabla}_k T^{i}_{\;jlm} = \overset{1}{\nabla}_k T^{i}_{\;jlm} + 2T^{i}_{\;nlm} S^{n}_{kj} + 2T^{i}_{\;jnm} S^{n}_{kl}, \tag{7.54f}$$

$$\overset{8}{\nabla}_k T^{i}_{\;jlm} = \overset{1}{\nabla}_k T^{i}_{\;jlm} + 2T^{i}_{\;nlm} S^{n}_{kj} + 2T^{i}_{\;jnm} S^{n}_{kl} + 2T^{i}_{\;jln} S^{n}_{km}, \tag{7.54g}$$

$$\overset{9}{\nabla}_k T^{i}_{\;jlm} = \overset{1}{\nabla}_k T^{i}_{\;jlm} + 2T^{n}_{\;jlm} S^{i}_{kn}, \tag{7.54h}$$

$$\overset{10}{\nabla}_k T^{i}_{\;jlm} = \overset{1}{\nabla}_k T^{i}_{\;jlm} - 2T^{n}_{\;jlm} S^{i}_{kn} + 2T^{i}_{\;jln} S^{n}_{km}, \tag{7.54i}$$

$$\overset{11}{\nabla}_k T^{i}_{\;jlm} = \overset{1}{\nabla}_k T^{i}_{\;jlm} - 2T^{n}_{\;jlm} S^{i}_{kn} + T^{i}_{\;jnm} S^{n}_{kl}, \tag{7.54j}$$

$$\overset{12}{\nabla}_k T^{i}_{\;jlm} = \overset{1}{\nabla}_k T^{i}_{\;jlm} - 2T^{n}_{\;jlm} S^{i}_{kn} + 2T^{i}_{\;jnm} S^{n}_{kl} + 2T^{i}_{\;jln} S^{n}_{km}, \tag{7.54k}$$

$$\overset{13}{\nabla}_k T^{i}_{\;jlm} = \overset{1}{\nabla}_k T^{i}_{\;jlm} - 2T^{n}_{\;jlm} S^{i}_{kn} + 2T^{i}_{\;nlm} S^{n}_{kj}, \tag{7.54l}$$

$$\overset{14}{\nabla}_k T^{i}_{\;jlm} = \overset{1}{\nabla}_k T^{i}_{\;jlm} - 2T^{n}_{\;jlm} S^{i}_{kn} + 2T^{i}_{\;nlm} S^{n}_{kj} + 2T^{i}_{\;jln} S^{n}_{km}, \tag{7.54m}$$

$$\overset{15}{\nabla}_k T^{i}_{\;jlm} = \overset{1}{\nabla}_k T^{i}_{\;jlm} - 2T^{n}_{\;jlm} S^{i}_{kn} + 2T^{i}_{\;nlm} S^{n}_{kj} + 2T^{i}_{\;jnm} S^{n}_{kl}, \tag{7.54n}$$

$$\overset{16}{\nabla}_k T^{i}_{\;jlm} = \overset{1}{\nabla}_k T^{i}_{\;jlm} - 2T^{n}_{\;jlm} S^{i}_{kn} + 2T^{i}_{\;nlm} S^{n}_{kj} + 2T^{i}_{\;jnm} S^{n}_{kl} + 2T^{i}_{\;jln} S^{n}_{km}. \tag{7.54o}$$

By the principles outlined in this section, the 2^m covariant derivatives of any tensor of valence m can be expanded uniquely, and the identities relating the covariant derivatives can be written. In the special case in which the linear connection is symmetric in its subscripts, the 2^m distinct covariant derivatives become identical and there is a single covariant derivative of each tensor.

7.06. Contracted covariant derivatives

Simpler derivatives that transform as tensors can be obtained from some of the covariant derivatives by the simple process of contraction. If a tensor has at least one contravariant index, this procedure is possible.

Consider the covariant derivatives of a contravariant vector v^i. Contraction of Equation (7.20) gives two scalars,

$$\overset{1}{\nabla}_k v^k = \frac{\partial v^k}{\partial x^k} + v^l \Gamma^k_{kl}, \tag{7.55a}$$

$$\overset{2}{\nabla}_k v^k = \frac{\partial v^k}{\partial x^k} + v^l \Gamma^k_{lk}. \tag{7.55b}$$

These scalars may be called the first and second kind of divergence of a contravariant vector. They differ by a constant,

$$\overset{2}{\nabla}_k v^k = \overset{1}{\nabla}_k v^k - 2v^l S^k_{kl}, \tag{7.56}$$

but become identical if the torsion tensor vanishes.

Similarly, contraction of the covariant derivative of a contravariant bivalent tensor[11] gives interesting results. Since there are four covariant derivatives of T^{ij} and two ways in which the contraction can be carried out, there are eight contravariant vectors that can be obtained by contraction of Equation (7.35).

$$\overset{1}{\nabla}_k T^{kj} = \frac{\partial T^{kj}}{\partial x^k} + T^{lj}\Gamma^k_{kl} + T^{kl}\Gamma^j_{kl}, \tag{7.57a}$$

$$\overset{2}{\nabla}_k T^{kj} = \frac{\partial T^{kj}}{\partial x^k} + T^{lj}\Gamma^k_{kl} + T^{kl}\Gamma^j_{lk}, \tag{7.57b}$$

$$\overset{3}{\nabla}_k T^{kj} = \frac{\partial T^{kj}}{\partial x^k} + T^{lj}\Gamma^k_{lk} + T^{kl}\Gamma^j_{kl}, \tag{7.57c}$$

$$\overset{4}{\nabla}_k T^{kj} = \frac{\partial T^{kj}}{\partial x^k} + T^{lj}\Gamma^k_{lk} + T^{kl}\Gamma^j_{lk}, \tag{7.57d}$$

$$\overset{1}{\nabla}_k T^{jk} = \frac{\partial T^{jk}}{\partial x^k} + T^{lk}\Gamma^j_{kl} + T^{jl}\Gamma^k_{kl}, \tag{7.57e}$$

$$\overset{2}{\nabla}_k T^{jk} = \frac{\partial T^{jk}}{\partial x^k} + T^{lk}\Gamma^j_{kl} + T^{jl}\Gamma^k_{lk}, \tag{7.57f}$$

$$\overset{3}{\nabla}_k T^{jk} = \frac{\partial T^{jk}}{\partial x^k} + T^{lk}\Gamma^j_{lk} + T^{jl}\Gamma^k_{kl}, \tag{7.57g}$$

$$\overset{4}{\nabla}_k T^{jk} = \frac{\partial T^{jk}}{\partial x^k} + T^{lk}\Gamma^j_{lk} + T^{jl}\Gamma^k_{lk}. \tag{7.57h}$$

According to Equation (7.36), these eight vectors are related by the equations

$$\overset{2}{\nabla}_k T^{kj} = \overset{1}{\nabla}_k T^{kj} - 2T^{kl}S^{j}_{kl}, \tag{7.58a}$$

$$\overset{3}{\nabla}_k T^{kj} = \overset{1}{\nabla}_k T^{kj} - 2T^{lj}S^{k}_{kl}, \tag{7.58b}$$

$$\overset{4}{\nabla}_k T^{kj} = \overset{1}{\nabla}_k T^{kj} - 2T^{kl}S^{j}_{kl} - 2T^{lj}S^{k}_{kl}, \tag{7.58c}$$

$$\overset{2}{\nabla}_k T^{jk} = \overset{1}{\nabla}_k T^{jk} - 2T^{jl}S^{k}_{kl}, \tag{7.58d}$$

$$\overset{3}{\nabla}_k T^{jk} = \overset{1}{\nabla}_k T^{jk} - 2T^{lk}S^{j}_{kl}, \tag{7.58e}$$

$$\overset{4}{\nabla}_k T^{jk} = \overset{1}{\nabla}_k T^{jk} - 2T^{jl}S^{k}_{jl} - 2T^{lk}S^{j}_{kl}. \tag{7.58f}$$

If the tensor T^{ij} is an alternating tensor $T^{ij} = -T^{ji}$, then Equation (7.58) simplifies somewhat. Since the contracted product of an alternating tensor (T^{il}) and a symmetric tensor (S^{j}_{il}) is always zero, half of the terms involving the torsion tensor in Equation (7.58) drop out and we are left with only

$$\overset{2}{\nabla}_k T^{kj} = \overset{1}{\nabla}_k T^{kj} = -\overset{2}{\nabla}_k T^{jk} = -\overset{1}{\nabla}_k T^{jk}, \tag{7.59a}$$

and

$$\begin{aligned} \overset{3}{\nabla}_k T^{kj} &= \overset{1}{\nabla}_k T^{kj} - 2T^{lj}S^{k}_{kl} = \overset{4}{\nabla}_k T^{kj} \\ &= -\overset{3}{\nabla}_k T^{jk} = -\overset{4}{\nabla}_k T^{jk}. \end{aligned} \tag{7.59b}$$

Similarly, from Equation (7.39), there are four distinct contracted covariant derivatives,

$$\overset{1}{\nabla}_k T^{k}_{j} = \frac{\partial T^{k}_{j}}{\partial x^k} + T^{l}_{j}\Gamma^{k}_{kl} - T^{k}_{l}\Gamma^{l}_{kj}, \tag{7.60a}$$

$$\overset{2}{\nabla}_k T^{k}_{j} = \frac{\partial T^{k}_{j}}{\partial x^k} + T^{l}_{j}\Gamma^{k}_{kl} - T^{k}_{l}\Gamma^{l}_{jk}, \tag{7.60b}$$

$$\overset{3}{\nabla}_k T^{k}_{j} = \frac{\partial T^{k}_{j}}{\partial x^k} + T^{l}_{j}\Gamma^{k}_{lk} - T^{k}_{l}\Gamma^{l}_{kj}, \tag{7.60c}$$

$$\overset{4}{\nabla}_k T^{k}_{j} = \frac{\partial T^{k}_{j}}{\partial x^k} + T^{l}_{j}\Gamma^{k}_{lk} - T^{k}_{l}\Gamma^{l}_{jk}. \tag{7.60d}$$

All of these transform as covariant vectors and are related according to Equation (7.40) by the relations

$$\overset{2}{\nabla}_k T^{k}_{j} = \overset{1}{\nabla}_k T^{k}_{j} + 2T^{k}_{l}S^{l}_{kj}, \tag{7.61a}$$

$$\overset{3}{\nabla}_k T^{k}_{j} = \overset{1}{\nabla}_k T^{k}_{j} - 2T^{l}_{j}S^{k}_{kl}, \tag{7.61b}$$

$$\overset{4}{\nabla}_k T^{k}_{j} = \overset{1}{\nabla}_k T^{k}_{j} + 2T^{k}_{l}S^{l}_{kj} - 2T^{l}_{j}S^{k}_{kl}. \tag{7.61c}$$

Many other types of tensor derivatives can readily be obtained from the covariant derivatives of higher-order tensors by the process of contraction.

7.07. Derivatives with respect to a scalar parameter

Suppose that a curve is defined in terms of a parameter $u : x^i(u)$. The parameter may be time t or a geometric parameter. Then the first derivative of x^i with respect to u is given by dx^i/du. Expressed in a different coordinate system,

$$\frac{dx^{i'}}{du} = \frac{\partial x^{i'}}{\partial x^i}\frac{dx^i}{du} \tag{7.62}$$

so dx^i/du transforms as a contravariant vector. If the parameter is time t, then the derivative is the velocity

$$v^i = \frac{dx^i}{dt}$$

and

$$v^{i'} = \frac{dx^{i'}}{dt} = \frac{\partial x^{i'}}{\partial x^i}\frac{dx^i}{dt} = \frac{\partial x^{i'}}{\partial x^i}v^i, \tag{7.62a}$$

so velocity always transforms as a contravariant vector.

However, the situation is entirely different if we consider the derivative of dx^i/du with respect to the parameter u. Differentiation of Equation (7.62) with respect to u gives

$$\frac{d^2x^{i'}}{du^2} = \frac{\partial x^{i'}}{\partial x^i}\frac{d^2x^i}{du^2} + \frac{\partial^2 x^{i'}}{\partial x^i \partial x^j}\frac{dx^i}{du}\frac{dx^j}{du}. \tag{7.63}$$

Because of the presence of the last term in Equation (7.63), we can conclude that d^2x^i/du^2 does not transform as a tensor. This result is particularly interesting when the scalar parameter is the time t. For this means that the ordinary definition of the acceleration as the time derivative of the velocity results in an "acceleration" that does not transform as a tensor!

Let us see if we can find a closely related tensor. To do so, substitute Equation (7.15) into (7.63). Changing names of indices, Equation (7.15) can be written

$$\frac{\partial^2 x^{i'}}{\partial x^i \partial x^j} = \frac{\partial x^{i'}}{\partial x^k}\Gamma^k_{ij} - \Gamma^{i'}_{m'n'}\frac{\partial x^{m'}}{\partial x^i}\frac{\partial x^{n'}}{\partial x^j} \tag{7.15a}$$

or

$$\frac{\partial^2 x^{i'}}{\partial x^i \partial x^j} = \frac{\partial x^{i'}}{\partial x^k}\Gamma^k_{ji} - \Gamma^{i'}_{n'm'}\frac{\partial x^{n'}}{\partial x^j}\frac{\partial x^{m'}}{\partial x^i}. \tag{7.15b}$$

Substitution of Equations (7.15a) and (7.15b) into (7.63) gives

$$\frac{d^2 x^{i'}}{du^2} = \frac{\partial x^{i'}}{\partial x^i}\frac{d^2 x^i}{du^2} + \left(\frac{\partial x^{i'}}{\partial x^k}\Gamma^k_{ij} - \Gamma^{i'}_{m'n'}\frac{\partial x^{m'}}{\partial x^i}\frac{\partial x^{n'}}{\partial x^j}\right)\frac{dx^i}{du}\frac{dx^j}{du} \tag{7.64a}$$

or

$$\frac{d^2 x^{i'}}{du^2} = \frac{\partial x^{i'}}{\partial x^i}\frac{d^2 x^i}{du^2} + \left(\frac{\partial x^{i'}}{\partial x^k}\Gamma^k_{ji} - \Gamma^{i'}_{n'm'}\frac{\partial x^{n'}}{\partial x^j}\frac{\partial x^{m'}}{\partial x^i}\right)\frac{dx^i}{du}\frac{dx^j}{du}. \tag{7.64b}$$

Now transpose the last term of Equation (7.64) to the left side of the equation and factor the remaining terms on the right side of the equation,

$$\frac{d^2 x^{i'}}{du^2} + \Gamma^{i'}_{m'n'}\left(\frac{\partial x^{m'}}{\partial x^i}\frac{dx^i}{du}\right)\left(\frac{\partial x^{n'}}{\partial x^j}\frac{dx^j}{du}\right) = \frac{\partial x^{i'}}{\partial x^k}\left(\frac{d^2 x^k}{du^2} + \Gamma^k_{ij}\frac{dx^i}{du}\frac{dx^j}{du}\right), \tag{7.65a}$$

$$\frac{d^2 x^{i'}}{du^2} + \Gamma^{i'}_{n'm'}\left(\frac{\partial x^{n'}}{\partial x^j}\frac{dx^j}{du}\right)\left(\frac{\partial x^{m'}}{\partial x^i}\frac{dx^i}{du}\right) = \frac{\partial x^{i'}}{\partial x^k}\left(\frac{d^2 x^k}{du^2} + \Gamma^k_{ji}\frac{dx^i}{du}\frac{dx^j}{du}\right). \tag{7.65b}$$

Next simplify the left side of Equation (7.65) by substitution of Equation (7.62) and change dummy indices so that

$$\frac{d^2 x^{i'}}{du^2} + \Gamma^{i'}_{j'k'}\frac{dx^{j'}}{du}\frac{dx^{k'}}{du} = \frac{\partial x^{i'}}{\partial x^i}\left(\frac{d^2 x^i}{du^2} + \Gamma^i_{jk}\frac{dx^j}{du}\frac{dx^k}{du}\right). \tag{7.66}$$

Two things about Equation (7.66) are particularly interesting: First, because the two coefficients of the linear connection are the same vector, both types of covariant derivative give the same expression. Second, it will be noted that the quantity

$$a^i = \frac{d^2x^i}{du^2} + \Gamma^i_{jk}\frac{dx^j}{du}\frac{dx^k}{du} \tag{7.67}$$

transforms as a contravariant vector. If the scalar parameter is the time t, this vector is the acceleration

$$a^i = \frac{d^2x^i}{du^2} + \Gamma^i_{jk}v^jv^k. \tag{7.68}$$

This means that even in a space with asymmetric linear connection, there is a single tensor that reduces to the ordinary idea of acceleration when $\Gamma^i_{jk} = 0$ and is therefore the only possible way of defining a *tensor acceleration*.

7.08. The covariant derivatives of akinetors

By the definition of a nilvalent akinetor, its transformation equation is

$$\tilde{S}' = \sigma(x^i)\tilde{S}. \tag{5.02a}$$

Differentiation of Equation (5.06) with respect to $x^{k'}$ gives

$$\frac{\partial\tilde{S}'}{\partial x^{k'}} = \sigma\frac{\partial x^k}{\partial x^{k'}}\frac{\partial\tilde{S}}{\partial x^k} + \tilde{S}\frac{\partial\sigma}{\partial x^{k'}}. \tag{7.69}$$

Because of the presence of the last term, $\partial\tilde{S}/\partial x^k$ does not transform as an akinetor.

However, there are two special cases in which akinetor derivatives can be obtained. If σ is a constant, then

$$\frac{\partial\tilde{S}'}{\partial x^{k'}} = \sigma\frac{\partial x^k}{\partial x^{k'}}\frac{\partial\tilde{S}}{\partial x^k}, \tag{7.70}$$

and $\partial\tilde{S}/\partial x^k$ transforms as an akinetor.

To derive the covariant derivative of the second important class of akinetors,[12] it is necessary to first obtain an expression for the partial derivative of a determinant.

In Chapter 1, it was shown that a determinant could be written compactly as a contracted product,

$$|a^i_j| = \delta^{12\cdots n}_{kl\cdots v}a^k_1a^l_2\cdots a^v_n. \tag{1.15}$$

Differentiating with respect to x^k,

$$\frac{\partial}{\partial x^k}|a^i_j| = \delta^{12\cdots n}_{kl\cdots v}\left(\frac{\partial a^k_1}{\partial x^k}a^l_2\cdots a^v_n + \frac{\partial a^l_2}{\partial x^k}a^k_1 a^m_3\cdots a^v_n + \cdots + \frac{\partial a^v_n}{\partial x^k}a^k_1\cdots a^u_{n-1}\right)$$

or

$$\frac{\partial}{\partial x^k}|a^i_j| = \frac{\partial a^i_j}{\partial x^k} = A^j_i, \tag{7.71}$$

where

$$A^i_j = \delta^{12\cdots n}_{kl\cdots v} a^k_1 a^l_2 \cdots a^{i-1}_{j-1} a^{i+1}_{j+1} \cdots a^v_n$$

is the cofactor of a^i_j.

In particular, if the matrix a^i_j is the transformation matrix

$$a^i_{i'} = \frac{\partial x^i}{\partial x^{i'}} \quad \text{and} \quad |a^i_{i'}| = \left|\frac{\partial x^i}{\partial x^{i'}}\right| = \Delta,$$

then Equation (7.71) becomes

$$\frac{\partial \Delta}{\partial x^{k'}} = \frac{\partial^2 x^i}{\partial x^{k'}\partial x^{i'}} = \frac{\partial^2 x^i}{\partial x^{k'}\partial x^{i'}}\frac{\partial x^{i'}}{\partial x^i}\Delta, \tag{7.72}$$

since the cofactor of $\partial x^i/\partial x^{i'}$ is readily shown to be

$$A^{i'}_i = \frac{\partial x^{i'}}{\partial x^i}\Delta. \tag{7.73}$$

Now consider the second and most important class of nilvalent akinetors in which the multiplicative coefficient σ is a power of Δ. Then the transformation equation of the nilvalent akinetor becomes

$$\tilde{S}' = \Delta^w \tilde{S}. \tag{5.06}$$

Differentiating with respect to $x^{k'}$ gives

$$\frac{\partial \tilde{S}'}{\partial x^{k'}} = \Delta^w \frac{\partial \tilde{S}}{\partial x^k}\frac{\partial x^k}{\partial x^{k'}} + w\tilde{S}\Delta^{w-1}\frac{\partial \Delta}{\partial x^{k'}}. \tag{7.74}$$

Substitution of Equation (7.72) into (7.74) gives

$$\frac{\partial \tilde{S}'}{\partial x^{k'}} = \Delta^w\left(\frac{\partial x^k}{\partial x^{k'}}\frac{\partial \tilde{S}}{\partial x^k} + w\tilde{S}\frac{\partial^2 x^i}{\partial x^{k'}\partial x^{i'}}\frac{\partial x^{i'}}{\partial x^i}\right). \tag{7.75}$$

But from Equation (7.15) we can express the second partial derivatives in terms of the linear connections as

$$\frac{\partial^2 x^i}{\partial x^{k'} \partial x^{i'}} = \frac{\partial x^i}{\partial x^{j'}} \Gamma^{j'}_{k'i'} - \Gamma^i_{mn} \frac{\partial x^m}{\partial x^{k'}} \frac{\partial x^n}{\partial x^{i'}}, \tag{7.15c}$$

or

$$\frac{\partial^2 x^i}{\partial x^{k'} \partial x^{i'}} = \frac{\partial x^i}{\partial x^{j'}} \Gamma^{j'}_{i'k'} - \Gamma^i_{nm} \frac{\partial x^n}{\partial x^{i'}} \frac{\partial x^m}{\partial x^{i'}}. \tag{7.15d}$$

Substitution of Equations (7.15c) and (7.15d) into (7.75) gives

$$\frac{\partial \tilde{S}'}{\partial x^{k'}} = \Delta^w \left(\frac{\partial x^k}{\partial x^{k'}} \frac{\partial \tilde{S}}{\partial x^k} + w\tilde{S} \frac{\partial x^{i'}}{\partial x^i} \frac{\partial x^i}{\partial x^{j'}} \Gamma^{j'}_{k'i'} - w\tilde{S} \frac{\partial x^{i'}}{\partial x^i} \frac{\partial x^m}{\partial x^{k'}} \frac{\partial x^n}{\partial x^{i'}} \Gamma^i_{mn} \right), \tag{7.76a}$$

or

$$\frac{\partial \tilde{S}'}{\partial x^{k'}} = \Delta^w \left(\frac{\partial x^k}{\partial x^{k'}} \frac{\partial \tilde{S}}{\partial x^k} + w\tilde{S} \frac{\partial x^{i'}}{\partial x^i} \frac{\partial x^i}{\partial x^{j'}} \Gamma^{j'}_{i'k'} - w\tilde{S} \frac{\partial x^{i'}}{\partial x^i} \frac{\partial x^n}{\partial x^{i'}} \frac{\partial x^m}{\partial x^{k'}} \Gamma^i_{nm} \right). \tag{7.76b}$$

The middle term on the right side of Equation (7.76) can be simplified since

$$\frac{\partial x^{i'}}{\partial x^i} \frac{\partial x^i}{\partial x^{j'}} = \delta^{i'}_{j'} \quad \text{and} \quad \tilde{S}' = \Delta^w \tilde{S},$$

so if we now transpose this simplified term to the left side of Equation (7.76), it becomes

$$\left(\frac{\partial \tilde{S}'}{\partial x^{k'}} - w\tilde{S}\Gamma^{i'}_{k'i'} \right) = \Delta^w \left(\frac{\partial x^k}{\partial x^{k'}} \frac{\partial \tilde{S}}{\partial x^k} - w\tilde{S} \frac{\partial x^{i'}}{\partial x^i} \frac{\partial x^m}{\partial x^{k'}} \frac{\partial x^n}{\partial x^{i'}} \Gamma^i_{mn} \right), \tag{7.77a}$$

$$\left(\frac{\partial \tilde{S}'}{\partial x^{k'}} - w\tilde{S}\Gamma^{i'}_{i'k'} \right) = \Delta^w \left(\frac{\partial x^k}{\partial x^{k'}} \frac{\partial \tilde{S}}{\partial x^k} - w\tilde{S} \frac{\partial x^{i'}}{\partial x^i} \frac{\partial x^n}{\partial x^{i'}} \frac{\partial x^m}{\partial x^{k'}} \Gamma^i_{nm} \right). \tag{7.77b}$$

The last term on the right side of Equation (7.77) can also be simplified since

$$\frac{\partial x^{i'}}{\partial x^i} \frac{\partial x^n}{\partial x^{i'}} = \delta^n_i.$$

Changing dummy indices, Equation (7.77) becomes

$$\left(\frac{\partial \tilde{S}'}{\partial x^{k'}} - w\tilde{S}\Gamma^{i'}_{k'i'}\right) = \Delta^w \frac{\partial x^k}{\partial x^{k'}}\left(\frac{\partial \tilde{S}}{\partial x^k} - w\tilde{S}\Gamma^i_{ki}\right), \tag{7.78a}$$

$$\left(\frac{\partial \tilde{S}'}{\partial x^{k'}} - w\tilde{S}\Gamma^{i'}_{i'k'}\right) = \Delta^w \frac{\partial x^k}{\partial x^{k'}}\left(\frac{\partial \tilde{S}}{\partial x^k} - w\tilde{S}\Gamma^i_{ik}\right). \tag{7.78b}$$

The bracketed quantity transforms as a covariant akinetor of weight w and will be defined as the covariant derivative of $\tilde{S}$. Thus, there are two equally important covariant derivatives of a nilvalent akinetor of weight w:

$$\overset{1}{\nabla}_k \tilde{S} = \frac{\partial \tilde{S}}{\partial x^k} - w\tilde{S}\Gamma^i_{ki}, \tag{7.79a}$$

$$\overset{2}{\nabla}_k \tilde{S} = \frac{\partial \tilde{S}}{\partial x^k} - w\tilde{S}\Gamma^i_{ik}. \tag{7.79b}$$

Their transformation equations are

$$\overset{1}{\nabla}_{k'} \tilde{S}' = \Delta^w \frac{\partial x^k}{\partial x^{k'}} \overset{1}{\nabla}_k \tilde{S}, \tag{7.80a}$$

$$\overset{2}{\nabla}_{k'} \tilde{S}' = \Delta^w \frac{\partial x^k}{\partial x^{k'}} \overset{2}{\nabla}_k \tilde{S}. \tag{7.80b}$$

The covariant derivative of a nilvalent akinetor of weight w always transforms as a covariant univalent akinetor of weight w. The two covariant derivatives are related by the equation

$$\overset{2}{\nabla}_k \tilde{S} = \overset{1}{\nabla}_k \tilde{S} + 2w\tilde{S}S^i_{ki} \tag{7.81}$$

and become identical if the torsion tensor vanishes.

Now let us consider how the covariant derivatives of higher-order akinetors can be defined. An akinetor of weight w and valence m, $\tilde{A}^{r_1 r_2 \cdots r_a}_{r_{a+1} r_{a+2} \cdots r_m}$ can always be expressed as the product of a nilvalent akinetor $\tilde{S}$ of weight w and a tensor of valence m, $T^{r_1 r_2 \cdots r_a}_{r_{a+1} r_{a+2} \cdots r_m}$,

$$\tilde{A}^{r_1 r_2 \cdots r_a}_{r_{a+1} r_{a+2} \cdots r_m} = \tilde{S} T^{r_1 r_2 \cdots r_a}_{r_{a+1} r_{a+2} \cdots r_m}. \tag{7.82}$$

The pth covariant derivative of the akinetor can be defined by assuming the usual product rule,

$$\overset{p}{\nabla}_k A^{r_1 r_2 \cdots r_a}_{r_{a+1} r_{a+2} \cdots r_m} = (\overset{i}{\nabla}_k \tilde{S}) T^{r_1 r_2 \cdots r_a}_{r_{a+1} r_{a+2} \cdots r_m} + \tilde{S}(\overset{p}{\nabla}_k T^{r_1 r_2 \cdots r_a}_{r_{a+1} r_{a+2} \cdots r_m}), \tag{7.83}$$

where $i = 1$ when $p = 1, 2, \ldots, 2^m$ and $i = 2$ when $p = 2^m + 1, 2^m + 2, \ldots, 2^m$.

As an example, consider the covariant derivatives of a univalent, contravariant akinetor of weight w,

$$\tilde{v}^i = \tilde{S} v^i. \tag{7.84}$$

From Equation (7.83) there are four possible covariant derivatives:

$$\overset{1}{\nabla}_k \tilde{v}^i = \frac{\partial \tilde{v}^i}{\partial x^k} + \tilde{v}^l \Gamma^i_{kl} - w \tilde{v}^i \Gamma^l_{kl}, \tag{7.85a}$$

$$\overset{2}{\nabla}_k \tilde{v}^i = \frac{\partial \tilde{v}^i}{\partial x^k} + \tilde{v}^l \Gamma^i_{lk} - w \tilde{v}^i \Gamma^l_{kl}, \tag{7.85b}$$

$$\overset{3}{\nabla}_k \tilde{v}^i = \frac{\partial \tilde{v}^i}{\partial x^k} + \tilde{v}^l \Gamma^i_{kl} - w \tilde{v}^i \Gamma^l_{lk}, \tag{7.85c}$$

$$\overset{4}{\nabla}_k \tilde{v}^i = \frac{\partial \tilde{v}^i}{\partial x^k} + \tilde{v}^l \Gamma^i_{lk} - w \tilde{v}^i \Gamma^l_{lk}, \tag{7.85d}$$

which are related by the equations,

$$\overset{2}{\nabla}_k \tilde{v}^i = \overset{1}{\nabla}_k \tilde{v}^i - 2 \tilde{v}^l S^i_{kl}, \tag{7.86a}$$

$$\overset{3}{\nabla}_k \tilde{v}^i = \overset{1}{\nabla}_k \tilde{v}^i + 2 w \tilde{v}^i \Gamma^l_{kl}, \tag{7.86b}$$

$$\overset{4}{\nabla}_k \tilde{v}^i = \overset{1}{\nabla}_k \tilde{v}^i - 2 \tilde{v}^l S^i_{kl} + 2 w \tilde{v}^i \Gamma^l_{kl}. \tag{7.86c}$$

An akinetor of valence m has twice as many distinct covariant derivatives as a tensor of valence m. The covariant derivatives of akinetors differ from the covariant derivatives of tensors of the same valence by a single additive term, which is the product of the weight of the akinetor, the akinetor itself, and a contracted linear connection Γ^l_{kl} or Γ^l_{lk}.

7.09. Summary

One might expect that differentiation of any akinetor would yield an akinetor. With a few exceptions, however, such is not the case. Differentiation of an akinetor usually results in an additive term that contains a second derivative.

To obtain derivatives that are akinetors, we introduce the linear connection Γ^i_{jk}, which is defined by its transformation equation

$$\Gamma^i_{jk} = \frac{\partial x^i}{\partial x^{i'}} \frac{\partial x^{j'}}{\partial x^j} \frac{\partial x^{k'}}{\partial x^k} \Gamma^{i'}_{j'k'} + \frac{\partial x^i}{\partial x^{i'}} \frac{\partial^2 x^{i'}}{\partial x^j \partial x^k}. \tag{7.07}$$

Though the linear connection is not a tensor, it can be combined with the additive second derivative terms to give a family of tensor derivatives. For example, we have the covariant derivative

$$\overset{1}{\nabla}_j v^k = \frac{\partial x^k}{\partial x^j} + v^l \Gamma^k_{jl}, \tag{7.20a}$$

which is a tensor.

The literature makes considerable use of this tensor where Γ^i_{jk} is symmetric in the lower indices. Recently the covariant derivatives have been expanded to include also the unsymmetric case, which gives

$$\overset{2}{\nabla}_j v^k = \frac{\partial x^k}{\partial x^j} + v^l \Gamma^k_{lj}. \tag{7.20b}$$

The chapter concludes with examples of covariant derivatives of various tensors and akinetors.

Problems

7.01 By combination of derivatives of the form $\partial v^i/\partial x^j$, attempt to obtain a tensor. Explain why your result is less satisfactory than with $\partial u_i/\partial x^j$.

7.02 Repeat Problem 7.01 for $\partial w^{ij}/\partial x^k$.

7.03

(a) By substitution, prove that if

$$\Gamma^i_{jk} = \frac{\partial x^i}{\partial x^{i'}} \frac{\partial x^{j'}}{\partial x^j} \frac{\partial x^{k'}}{\partial x^k} \Gamma^{i'}_{j'k'} + \frac{\partial^2 x^{i'}}{\partial x^j \partial x^k} \frac{\partial x^i}{\partial x^{i'}},$$

then

$$\Gamma^{i'}_{j'k'} = \frac{\partial x^{i'}}{\partial x^i} \frac{\partial x^j}{\partial x^{j'}} \frac{\partial x^k}{\partial x^{k'}} \Gamma^i_{jk} + \frac{\partial^2 x^i}{\partial x^{j'} \partial x^{k'}} \frac{\partial x^{i'}}{\partial x^i}.$$

(b) Write the transformation equations for Γ^j_{ki} and Γ^k_{ij}.

7.04 Derive Equation (7.29) from the product $u_i v^i = S$ and Equation (7.20).

7.05 Suppose that $\overset{S}{\nabla}_k w^{ij}$ is to be obtained from the product

$$w^{ij} u_i = v^j$$

and Equations (7.20) and (7.29). Derive the general expressions for $\overset{S}{\nabla}_k w^{ij}$ and compare with Equation (7.35).

7.06 Derive Equation (7.36).

7.07 Derive Equation (7.37).

7.08 Derive Equation (7.38).

7.09 Derive Equation (7.39).

7.10 Derive Equation (7.40).

7.11 Derive an equation for all of the covariant derivatives $\overset{s}{\nabla}_l v^i_{\ jk}$, where $v^i_{\ jk}$ is a tensor.

7.12 Find all of the linear relations between the covariant derivatives of Problem 7.11.

7.13 Derive equations for all of the covariant derivatives $\overset{s}{\nabla}_l v^{ijk}$, where v^{ijk} is a tensor.

7.14 Find all of the linear relations between the covariant derivatives of Problem 7.13.

7.15 A linear connection is given by the matrices

$$\Gamma^1_{jk} = \begin{bmatrix} 1 & 2 \\ -1 & 0 \end{bmatrix}, \qquad \Gamma^2_{jk} = \begin{bmatrix} 0 & 1 \\ 0 & 0 \end{bmatrix}.$$

(a) Find the torsion tensor $S^i_{\ jk}$.
(b) Find $\overset{1}{\nabla}_j v^k$ and $\overset{2}{\nabla}_j v^k$.

7.16 A linear connection is defined by the equations

$$\Gamma^1_{jk} = \Delta \begin{bmatrix} A & B \\ -B & A \end{bmatrix}, \qquad \Gamma^2_{jk} = -\Delta \begin{bmatrix} B & -A \\ A & B \end{bmatrix}.$$

(a) Find the torsion tensor $S^i_{\ jk}$.
(b) Find $\overset{1}{\nabla}_j u_k$ and $\overset{2}{\nabla}_j u_k$.

7.17 For the linear connection of Problem 7.15, find
(a) $\overset{3}{\nabla}_k T^{ij}$ and
(b) $\overset{3}{\nabla}_k T^i_{\ j}$.

7.18 For the linear connection of Problem 7.16, find
(a) $\overset{4}{\nabla}_k T_{ij}$ and
(b) $\overset{4}{\nabla}_k T^{ij}$.

7.19 Find the contracted covariant derivatives of a trivalent tensor $T^i_{\ jk}$.

7.20 Find all of the contracted covariant derivatives of a quadrivalent tensor $T^i_{\ jkl}$.

7.21 Find the acceleration tensor a^i in a region in which the linear connection is that given in Problem 7.15.

7.22 For the linear connection of Problem 7.15,
(a) evaluate $\Gamma^i_{\ ki}$ and $\Gamma^i_{\ ik}$, and
(b) find $\overset{1}{\nabla}_k \tilde{S}$ and $\overset{2}{\nabla}_k \tilde{S}$.

7.23 For the linear connection of Problem 7.16, find
(a) $\Gamma^i_{\ ki}$,
(b) $\Gamma^i_{\ ik}$, and
(c) all of the covariant derivatives $\overset{s}{\nabla}_k \tilde{v}^i$.

8

The Riemann–Christoffel tensors

This chapter deals with two important questions whose answer can be expressed neatly in terms of two quadrivalent tensors. The first of these, R^l_{ijk}, has its roots in the work of Riemann and Christoffel in nineteenth-century differential geometry.[1] The second, T^l_{ijk}, has been introduced only recently[2] but is essential to a general treatment of the subject.

The two questions are:

(1) Does $\overset{(S)}{\nabla}_i \overset{(T)}{\nabla}_j \tilde{U}^{lm\cdots n}_{pq\cdots r} = \overset{(S)}{\nabla}_j \overset{(T)}{\nabla}_i \tilde{U}^{lm\cdots n}_{pq\cdots r}$?

(2) If the covariant derivative of an akinetor $\tilde{U}^{lm\cdots n}_{pq\cdots r}$ is equated to zero,

$$\overset{(S)}{\nabla}_i \tilde{U}^{lm\cdots n}_{pq\cdots r} = 0$$

and if the akinetor is known at a single point, is it possible to determine $\tilde{U}^{lm\cdots n}_{pq\cdots r}$ uniquely at any other point by integration?

8.01. Second covariant derivatives of scalars

In Chapter 7, it was shown that the covariant derivatives of all tensors transform as tensors. Therefore, second and higher covariant derivatives of all tensors will also transform as tensors.

An interesting question is how the various possible second derivatives of tensors are related to one another. We know[3] that

$$\frac{\partial^2 \phi}{\partial x^i \partial x^j} = \frac{\partial^2 \phi}{\partial x^j \partial x^i}. \tag{8.01}$$

If instead of the partial derivatives of a scalar we consider the covariant derivatives of a scalar, does the same simple equation hold true? According to Equation (7.23), there is only one covariant derivative of a scalar

$$\nabla_j \phi = \frac{\partial \phi}{\partial x^j} \tag{7.23}$$

that transforms as a covariant vector. Since there are two possible covariant derivatives of a covariant vector, Equation (7.28),

$$\overset{1}{\nabla}_i \nabla_j \phi = \frac{\partial^2 \phi}{\partial x^i \partial x^j} - \frac{\partial \phi}{\partial x^l} \Gamma^l_{ij}, \tag{8.02a}$$

$$\overset{2}{\nabla}_i \nabla_j \phi = \frac{\partial^2 \phi}{\partial x^i \partial x^j} - \frac{\partial \phi}{\partial x^l} \Gamma^l_{ji}. \tag{8.02b}$$

Interchanging indices i and j in Equation (8.02),

$$\overset{1}{\nabla}_j \nabla_i \phi = \frac{\partial^2 \phi}{\partial x^j \partial x^i} - \frac{\partial \phi}{\partial x^l} \Gamma^l_{ji}, \tag{8.03a}$$

$$\overset{2}{\nabla}_j \nabla_i \phi = \frac{\partial^2 \phi}{\partial x^i \partial x^j} - \frac{\partial \phi}{\partial x^l} \Gamma^l_{ij}. \tag{8.03b}$$

Subtraction of Equations (8.02) and (8.03) gives

$$\overset{1}{\nabla}_{[i} \nabla_{j]} \phi = \overset{1}{\nabla}_i \nabla_j \phi - \overset{1}{\nabla}_j \nabla_i \phi = 2 \frac{\partial \phi}{\partial x^l} S^l_{ji}, \tag{8.04a}$$

$$\overset{2}{\nabla}_{[i} \nabla_{j]} \phi = \overset{2}{\nabla}_i \nabla_j \phi - \overset{2}{\nabla}_j \nabla_i \phi = 2 \frac{\partial \phi}{\partial x^l} S^l_{ij}. \tag{8.04b}$$

Thus, the necessary and sufficient condition for the equality of the two second covariant derivatives of a scalar is the vanishing of the torsion tensor S^l_{ij}. If the linear connection is symmetric and the torsion tensor S^l_{ij} is zero, then

$$\nabla_{[i} \nabla_{j]} \phi = 0, \tag{8.05}$$

which is equivalent to Equation (8.01).

8.02. Second covariant derivatives of contravariant vectors

Now consider the extension to the second covariant derivatives of a contravariant vector v^k. By Equation (7.20) there are two covariant derivatives of a contravariant vector,

$$\overset{1}{\nabla}_j v^k = \frac{\partial v^k}{\partial x^j} + v^l \Gamma^k_{jl} \tag{7.20a}$$

and

$$\overset{2}{\nabla}_j v^k = \frac{\partial v^k}{\partial x^j} + v^l \Gamma^k_{lj}, \tag{7.20b}$$

which transform as mixed bivalent tensors. According to Equation (7.39), there are four possible covariant derivatives of a mixed bivalent tensor T^k_j:

$$\overset{1}{\nabla}_i T^k_j = \frac{\partial T^k_j}{\partial x^i} + T^l_j \Gamma^k_{il} - T^k_l \Gamma^l_{ij}, \tag{7.39a}$$

$$\overset{2}{\nabla}_i T^k_j = \frac{\partial T^k_j}{\partial x^i} + T^l_j \Gamma^k_{il} - T^k_l \Gamma^l_{ji}, \tag{7.39b}$$

$$\overset{3}{\nabla}_i T^k_j = \frac{\partial T^k_j}{\partial x^i} + T^l_j \Gamma^k_{li} - T^k_l \Gamma^l_{ij}, \tag{7.39c}$$

$$\overset{4}{\nabla}_i T^k_j = \frac{\partial T^k_j}{\partial x^i} + T^l_j \Gamma^k_{li} - T^k_l \Gamma^l_{ji}. \tag{7.39d}$$

Substitution of Equation (7.20a) into (7.39a) gives

$$\overset{1}{\nabla}_i \overset{1}{\nabla}_j v^k = \frac{\partial^2 v^k}{\partial x^i \partial x^j} + \frac{\partial v^l}{\partial x^i}\Gamma^k_{jl} + v^l \frac{\partial \Gamma^k_{jl}}{\partial x^i} + \frac{\partial v^l}{\partial x^j}\Gamma^k_{il} + v^l \Gamma^m_{jl}\Gamma^k_{im} - \frac{\partial v^k}{\partial x^l}\Gamma^l_{ij} - v^l \Gamma^k_{lm}\Gamma^m_{ij}. \tag{8.06}$$

Now interchange indices i and j in Equation (8.06):

$$\overset{1}{\nabla}_j \overset{1}{\nabla}_i v^k = \frac{\partial^2 v^k}{\partial x^j \partial x^i} + \frac{\partial v^l}{\partial x^j}\Gamma^k_{il} + v^l \frac{\partial \Gamma^k_{il}}{\partial x^j} + \frac{\partial v^l}{\partial x^i}\Gamma^k_{jl} + v^l \Gamma^m_{il}\Gamma^k_{jm} - \frac{\partial v^k}{\partial x^l}\Gamma^l_{ji} - v^l \Gamma^k_{lm}\Gamma^m_{ji}. \tag{8.07}$$

Subtraction of Equation (8.07) from (8.06) gives

$$\overset{1}{\nabla}_{[i} \overset{1}{\nabla}_{j]} v^k = 2(\overset{1}{\nabla}_l v^k) S^l_{ji} + v^l R^k_{ijl} \tag{8.08a}$$

where the first Riemann–Christoffel tensor (the one that occurs in classical differential geometry[4]) is defined by the equation

$$R^k_{ijl} = \frac{\partial \Gamma^k_{jl}}{\partial x^i} - \frac{\partial \Gamma^k_{il}}{\partial x^j} + \Gamma^m_{jl}\Gamma^k_{im} - \Gamma^m_{il}\Gamma^k_{jm}. \tag{8.09a}$$

Similarly, from direct expansion of Equations (7.20) and (7.39), it can readily be shown that

$$\overset{2}{\nabla}_{[i}\, \overset{1}{\nabla}_{j]}\, v^k = 2(\overset{1}{\nabla}_l v^k) S^l_{ij} + v^l R^k_{ijl}, \tag{8.08b}$$

$$\overset{3}{\nabla}_{[i}\, \overset{1}{\nabla}_{j]}\, v^k = 2[(\overset{1}{\nabla}_i v^l) S^k_{jl} + (\overset{1}{\nabla}_j v^l) S^k_{li} + (\overset{1}{\nabla}_l v^k) S^l_{ji}] + v^l R^k_{ijl}, \tag{8.08c}$$

$$\overset{4}{\nabla}_{[i}\, \overset{1}{\nabla}_{j]}\, v^k = 2[(\overset{1}{\nabla}_i v^l) S^k_{jl} + (\overset{1}{\nabla}_j v^l) S^k_{li} + (\overset{1}{\nabla}_l v^k) S^l_{ij}] + v^l R^k_{ijl}. \tag{8.08d}$$

So the differences of these various second covariant derivatives of a contravariant vector can always be expressed as a linear combination of the torsion tensor S^l_{ij} and the first Riemann–Christoffel tensor R^k_{ijl}.

However, if we consider those second covariant derivatives of a contravariant vector in which we start with the second kind of covariant derivative of a contravariant vector, it becomes necessary to introduce a second type of Riemann–Christoffel tensor[2] T^k_{ijl},

$$T^k_{ijl} = \frac{\partial \Gamma^k_{lj}}{\partial x^i} - \frac{\partial \Gamma^k_{li}}{\partial x^j} + \Gamma^m_{lj}\Gamma^k_{mi} - \Gamma^m_{li}\Gamma^k_{mj}. \tag{8.09b}$$

In terms of this tensor, direct expansion of Equations (7.20) and (7.39) gives

$$\overset{1}{\nabla}_{[i}\, \overset{2}{\nabla}_{j]}\, v^k = 2[(\overset{2}{\nabla}_i v^l) S^k_{lj} + (\overset{2}{\nabla}_j v^l) S^k_{il} + (\overset{2}{\nabla}_l v^k) S^l_{ji}] + v^l T^k_{ijl}, \tag{8.10a}$$

$$\overset{2}{\nabla}_{[i}\, \overset{2}{\nabla}_{j]}\, v^k = 2[(\overset{2}{\nabla}_i v^l) S^k_{lj} + (\overset{2}{\nabla}_j v^l) S^k_{il} + (\overset{2}{\nabla}_l v^k) S^l_{ij}] + v^l T^k_{ijl}, \tag{8.10b}$$

$$\overset{3}{\nabla}_{[i}\, \overset{2}{\nabla}_{j]}\, v^k = 2(\overset{2}{\nabla}_l v^k) S^l_{ji} + v^l T^k_{ijl}, \tag{8.10c}$$

$$\overset{4}{\nabla}_{[i}\, \overset{2}{\nabla}_{j]}\, v^k = 2(\overset{2}{\nabla}_l v^k) S^l_{ij} + v^l T^k_{ijl}. \tag{8.10d}$$

In the important special case of a symmetric linear connection,

$$R^k_{ijl} = T^k_{ijl} \tag{8.11}$$

and Equations (8.08) and (8.10) reduce to

$$\nabla_{[i}\, \nabla_{j]}\, v^k = v^l R^k_{ijl}. \tag{8.12}$$

If the Riemann–Christoffel tensor R^i_{jkl} vanishes,

$$\nabla_{[i}\, \nabla_{j]}\, v^k = 0$$

or

$$\nabla_i \nabla_j v^k = \nabla_j \nabla_i v^k \tag{8.13}$$

and the covariant derivative of a contravariant vector becomes commutative.

8.03. Second covariant derivatives of covariant vectors

Similar results can also be obtained for a covariant vector u_k by direct expansion of Equations (7.28) and (7.37).

$$\overset{1}{\nabla}_{[i}\overset{1}{\nabla}_{j]}u_k=2(\overset{1}{\nabla}_l u_k)S^l_{ji}-u_l R^l_{ijk}, \tag{8.14a}$$

$$\overset{2}{\nabla}_{[i}\overset{1}{\nabla}_{j]}u_k=2[(\overset{1}{\nabla}_i u_l)S^l_{kj}+(\overset{1}{\nabla}_j u_l)S^l_{ik}+(\overset{1}{\nabla}_l u_k)S^l_{ji}]-u_l R^l_{ijk}, \tag{8.14b}$$

$$\overset{3}{\nabla}_{[i}\overset{1}{\nabla}_{j]}u_k=2(\overset{1}{\nabla}_l u_k)S^l_{ij}-u_l R^l_{ijk}, \tag{8.14c}$$

$$\overset{4}{\nabla}_{[i}\overset{1}{\nabla}_{j]}u_k=2[(\overset{1}{\nabla}_i u_l)S^l_{kj}+(\overset{1}{\nabla}_j u_l)S^l_{ik}+(\overset{1}{\nabla}_l u_k)S^l_{ij}]-u_l R^l_{ijk}, \tag{8.14d}$$

and

$$\overset{1}{\nabla}_{[i}\overset{2}{\nabla}_{j]}u_k=2[(\overset{2}{\nabla}_i u_l)S^l_{jk}+(\overset{2}{\nabla}_j u_l)S^l_{ki}+(\overset{2}{\nabla}_l u_k)S^l_{ji}]-u_l T^l_{ijk}, \tag{8.15a}$$

$$\overset{2}{\nabla}_{[i}\overset{2}{\nabla}_{j]}u_k=2(\overset{2}{\nabla}_l u_k)S^l_{ji}-u_l T^l_{ijk}, \tag{8.15b}$$

$$\overset{3}{\nabla}_{[i}\overset{2}{\nabla}_{j]}u_k=2[(\overset{2}{\nabla}_i u_l)S^l_{jk}+(\overset{2}{\nabla}_j u_l)S^l_{ki}+(\overset{2}{\nabla}_l u_k)S^l_{ij}]-u_l T^l_{ijk}, \tag{8.15c}$$

$$\overset{4}{\nabla}_{[i}\overset{2}{\nabla}_{j]}u_k=2(\overset{2}{\nabla}_l u_k)S^l_{ij}-u_l T'^{\,l}_{ijk}. \tag{8.15d}$$

If the linear connection is symmetric, Equations (8.14) and (8.15) reduce to

$$\nabla_{[i}\nabla_{j]}u_k=-u_l R^l_{ijk}. \tag{8.16}$$

If the Riemann–Christoffel tensor R^l_{ijk} vanishes, then

$$\nabla_{[i}\nabla_{j]}u_k=0$$

or

$$\nabla_i\nabla_j u_k=\nabla_j\nabla_i u_k. \tag{8.17}$$

8.04. Second covariant derivatives of tensors

The foregoing results can readily be extended to tensors of any valence. For example,

$$\overset{1}{\nabla}_{[i}\overset{1}{\nabla}_{j]}U^k=2(\overset{1}{\nabla}_m U^k_l)S^m_{ji}+U^m_l R^k_{ijm}-U^k_m R^m_{ijl}. \tag{8.18}$$

In general, the expansion for

$$\overset{p}{\nabla}_{[i}\overset{q}{\nabla}_{j]}U^{kl\cdots m}_{pq\cdots r}\neq 0 \tag{8.19}$$

and contains three types of terms:

(1) contracted products of $\overset{q}{\nabla}_j U^{kl\cdots m}_{pq\cdots r}$ and the torsion tensor S^{u}_{vw};

(2) a set of positive products of $U^{kl\cdots m}_{pq\cdots r}$ and the Riemann–Christoffel tensor in which contraction takes place successively on each contravariant index of the tensor and the last subscript of the Riemann–Christoffel tensors; and

(3) a set of negative products of $U^{kl\cdots m}_{pq\cdots r}$ and the Riemann–Christoffel tensors in which contraction takes place successively on each subscript of the tensor and the superscript of the Riemann–Christoffel tensors.

If the linear connection is symmetric, these equations reduce to the relatively simple equation

$$\nabla_{[i}\nabla_{j]} U^{kl\cdots m}_{pq\cdots r} = U^{sl\cdots m}_{pq\cdots r} R^{k}_{ijs} + U^{ks\cdots m}_{pq\cdots r} R^{l}_{ijs} + \cdots + U^{kl\cdots s}_{pq\cdots r} R^{m}_{ijs} - U^{kl\cdots m}_{sq\cdots r} R^{s}_{ijp} - U^{kl\cdots m}_{ps\cdots r} R^{s}_{ijq} - \cdots - U^{kl\cdots m}_{pq\cdots s} R^{s}_{ijr}. \qquad (8.20)$$

If the Riemann–Christoffel tensor vanishes, then

$$\nabla_{[i}\nabla_{j]} U^{kl\cdots m}_{pq\cdots r} = 0. \qquad (8.21)$$

Thus, it is possible to draw very general conclusions for all types of tensors:

Theorem XXVIII. *If the torsion tensor* $S^{i}_{jk} \not\equiv 0$,

$$\overset{s}{\nabla}_i \overset{t}{\nabla}_j U^{kl\cdots m}_{pq\cdots r} \not\equiv \overset{s}{\nabla}_j \overset{t}{\nabla}_i U^{kl\cdots m}_{pq\cdots r}$$

for any arbitrary tensor $U^{kl\cdots m}_{pq\cdots r}$ *of any valence.*

Theorem XXIX. *If the torsion tensor* $S^{i}_{jk} \equiv 0$,

(A)
$$\nabla_i \nabla_j \phi = \nabla_j \nabla_i \phi$$

for any scalar ϕ,

(B)
$$\nabla_i \nabla_j U^{kl\cdots m}_{pq\cdots r} \not\equiv \nabla_j \nabla_i U^{kl\cdots m}_{pq\cdots r}$$

for an arbitrary tensor $U^{kl\cdots m}_{pq\cdots r}$ *of valence greater than or equal to* 1.

Theorem XXX. *If the torsion tensor* $S^{i}_{jk} \equiv 0$ *and the Riemann–Christoffel tensor* $R^{l}_{ijk} \equiv 0$, *then*

$$\nabla_i \nabla_j U^{kl\cdots m}_{pq\cdots r} = \nabla_j \nabla_i U^{kl\cdots m}_{pq\cdots r}$$

for all tensors $U^{kl\cdots m}_{pq\cdots r}$ *of any valence.*

8.05. Second covariant derivatives of nilvalent akinetors

Next let us extend these considerations to a nilvalent akinetor $\tilde{S}$ of weight w. By Equation (7.79) there are two possible covariant derivatives[5] of $\tilde{S}$,

$$\overset{1}{\nabla}_j \tilde{S} = \frac{\partial \tilde{S}}{\partial x^j} - w\tilde{S}\Gamma^l_{jl}, \tag{7.79a}$$

and

$$\overset{2}{\nabla}_j \tilde{S} = \frac{\partial \tilde{S}}{\partial x^j} - w\tilde{S}\Gamma^l_{lj}, \tag{7.79b}$$

both of which transform as covariant akinetors of valence 1 and weight 1. By Section 7.08 there are four covariant derivatives of a covariant akinetor u_j:

$$\overset{1}{\nabla}_i \tilde{u}_j = \frac{\partial \tilde{u}_j}{\partial x^i} - \tilde{u}_m \Gamma^m_{ij} - w\tilde{u}_j \Gamma^m_{im}, \tag{8.22a}$$

$$\overset{2}{\nabla}_i \tilde{u}_j = \frac{\partial \tilde{u}_j}{\partial x^i} - \tilde{u}_m \Gamma^m_{ji} - w\tilde{u}_j \Gamma^m_{im}, \tag{8.22b}$$

$$\overset{3}{\nabla}_i \tilde{u}_j = \frac{\partial \tilde{u}_j}{\partial x^i} - \tilde{u}_m \Gamma^m_{ij} - w\tilde{u}_j \Gamma^m_{mi}, \tag{8.22c}$$

$$\overset{4}{\nabla}_i \tilde{u}_j = \frac{\partial \tilde{u}_j}{\partial x^i} - \tilde{u}_m \Gamma^m_{ji} - w\tilde{u}_j \Gamma^m_{mi}. \tag{8.22d}$$

Substitution of Equation (7.79) into (8.22) gives

$$\overset{1}{\nabla}_{[i} \overset{1}{\nabla}_{j]} \tilde{S} = 2(\overset{1}{\nabla}_l \tilde{S}) S^l_{ji} - w\tilde{S}R_{ij}, \tag{8.23a}$$

$$\overset{2}{\nabla}_{[i} \overset{1}{\nabla}_{j]} \tilde{S} = 2(\overset{1}{\nabla}_l \tilde{S}) S^l_{ij} - w\tilde{S}R_{ij}, \tag{8.23b}$$

$$\overset{3}{\nabla}_{[i} \overset{1}{\nabla}_{j]} \tilde{S} = 2w[(\overset{1}{\nabla}_i \tilde{S}) S^l_{lj} + (\overset{1}{\nabla}_j \tilde{S}) S^l_{il}] + 2(\overset{1}{\nabla}_l \tilde{S}) S^l_{ji} - w\tilde{S}R_{ij}, \tag{8.23c}$$

$$\overset{4}{\nabla}_{[i} \overset{1}{\nabla}_{j]} \tilde{S} = 2w[(\overset{1}{\nabla}_i \tilde{S}) S^l_{lj} + (\overset{1}{\nabla}_j \tilde{S}) S^l_{il}] + 2(\overset{1}{\nabla}_l \tilde{S}) S^l_{ij} - w\tilde{S}R_{ij}, \tag{8.23d}$$

where

$$R_{ij} = R^l_{ijl} = \frac{\partial \Gamma^l_{jl}}{\partial x^i} - \frac{\partial \Gamma^l_{il}}{\partial x^j}. \tag{8.24}$$

The contracted Riemann-Christoffel tensor of the first kind is called[5] the Ricci tensor of the first kind. Also,

$$\overset{1}{\nabla}_{[i} \overset{2}{\nabla}_{j]} \tilde{S} = 2w[(\overset{2}{\nabla}_i \tilde{S}) S^l_{jl} + (\overset{2}{\nabla}_j \tilde{S}) S^l_{li}] + 2(\overset{2}{\nabla}_l \tilde{S}) S^l_{ji} - w\tilde{S}T_{ij}, \tag{8.25a}$$

$$\overset{2}{\nabla}_{[i} \overset{2}{\nabla}_{j]} \tilde{S} = 2w[(\overset{2}{\nabla}_i \tilde{S}) S^l_{jl} + (\overset{2}{\nabla}_j \tilde{S}) S^l_{li}] + 2(\overset{2}{\nabla}_l \tilde{S}) S^l_{ij} - w\tilde{S}T_{ij}, \tag{8.25b}$$

$$\overset{3}{\nabla}_{[i} \overset{2}{\nabla}_{j]} \tilde{S} = 2(\overset{2}{\nabla}_l \tilde{S}) S^l_{ji} - w\tilde{S}T_{ij}, \tag{8.25c}$$

$$\overset{4}{\nabla}_{[i} \overset{2}{\nabla}_{j]} \tilde{S} = 2(\overset{2}{\nabla}_l \tilde{S}) S^l_{ij} - w\tilde{S}T_{ij}, \tag{8.25d}$$

where

$$T_{ij} = T^l_{ijl} = \frac{\partial \Gamma^l_{lj}}{\partial x^i} - \frac{\partial \Gamma^l_{li}}{\partial x^j}. \tag{8.26}$$

The contracted Riemann–Christoffel tensor of the second kind is called[5] the Ricci tensor of the second kind.

If the linear connection is symmetric, $S^i_{jk} = 0$ and

$$R_{ij} = T_{ij}, \tag{8.27}$$

so Equations (8.23) and (8.25) reduce to the single equation

$$\nabla_{[i} \nabla_{j]} \tilde{S} = -w\tilde{S}R_{ij}. \tag{8.28}$$

If the linear connection is symmetric and the Ricci tensor $R_{ij} \equiv 0$, then

$$\nabla_{[i} \nabla_{j]} \tilde{S} = 0$$

or

$$\nabla_i \nabla_j \tilde{S} = \nabla_j \nabla_i \tilde{S}. \tag{8.29}$$

8.06. Second covariant derivatives of akinetors

The procedure can readily be extended to akinetors of higher valence. For example,

$$\overset{1}{\nabla}_{[i} \overset{1}{\nabla}_{j]} \tilde{v}^k = 2(\nabla_l \tilde{v}^k) S^l_{ji} + \tilde{v}^l R^k_{ijl} - w\tilde{v}^k R_{ij}. \tag{8.30}$$

The akinetor expressions for $\overset{s}{\nabla}_{[i} \overset{t}{\nabla}_{j]} \tilde{U}^{kl\cdots m}_{pq\cdots r}$ differ from the corresponding expressions for a tensor of the same valence only in the presence of a subtractive term containing the product of the weight of the akinetor, the akinetor itself, and one of the Ricci tensors.

For a symmetric linear connection and an akinetor of weight w, $\tilde{U}^{kl\cdots m}_{pq\cdots n}$,

$$\begin{aligned} \nabla_{[i} \nabla_{j]} \tilde{U}^{kl\cdots m}_{pq\cdots r} = {} & \tilde{U}^{sl\cdots m}_{pq\cdots r} R^k_{ijs} + \tilde{U}^{ks\cdots m}_{pq\cdots r} R^l_{ijs} + \cdots + \tilde{U}^{kl\cdots s}_{pq\cdots r} R^m_{ijs} \\ & - \tilde{U}^{kl\cdots m}_{sq\cdots r} R^s_{ijp} - \tilde{U}^{kl\cdots m}_{ps\cdots r} R^s_{ijq} - \cdots \\ & - \tilde{U}^{kl\cdots m}_{pq\cdots s} R^s_{ijr} - w\tilde{U}^{kl\cdots m}_{pq\cdots r} R_{ij}. \end{aligned} \tag{8.31}$$

If the Ricci tensor vanishes, then Equation (8.31) has exactly the same number of terms as Equation (8.20), and if the Riemann–Christoffel tensor and the Ricci tensor both vanish,

$$\nabla_{[i}\nabla_{j]}\tilde{U}^{kl\cdots m}_{pq\cdots r}=0. \tag{8.32}$$

The final conclusions for second covariant derivatives of akinetors may be summarized[5] in four theorems:

Theorem XXXI. *If the torsion tensor* $S^{i}_{\ jk}\not\equiv 0$,

$$\overset{(s)}{\nabla}_i\overset{(t)}{\nabla}_j\tilde{U}^{kl\cdots m}_{pq\cdots r}\not\equiv\overset{(s)}{\nabla}_j\overset{(t)}{\nabla}_i\tilde{U}^{kl\cdots m}_{pq\cdots r}$$

for any arbitrary akinetor $U^{kl\cdots m}_{pq\cdots r}$ *of any weight and any valence.*

Theorem XXXII. *If the torsion tensor* $S^{i}_{\ jk}\equiv 0$,

(A) $$\nabla_i\nabla_j\phi=\nabla_j\nabla_i\phi$$

for any scalar ϕ,

(B) $$\nabla_i\nabla_j\tilde{U}^{kl\cdots m}_{pq\cdots r}\not\equiv\nabla_j\nabla_i\tilde{U}^{kl\cdots m}_{pq\cdots r}$$

for any arbitrary akinetor of weight or valence greater than 0.

Theorem XXXIII. *If the torsion tensor* $S^{i}_{\ jk}\equiv 0$ *and the Ricci tensor* $R_{ij}\equiv 0$,

(A) $$\nabla_i\nabla_j\phi=\nabla_j\nabla_i\phi$$

for any scalar ϕ,

(B) $$\nabla_i\nabla_j\tilde{S}=\nabla_j\nabla_i\tilde{S}$$

for any nilvalent akinetor $\tilde{S}$ *of weight* w,

(C) $$\nabla_i\nabla_j\tilde{U}^{kl\cdots m}_{pq\cdots r}\not\equiv\nabla_j\nabla_i\tilde{U}^{kl\cdots m}_{pq\cdots r}$$

for any arbitrary akinetor $\tilde{U}^{kl\cdots m}_{pq\cdots r}$ *of valence greater than or equal to* 1.

Theorem XXXIV. *If the torsion tensor* $S^{i}_{\ jk}\equiv 0$, *the Ricci tensor* $R_{ij}\equiv 0$, *and the Riemann–Christoffel tensor* $R^{i}_{\ jkl}\equiv 0$, *then*

$$\nabla_i\nabla_j\tilde{U}^{kl\cdots m}_{pq\cdots r}=\nabla_j\nabla_i\tilde{U}^{kl\cdots m}_{pq\cdots r}$$

for all akinetors $\tilde{U}^{kl\cdots m}_{pq\cdots r}$ *of any valence.*

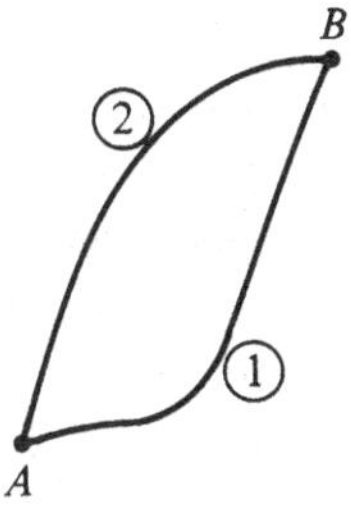

Figure 8.01. Line integral from A to B.

8.07. Integrability of scalar fields

Consider a scalar field $\phi(x^i)$ that is defined at every point of space. There is only one covariant derivative of this scalar:

$$f_i = \nabla_i \phi = \frac{\partial \phi}{\partial x^i}. \tag{8.33}$$

This covariant derivative of ϕ defines a covariant vector field f_i at every point.

Suppose that the covariant vector f_i is known at every point of space and the value of the scalar potential $(\phi)_A$ is known at a single point A. Is it possible to determine the scalar potential uniquely at every other point of space?

The potential $(\phi)_B$ at any other point B might be defined by the equation

$$(\phi)_B = (\phi)_A + \int_A^B \frac{\partial \phi}{\partial x^i}\, dx^i = (\phi)_A + \int_A^B f_i\, dx^i. \tag{8.34}$$

The expression $\int_A^B f_i\, dx^i$ is a line integral. Its value depends on the end points A and B and on the path along which the integral is evaluated (Figure 8.01).

However, there is one important special case in which the integral is independent of path. If the same result is obtained for all possible paths joining A and B, then the integral will be uniquely defined. Consider any two paths joining points A and B in Figure 8.01. If the integral is independent of path

$$\int_{A\,①}^{B} f_i\, dx^i = \int_{A\,②}^{B} f_i\, dx^i = -\int_{B\,②}^{A} f_i\, dx^i$$

or

$$\oint_C f_i\, dx^i = 0, \tag{8.35}$$

where C is a closed path containing points A and B. Therefore, if Equation (8.34) can be used to define $(\phi)_B$ uniquely, then the line integral of f_i about every closed path in the region must vanish.

But according to the generalized Stokes theorem [Equation (5.48)],

$$\oint_C f_i\, dx^i = \int_{\tilde{\alpha}} \delta_{123}^{ijk} \frac{\partial f_k}{\partial x^j}\, d\tilde{\alpha}_i, \tag{8.36}$$

where $\tilde{\alpha}$ is any pseudoarea bounded by the closed path C. If the line integral is to vanish for all possible paths in the region under consideration, then the integrand of the right side of Equation (8.36) must also vanish throughout the region,

$$\delta_{123}^{ijk} \frac{\partial f_k}{\partial x^j} = 0. \tag{8.37}$$

This is the integrability condition for a covariant vector field. Expansion of Equation (8.37) gives

$$\frac{\partial f_2}{\partial x^3} - \frac{\partial f_3}{\partial x^2} = 0, \qquad \frac{\partial f_1}{\partial x^3} - \frac{\partial f_3}{\partial x^1} = 0, \qquad \frac{\partial f_2}{\partial x^1} - \frac{\partial f_1}{\partial x^2} = 0. \tag{8.37a}$$

An arbitrary covariant vector field f_i does not in general satisfy this condition.

However, if the covariant vector f_i is defined as the gradient of a scalar [Equation (8.33)], then Equation (8.37a) becomes

$$\frac{\partial^2 \phi}{\partial x^3 \partial x^2} - \frac{\partial^2 \phi}{\partial x^2 \partial x^3} = 0, \qquad \frac{\partial^2 \phi}{\partial x^3 \partial x^1} - \frac{\partial^2 \phi}{\partial x^1 \partial x^3} = 0,$$

$$\frac{\partial^2 \phi}{\partial x^1 \partial x^2} - \frac{\partial^2 \phi}{\partial x^2 \partial x^1} = 0, \tag{8.37b}$$

which is satisfied identically[6] for all scalars ϕ.

To summarize:

Theorem XXXV. *If a covariant vector field is defined as the gradient of a scalar potential* ϕ,

$$f_i = \nabla_i \phi = \frac{\partial \phi}{\partial x^i},$$

and the scalar potential $(\phi)_A$ is known at a single point A, it is always possible to define the potential at any other point B by the equation

$$(\phi)_B = (\phi)_A + \int_A^B \frac{\partial \phi}{\partial x^i}\, dx^i.$$

Such a field is always integrable.

An interesting special case is that in which

$$\nabla_i \phi = 0$$

throughout the region. Then $f_i = 0$ throughout the region, the line integral is everywhere zero, and $(\phi)_B = (\phi)_A$. If the covariant derivative of the scalar potential is everywhere zero, the scalar potential is a constant.

8.08. Integrability of contravariant vector fields

Recall that there are two covariant derivatives of a contravariant vector,

$$\overset{1}{\nabla}_j v^k = \frac{\partial v^k}{\partial x^j} + v^l \Gamma^k_{jl}, \tag{8.38a}$$

$$\overset{2}{\nabla}_j v^k = \frac{\partial v^k}{\partial x^j} + v^l \Gamma^k_{lj}. \tag{8.38b}$$

We will consider vector fields generated in such a way that the covariant derivative of the vector field is zero. Two such fields exist in general simultaneously for we can require

$$\overset{1}{\nabla}_j v^k = 0 \quad \text{or} \quad \overset{2}{\nabla}_j v^k = 0.$$

In the first case,

$$\frac{\partial v^k}{\partial x^j} = -v^l \Gamma^k_{jl}, \tag{8.39a}$$

and in the second,

$$\frac{\partial v^k}{\partial x^j} = -v^l \Gamma^k_{lj}. \tag{8.39b}$$

If the partial derivatives of the contravariant vector, $\partial v^k / \partial x^j$, are known at every point in the region and if the vector $(v^k)_A$ is known at point A, then it is possible to determine the vector at any other point B by integration,

$$(v^k)_B = (v^k)_A + \int_A^B \frac{\partial v^k}{\partial x^j} dx^j. \tag{8.40}$$

Two distinct vector fields can be defined in this way, depending on whether the first or second covariant derivative of v^k is set equal to zero.

Are these fields uniquely defined? The integral in Equation (8.40) depends on both the end points A and B and the path of integration joining A and B. However, the integral will be independent of path if

$$\oint_C \frac{\partial v^k}{\partial x^j} dx^j = 0. \tag{8.41}$$

However, by the generalized Stokes theorem for a mixed bivalent holor [Equation (5.48a)],

$$\oint_C \frac{\partial v^k}{\partial x^j} dx^j = \int_{\tilde{\mathfrak{a}}} \delta^{lij}_{123} \frac{\partial^2 v^k}{\partial x^i \partial x^j} d\tilde{\mathfrak{a}}_l, \tag{8.42}$$

where C is a contour bounding the pseudoarea $\tilde{\mathfrak{a}}$. The contour integral about C will vanish for all possible paths in the region only if the integrand of the surface integral is zero. Thus,

$$\delta^{lij}_{123} \frac{\partial^2 v^k}{\partial x^i \partial x^j} = 0 \tag{8.43}$$

or on expansion

$$\frac{\partial^2 v^k}{\partial x^i \partial x^j} = \frac{\partial^2 v^k}{\partial x^j \partial x^i}. \tag{8.44}$$

But the first partial derivatives of the vector are restricted by Equation (8.39). Substitution into Equation (8.44) gives

$$\frac{\partial}{\partial x^i}(v^l \Gamma^k_{jl}) = \frac{\partial}{\partial x^j}(v^l \Gamma^k_{il}) \tag{8.45a}$$

or

$$\frac{\partial}{\partial x^i}(v^l \Gamma^k_{lj}) = \frac{\partial}{\partial x^j}(v^l \Gamma^k_{li}). \tag{8.45b}$$

Expanding,

$$\frac{\partial v^l}{\partial x^i} \Gamma^k_{jl} + v^l \frac{\partial \Gamma^k_{jl}}{\partial x^i} = \frac{\partial v^l}{\partial x^j} \Gamma^k_{il} + v^l \frac{\partial \Gamma^k_{il}}{\partial x^j}, \tag{8.46a}$$

or

$$\frac{\partial v^l}{\partial x^i}\Gamma^k_{lj}+v^l\frac{\partial \Gamma^k_{lj}}{\partial x^i}=\frac{\partial v^l}{\partial x^j}\Gamma^k_{li}+v^l\frac{\partial \Gamma^k_{li}}{\partial x^j}. \tag{8.46b}$$

Substitution of Equation (8.39) for the partial derivative of v^l gives

$$-v^m\Gamma^l_{im}\Gamma^k_{jl}+v^l\frac{\partial \Gamma^k_{jl}}{\partial x^i}=-v^m\Gamma^l_{jm}\Gamma^k_{il}+v^l\frac{\partial \Gamma^k_{il}}{\partial x^j}, \tag{8.47a}$$

or

$$-v^m\Gamma^l_{im}\Gamma^k_{lj}+v^l\frac{\partial \Gamma^k_{lj}}{\partial x^i}=-v^m\Gamma^l_{mj}\Gamma^k_{li}+v^l\frac{\partial \Gamma^k_{li}}{\partial x^j}. \tag{8.47b}$$

Changing dummy indices and rearranging terms,

$$v^l\left(\frac{\partial \Gamma^k_{jl}}{\partial x^i}-\frac{\partial \Gamma^k_{il}}{\partial x^j}+\Gamma^m_{jl}\Gamma^k_{im}-\Gamma^m_{il}\Gamma^k_{jm}\right)=0, \tag{8.48a}$$

or

$$v^l\left(\frac{\partial \Gamma^k_{lj}}{\partial x^i}-\frac{\partial \Gamma^k_{li}}{\partial x^j}+\Gamma^m_{lj}\Gamma^k_{mi}-\Gamma^m_{li}\Gamma^k_{mj}\right)=0, \tag{8.48b}$$

which can be rewritten as

$$v^lR^k_{ijl}=0, \tag{8.49a}$$

or

$$v^lT^k_{ijl}=0. \tag{8.49b}$$

These are the integrability conditions for the two vector fields defined by Equation (8.39).

To summarize:

Theorem XXXVI. (A) *If a contravariant vector field is defined by the condition*

$$\nabla_j v^k=0,$$

and the value of the vector field $(v^k)_A$ *is known at point* A*, then the vector field is defined uniquely at any other point* B *by the equation*

$$(v^k)_B=(v^k)_A+\int_A^B\frac{\partial v^k}{\partial x^j}dx^j$$

for any arbitrary vector v^k *if*

$$R^k_{ijl}=0.$$

(B) *If the contravariant vector field is defined instead by the condition*

$$\overset{2}{\nabla}_j v^k = 0,$$

then the integrability condition is

$$T^k_{ijl} = 0.$$

Theorem XXXVII. *If the linear connection Γ^i_{jk} is symmetric, then there is a single contravariant vector field defined by the condition*

$$\nabla_j v^k = 0,$$

and the integrability condition is

$$R^k_{ijl} = 0.$$

8.09. Integrability of covariant vector fields

Similarly, there are two covariant derivatives of a covariant vector,

$$\overset{1}{\nabla}_j u_k = \frac{\partial u_k}{\partial x^j} - u_l \Gamma^l_{jk}, \tag{8.50a}$$

$$\overset{2}{\nabla}_j u_k = \frac{\partial u_k}{\partial x^j} - u_l \Gamma^l_{kj}. \tag{8.50b}$$

Two distinct covariant vector fields are defined by the conditions

$$\overset{1}{\nabla}_j u_k = 0 \quad \text{and} \quad \overset{2}{\nabla}_j u_k = 0.$$

In the first case,

$$\frac{\partial u_k}{\partial x^j} = +u_l \Gamma^l_{jk}, \tag{8.51a}$$

or

$$\frac{\partial u_k}{\partial x^j} = +u_l \Gamma^l_{kj}. \tag{8.51b}$$

If the partial derivatives of the covariant vector $\partial u_k / \partial x^j$ are known at every point in the region and if $(u_k)_A$ is known at point A, then it is possible to determine $(u_k)_B$ at any other point B by integration,

$$(u_k)_B = (u_k)_A + \int_A^B \frac{\partial u_k}{\partial x^j}\, dx^j. \tag{8.52}$$

Two distinct vector fields are defined by Equation (8.52), depending on whether we let $\overset{1}{\nabla}_j u_k = 0$ or $\overset{2}{\nabla}_j u_k = 0$.

These fields will be uniquely defined if

$$\oint_C \frac{\partial u_k}{\partial x^j}\, dx^j = 0. \tag{8.53}$$

But by the generalized Stokes theorem for a bivalent holor with two subscripts [Equation (5.48a)],

$$\oint_C \frac{\partial u_k}{\partial x^j}\, dx^j = \int_{\tilde{\mathfrak{a}}} \delta^{lij}_{123} \frac{\partial^2 u_k}{\partial x^i \partial x^j}\, d\tilde{\mathfrak{a}}_l, \tag{8.54}$$

where contour C bounds $\tilde{\mathfrak{a}}$. Since this equation holds for all contours in the region under consideration,

$$\delta^{lij}_{123} \frac{\partial^2 u_k}{\partial x^i \partial x^j} = 0,$$

or

$$\frac{\partial^2 u_k}{\partial x^i \partial x^j} = \frac{\partial^2 u_k}{\partial x^j \partial x^i}. \tag{8.55}$$

Substitution of Equation (8.51) into (8.55) gives

$$u_l\left(\frac{\partial \Gamma^l_{jk}}{\partial x^i} - \frac{\partial \Gamma^l_{ik}}{\partial x^j} + \Gamma^l_{im}\Gamma^m_{jk} - \Gamma^l_{jm}\Gamma^m_{ik}\right) = 0, \tag{8.55a}$$

or

$$u_l\left(\frac{\partial \Gamma^l_{kj}}{\partial x^i} - \frac{\partial \Gamma^l_{ki}}{\partial x^j} + \Gamma^l_{mi}\Gamma^m_{kj} - \Gamma^l_{mj}\Gamma^m_{ki}\right) = 0, \tag{8.55b}$$

which can be abbreviated into

$$u_l R^l_{ijk} = 0, \tag{8.56a}$$

or

$$u_l T^l_{ijk} = 0. \tag{8.56b}$$

These are the integrability conditions[2] for a covariant vector field.

Thus, we can conclude:

Theorem XXXVIII. (A) *If a covariant vector field is defined by the condition*

$$\overset{1}{\nabla}_j u_k = 0,$$

and the value of the vector field $(u_k)_A$ *is known at point* A, *then the vector field is defined uniquely at any other point* B *by the equation*

$$(u_k)_B = (u_k)_A + \int_A^B \frac{\partial u_k}{\partial x^j}\, dx^j$$

for any arbitrary covariant vector u_k *if*

$$R^l_{ijk} = 0.$$

(B) *If the covariant vector field is defined by the condition*

$$\overset{2}{\nabla}_j u_k = 0,$$

then the integrability condition is

$$T^l_{ijk} = 0.$$

Theorem XXXIX. *If the linear connection is symmetric, there is a single covariant vector field defined by the condition*

$$\nabla_j u_k = 0$$

and the integrability condition is

$$R^l_{ijk} = 0.$$

8.10. Integrability of tensor fields

Integrability studies can readily be extended to tensors of higher valence. For example, consider the mixed tensor U^i_j. According to Equation (7.39), there are four distinct covariant derivatives,

$$\overset{1}{\nabla}_k U^i_j = \frac{\partial U^i_j}{\partial x^k} + U^l_j \Gamma^i_{kl} - U^i_l \Gamma^l_{kj}, \tag{7.39a}$$

$$\overset{2}{\nabla}_k U^i_j = \frac{\partial U^i_j}{\partial x^k} + U^l_j \Gamma^i_{kl} - U^i_l \Gamma^l_{jk}, \tag{7.39b}$$

$$\overset{3}{\nabla}_k U^i_j = \frac{\partial U^i_j}{\partial x^k} + U^l_j \Gamma^i_{lk} - U^i_l \Gamma^l_{kj}, \tag{7.39c}$$

$$\overset{4}{\nabla}_k U^i_j = \frac{\partial U^i_j}{\partial x^k} + U^l_j \Gamma^i_{lk} - U^i_l \Gamma^l_{jk}. \tag{7.39d}$$

If the partial derivatives $\partial U^i_j/\partial x^k$ are known throughout a region and the value of the tensor $(U^i_j)_A$ is known at a single point A, then it is possible to define the tensor at any other point B by the equation

$$(U^i_j)_B = (U^i_j)_A + \int_A^B \frac{\partial U^i_j}{\partial x^k}\, dx^k. \tag{8.57}$$

Four distinct fields can be defined in this way depending on which of the covariant derivatives is equivalent to zero. For each of these fields, a unique field will be defined at every point if

$$\oint_C \frac{\partial U^i_j}{\partial x^k}\, dx^k = 0. \tag{8.58}$$

Application of the generalized Stokes theorem for a trivalent holor gives, from Equation (5.48a),

$$\oint_C \frac{\partial U^i_j}{\partial x^k}\, dx^k = \int_{\tilde{\mathfrak{a}}} \delta^{lmk}_{123} \frac{\partial^2 U^i_j}{\partial x^m \partial x^k}\, d\tilde{\mathfrak{a}}_l. \tag{8.59}$$

Consequently, from Equations (8.58) and (8.59), we have the general integrability condition

$$\frac{\partial^2 U^i_j}{\partial x^m \partial x^k} = \frac{\partial^2 U^i_j}{\partial x^k \partial x^m}. \tag{8.60}$$

If we wish to study fields in which one of the four covariant derivatives is equated to zero, we have four possible conditions from Equation (7.39),

$$\frac{\partial U^i_j}{\partial x^k} = -U^l_j \Gamma^i_{kl} + U^i_l \Gamma^l_{kj}, \tag{8.61a}$$

$$\frac{\partial U^i_j}{\partial x^k} = -U^l_j \Gamma^i_{kl} + U^i_l \Gamma^l_{jk}, \tag{8.61b}$$

$$\frac{\partial U^i_j}{\partial x^k} = -U^l_j \Gamma^i_{lk} + U^i_l \Gamma^l_{kj}, \tag{8.61c}$$

$$\frac{\partial U^i_j}{\partial x^k} = -U^l_j \Gamma^i_{lk} + U^i_l \Gamma^l_{jk}. \tag{8.61d}$$

Substitution of Equation (8.61) into (8.60) gives the integrability conditions

$$U^m_j R^i_{klm} - U^i_m R^m_{klj} = 0, \tag{8.62a}$$

$$U^m_j R^i_{klm} - U^i_m T^m_{klj} = 0, \tag{8.62b}$$

$$U^m_j T^i_{klm} - U^i_m R^m_{klj} = 0, \tag{8.62c}$$

$$U^m_j T^i_{klm} - U^i_m T^m_{klj} = 0. \tag{8.62d}$$

The integrability conditions[2] can be readily extended to tensors of any valence:

Theorem XL. *If a tensor field of valence m is defined by the condition*

$$\overset{s}{\nabla}_i U^{jk\cdots l}_{pq\cdots r} = 0$$

and the value of the tensor is given at a single point A, the tensor is defined uniquely at any other point B by the equation

$$(U^{jk\cdots l}_{pq\cdots r})_B = (U^{jk\cdots l}_{pq\cdots r})_A + \int_A^B \frac{\partial U^{jk\cdots l}_{pq\cdots r}}{\partial x^i}\, dx^i$$

if (a) $s=1$, $R^i_{\ klm}=0$; (b) $1<s<2^m$, $R^i_{\ klm}=0$, *and* $T^i_{\ klm}=0$; *and* (c) $s=2^m$, $T^i_{\ klm}=0$.

Theorem XLI. *If the linear connection is symmetric, $S^i_{\ jk}=0$, the integrability condition for all tensor fields defined by the condition*

$$\nabla_i U^{jk\cdots l}_{pq\cdots r} = 0$$

is

$$R^i_{\ klm} = 0.$$

8.11. Integrability of nilvalent akinetor fields

Now consider the extension of the integrability question to include akinetors. The key to the study[5] is the behavior of a nilvalent akinetor $\tilde{S}$ of weight w. According to Equation (7.79), there are two covariant derivatives,

$$\overset{1}{\nabla}_k \tilde{S} = \frac{\partial \tilde{S}}{\partial x^k} - w\tilde{S}\Gamma^i_{\ ki}, \tag{7.79a}$$

and

$$\overset{2}{\nabla}_k \tilde{S} = \frac{\partial \tilde{S}}{\partial x^k} - w\tilde{S}\Gamma^i_{\ ik}. \tag{7.79b}$$

Two akinetor fields of particular interest are generated by the two conditions

$$\overset{1}{\nabla}_k \tilde{S} = 0 \quad \text{or} \quad \overset{2}{\nabla}_k \tilde{S} = 0,$$

which are equivalent to

$$\frac{\partial \tilde{S}}{\partial x^k} = w\tilde{S}\Gamma^i_{\ ki}, \tag{8.63a}$$

or

$$\frac{\partial \tilde{S}}{\partial x^k} = w\tilde{S}\Gamma^i_{ik}. \tag{8.63b}$$

If the partial derivatives $\partial \tilde{S}/\partial x^k$ are known throughout the region and the value of $\tilde{S}$ is known at point A, the nilvalent akinetor field can be defined at any other point B by the equation

$$(\tilde{S})_B = (\tilde{S})_A + \int_A^B \frac{\partial \tilde{S}}{\partial x^k}\, dx^k. \tag{8.64}$$

This integral will give a uniquely determined field if

$$\oint_C \frac{\partial \tilde{S}}{\partial x^k}\, dx^k = 0 \tag{8.65}$$

for any closed path C in the region. But by the generalized Stokes theorem [Equation (5.48a)],

$$\oint_C \frac{\partial \tilde{S}}{\partial x^k}\, dx^k = \int_{\tilde{\mathfrak{a}}} \delta^{ijk}_{123} \frac{\partial^2 S}{\partial x^j \partial x^k}\, d\tilde{\mathfrak{a}}_i, \tag{8.66}$$

where C bounds $\tilde{\mathfrak{a}}$. Since the contour integral must vanish for all closed paths in the region, the integrand of the surface integral must vanish throughout the region, or

$$\frac{\partial^2 \tilde{S}}{\partial x^j \partial x^k} = \frac{\partial^2 \tilde{S}}{\partial x^k \partial x^j}. \tag{8.67}$$

This is the integrability condition for a nilvalent akinetor.

Substitution of Equation (8.63) into (8.67) gives

$$\tilde{S}\left[\frac{\partial \Gamma^i_{ki}}{\partial x^j} - \frac{\partial \Gamma^i_{ji}}{\partial x^k}\right] = 0, \tag{8.68a}$$

or

$$\tilde{S}\left[\frac{\partial \Gamma^i_{ik}}{\partial x^j} - \frac{\partial \Gamma^i_{ij}}{\partial x^k}\right] = 0. \tag{8.68b}$$

But according to Equations (8.21) and (8.23) the integrability conditions[5] for nilvalent akinetors can be written

$$\tilde{S}R_{jk} = 0, \tag{8.69a}$$

or

$$\tilde{S}T_{jk} = 0. \tag{8.69b}$$

To summarize:

Theorem XLII. (A) *If a nilvalent akinetor field is defined by the condition*

$$\overset{1}{\nabla}_k \tilde{S} = 0$$

and $(\tilde{S})_A$ *is known at point A, then* $(\tilde{S})_B$ *can be defined at any other point B by the equation*

$$(\tilde{S})_B = (\tilde{S})_A + \int_A^B \frac{\partial \tilde{S}}{\partial x^k}\, dx^k$$

if

$$R_{ij} = 0.$$

(B) *If a nilvalent akinetor field is defined by the condition*

$$\overset{2}{\nabla}_k \tilde{S} = 0,$$

then the integrability condition is

$$T_{ij} = 0.$$

Theorem XLIII. *If the linear connection is symmetric, a single field* $\tilde{S}$ *is defined by the condition*

$$\nabla_k \tilde{S} = 0,$$

and the integrability condition is

$$R_{ij} = 0.$$

8.12. Integrability of akinetor fields

According to Equation (7.85), a contravariant univalent akinetor has four distinct covariant derivatives:

$$\overset{1}{\nabla}_k \tilde{v}^i = \frac{\partial \tilde{v}^i}{\partial x^k} + \tilde{v}^l \Gamma^i_{kl} - w\tilde{v}^i \Gamma^l_{kl}, \tag{7.85a}$$

$$\overset{2}{\nabla}_k \tilde{v}^i = \frac{\partial \tilde{v}^i}{\partial x^k} + \tilde{v}^l \Gamma^i_{lk} - w\tilde{v}^i \Gamma^l_{kl}, \tag{7.85b}$$

$$\overset{3}{\nabla}_k \tilde{v}^i = \frac{\partial \tilde{v}^i}{\partial x^k} + \tilde{v}^l \Gamma^i_{kl} - w\tilde{v}^i \Gamma^l_{lk}, \tag{7.85c}$$

$$\overset{4}{\nabla}_k \tilde{v}^i = \frac{\partial \tilde{v}^i}{\partial x^k} + \tilde{v}^l \Gamma^i_{lk} - w\tilde{v}^i \Gamma^l_{lk}. \tag{7.85d}$$

Four distinct akinetor fields can be generated by setting each of the four covariant derivatives equal to zero in turn:

$$\frac{\partial \tilde{v}^i}{\partial x^k} = -\tilde{v}^l \Gamma^i_{kl} + w\tilde{v}^i \Gamma^l_{kl}, \tag{8.70a}$$

or

$$\frac{\partial \tilde{v}^i}{\partial x^k} = -\tilde{v}^l \Gamma^i_{lk} + w\tilde{v}^i \Gamma^l_{kl}, \tag{8.70b}$$

or

$$\frac{\partial \tilde{v}^i}{\partial x^k} = -\tilde{v}^l \Gamma^i_{kl} + w\tilde{v}^i \Gamma^l_{lk}, \tag{8.70c}$$

or

$$\frac{\partial \tilde{v}^i}{\partial x^k} = -\tilde{v}^l \Gamma^i_{lk} + w\tilde{v}^i \Gamma^l_{lk}. \tag{8.70d}$$

If the partial derivatives $\partial \tilde{v}^i / \partial x^k$ are known throughout a given region and the value of $(\tilde{v}^i)_A$ is given, $(\tilde{v}^i)_B$ can always be defined by the equation

$$(\tilde{v}^i)_B = (\tilde{v}^i)_A + \int_A^B \frac{\partial \tilde{v}^i}{\partial x^k} dx^k. \tag{8.71}$$

This integral will be defined uniquely only if

$$\oint_C \frac{\partial \tilde{v}^i}{\partial x^k} dx^k = 0$$

throughout the region. By the generalized Stokes theorem [Equation (5.48a)],

$$\oint_C \frac{\partial \tilde{v}^i}{\partial x^k} dx^k = \int_{\tilde{a}} \delta^{ljk}_{123} \frac{\partial^2 \tilde{v}^i}{\partial x^j \partial x^k}. \tag{8.72}$$

Since the contour integral must vanish for all closed paths in the region, the integrand of the surface integral must vanish, so the integrability condition becomes

$$\frac{\partial^2 \tilde{v}^i}{\partial x^j \partial x^k} = \frac{\partial^2 \tilde{v}^i}{\partial x^k \partial x^j}. \tag{8.73}$$

Substitution of Equations (8.70) into (8.73) gives the integrability conditions

$$\tilde{v}^l R^i_{kjl} - w\tilde{v}^i R_{kj} = 0, \tag{8.74a}$$

$$\tilde{v}^l T^i_{kjl} - w\tilde{v}^i R_{kj} = 0, \tag{8.74b}$$

$$\tilde{v}^l R^i_{kjl} - w\tilde{v}^i T_{kj} = 0, \tag{8.74c}$$

$$\tilde{v}^l T^i_{kjl} - w\tilde{v}^i T_{kj} = 0. \tag{8.74d}$$

The results for all akinetors may be summarized[5] in the following theorem:

Theorem XLIV. *If an akinetor field of valence $m \geq 0$ and weight $w > 0$ is defined by the condition that*

$$\overset{s}{\nabla}_i \tilde{U}^{jk\cdots l}_{pq\cdots r} = 0$$

and the value of the akinetor is given at a single point A, the akinetor is defined uniquely at any other point B by the equation

$$(\tilde{U}^{jk\cdots l}_{pq\cdots r})_B = (\tilde{U}^{jk\cdots l}_{pq\cdots r})_A + \int_A^B \frac{\partial \tilde{U}^{jk\cdots l}_{pq\cdots r}}{\partial x^i}\, dx^i$$

if (a) $s=1$, $R^i_{kjl}=0$, *and* $R_{kj}=0$; (b) $1<s<2^m$, $R^i_{kjl}=0$, $T^i_{kjl}=0$, *and* $R_{kj}=0$; (c) $2^{m+1} \leq s \leq 2^{m+1}-1$, $R^i_{kjl}=0$, $T^i_{kjl}=0$, *and* $T_{kj}=0$; *and* (d) $s=2^{m+1}$, $T^i_{kjl}=0$, *and* $T_{kj}=0$.

Theorem XLV. *If the linear connection is symmetric,* $S^i_{jk}=0$, *and the akinetor field of valence $m>0$ and weight $w>0$ is defined by the condition*

$$\nabla_i \tilde{U}^{jk\cdots l}_{pq\cdots r} = 0,$$

the integrability condition is

$$R^i_{kjl}=0 \quad \textit{and} \quad R_{kj}=0.$$

8.13. Properties of the Riemann–Christoffel tensors

We have defined two quadrivalent tensors

$$R^k_{ijl} = \frac{\partial \Gamma^k_{jl}}{\partial x^i} - \frac{\partial \Gamma^k_{il}}{\partial x^j} + \Gamma^m_{jl}\Gamma^k_{im} - \Gamma^m_{il}\Gamma^k_{jm} \tag{8.09a}$$

and

$$T^k_{ijl} = \frac{\partial \Gamma^k_{lj}}{\partial x^i} - \frac{\partial \Gamma^k_{li}}{\partial x^j} + \Gamma^m_{lj}\Gamma^k_{mi} - \Gamma^m_{li}\Gamma^k_{mj}. \tag{8.09b}$$

By direct expansion, two important identities are readily proved:

$$R^k_{(ij)l} = R^k_{ijl} + R^k_{jil} = 0, \tag{8.75a}$$

$$T^k_{(ij)l} = T^k_{ijl} + T^k_{jil} = 0. \tag{8.75b}$$

A practical difficulty in the use of the Riemann–Christoffel tensors is the large number of independent merates. In n-space, there are n^4 distinct merates. Thus, in 2-space the Riemann–Christoffel tensors each have 16 merates, in 3-space 81 merates, and in 4-space 256 distinct merates.

However, because of Equation (8.75), the number of independent merates is much less than n^4. From Equation (8.75),

$$R^k_{ijl} = -R^k_{jil}, \tag{8.76a}$$

$$T^k_{ijl} = -T^k_{jil}. \tag{8.76b}$$

In the special case in which $i = j$,

$$R^k_{iil} = -R^k_{iil} = 0, \tag{8.77a}$$

$$T^k_{iil} = -T^k_{iil} = 0. \tag{8.77b}$$

Consequently n^3 of the n^4 merates of the quadrivalent tensors are zero. This leaves a possibility of

$$n^4 - n^3 = n^3(n-1)$$

distinct nonvanishing merates of each of the Riemann–Christoffel tensors. However, since Equation (8.76) shows that the tensors are alternating in indices i and j, half of these remaining merates are the negative of the corresponding terms in the other half. So the number of linearly independent merates of R^k_{ijl} and T^k_{ijl} that are not necessarily zero is

$$N = \tfrac{1}{2}[n^3(n-1)]. \tag{8.78}$$

There is another family of relations between the merates of the Riemann–Christoffel tensors[2]:

$$R^k_{ijl} + R^k_{jli} + R^k_{lij} = 2\left[\frac{\partial S^k_{ij}}{\partial x^l} + \frac{\partial S^k_{jl}}{\partial x^i} + \frac{\partial S^k_{li}}{\partial x^j} + \Gamma^k_{lm} S^m_{ij} + \Gamma^k_{im} S^m_{jl} + \Gamma^k_{jm} S^m_{li}\right], \tag{8.79a}$$

$$T^k_{ijl} + T^k_{jli} + T^k_{lij} = 2\left[\frac{\partial S^k_{ji}}{\partial x^l} + \frac{\partial S^k_{lj}}{\partial x^i} + \frac{\partial S^k_{il}}{\partial x^j} + \Gamma^k_{ml} S^m_{ji} + \Gamma^k_{mi} S^m_{lj} + \Gamma^k_{mj} S^m_{il}\right]. \tag{8.79b}$$

These equations further reduce the number of independent merates of the Riemann–Christoffel tensor. If any two of the three subscripts in Equation (8.79) are equal, Equation (8.79) reduces to Equation (8.75) and yields no new information. Thus, for $n = 2$, no new information results from this equation.

In an n-dimensional space, the subscripts i, j, and l can be chosen in C_3^n ways,

$$C_3^n = \frac{n!}{3!(n-3)!} = \frac{n}{6}(n-1)(n-2).$$

For each of these the superscript k can be chosen in n ways so Equation (8.79) yields

$$\tfrac{1}{6}(n)^2(n-1)(n-2)$$

distinct equations for each of the Riemann–Christoffel tensors. Each of these introduces a linear relation between the merates of each Riemann–Christoffel tensor. So the number of linearly independent merates of each Riemann–Christoffel tensor that are not necessarily zero is

$$M = \tfrac{1}{2}(n)^3(n-1) - \tfrac{1}{6}(n)^2(n-1)(n-2),$$

or

$$M = \tfrac{1}{3}(n)^2(n^2-1). \tag{8.80}$$

Thus, in 2-space there are four distinct linearly independent merates that are not necessarily equal to zero for each Riemann–Christoffel tensor,

$$R^k_{12l} = \begin{bmatrix} R^1_{121} & R^1_{122} \\ R^2_{121} & R^2_{122} \end{bmatrix}, \qquad T^k_{12l} = \begin{bmatrix} T^1_{121} & T^1_{122} \\ T^2_{121} & T^2_{122} \end{bmatrix}.$$

In 3-space, according to Equation (8.78), there are $N = 27$ merates that can be arranged into the three matrices R^k_{12l}, R^k_{13l}, and R^k_{23l}, but there are also three distinct linear equations from Equation (8.79a). So there are $M = 24$ distinct linearly independent merates of R^k_{ijl} that do not necessarily vanish. Similarly, there are 24 distinct linearly independent merates of T^k_{ijl} that do not necessarily vanish.

In 4-space, according to Equation (8.78), there are $N = 96$ merates, but by Equation (8.79), these are related by 16 linear relations from Equation (8.79). So $M = 80$. For each of the Riemann–Christoffel tensors in 4-space, there are 80 distinct, linearly independent merates that do not necessarily vanish.

8.14. The Bianchi identities

The Riemann–Christoffel tensors can be differentiated covariantly in 16 different ways (Section 7.05). Consequently, it is possible by expansion of Equation (7.54) to derive a family of 32 equations that are generalizations of the familiar Bianchi identity.[7] These can be written for the first Riemann–Christoffel tensor:

$$\overset{1}{\nabla}_{[i} R^m_{jk]l} = 2S^n_{[ij} R^m_{k]nl}, \tag{8.81a}$$

$$\overset{2}{\nabla}_{[i} R^m_{jk]l} = 2S^n_{[ij} R^m_{k]nl} + 2S^n_{l[i} R^m_{kj]n}, \tag{8.81b}$$

$$\overset{3}{\nabla}_{[i} R^m_{jk]l} = 0, \tag{8.81c}$$

$$\overset{4}{\nabla}_{[i} R^m_{jk]l} = 2S^n_{l[i} R^m_{kj]n}, \tag{8.81d}$$

$$\overset{5}{\nabla}_{[i} R^m_{jk]l} = 0, \tag{8.81e}$$

$$\overset{6}{\nabla}_{[i} R^m_{jk]l} = 2S^n_{l[i} R^m_{kj]n}, \tag{8.81f}$$

$$\overset{7}{\nabla}_{[i} R^m_{jk]l} = 2S^n_{[ik} R^m_{j]nl}, \tag{8.81g}$$

$$\overset{8}{\nabla}_{[i} R^m_{jk]l} = 2S^n_{[ik} R^m_{j]nl} + 2S^n_{l[i} R^m_{kj]n}, \tag{8.81h}$$

$$\overset{9}{\nabla}_{[i} R^m_{jk]l} = 2S^m_{n[i} R^n_{jk]l} + 2S^n_{[ij} R^m_{k]nl}, \tag{8.81i}$$

$$\overset{10}{\nabla}_{[i} R^m_{jk]l} = 2S^m_{n[i} R^n_{jk]l} + 2S^n_{[ij} R^m_{k]nl} + 2S^n_{l[i} R^m_{kj]n}, \tag{8.81j}$$

$$\overset{11}{\nabla}_{[i} R^m_{jk]l} = 2S^m_{n[i} R^n_{jk]l}, \tag{8.81k}$$

$$\overset{12}{\nabla}_{[i} R^m_{jk]l} = 2S^m_{n[i} R^n_{jk]l} + 2S^n_{l[i} R^m_{kj]n}, \tag{8.81l}$$

$$\overset{13}{\nabla}_{[i} R^m_{jk]l} = 2S^m_{n[i} R^n_{jk]l}, \tag{8.81m}$$

$$\overset{14}{\nabla}_{[i} R^m_{jk]l} = 2S^m_{n[i} R^n_{jk]l} + 2S^n_{l[i} R^m_{kj]n}, \tag{8.81n}$$

$$\overset{15}{\nabla}_{[i} R^m_{jk]l} = 2S^m_{n[i} R^n_{jk]l} + 2S^n_{[ik} R^m_{j]nl}, \tag{8.81o}$$

$$\overset{16}{\nabla}_{[i} R^m_{jk]l} = 2S^n_{[ik} R^m_{j]nl} + 2S^m_{n[i} R^n_{jk]l} + 2S^n_{l[i} R^m_{kj]n}. \tag{8.81p}$$

For the second Riemann–Christoffel tensor, there is a similar set of 16 identities:

$$\overset{1}{\nabla}_{[i} T^m_{jk]l} = 2S^n_{[ji} T^m_{k]nl}, \tag{8.82a}$$

$$\overset{2}{\nabla}_{[i} T^m_{jk]l} = 2S^n_{[ji} T^m_{k]nl} + 2S^n_{i[l} T^m_{kj]n}, \tag{8.82b}$$

$$\overset{3}{\nabla}_{[i} T^m_{jk]l} = 0, \tag{8.82c}$$

$$\overset{4}{\nabla}_{[i} T^m_{jk]l} = 2S^n_{i[l} T^m_{kj]n}, \tag{8.82d}$$

$$\overset{5}{\nabla}_{[i} T^m_{jk]l} = 0, \tag{8.82e}$$

$$\overset{6}{\nabla}_{[i} T^m_{jk]l} = 2S^n_{i[l} T^m_{kj]n}, \tag{8.82f}$$

$$\overset{7}{\nabla}_{[i} T^m_{\ jk]l} = 2S^n_{\ [ki} T^m_{\ j]nl}, \tag{8.82g}$$

$$\overset{8}{\nabla}_{[i} T^m_{\ jk]l} = 2S^n_{\ [ki} T^m_{\ j]nl} + 2S^n_{\ i[l} T^m_{\ kj]n}, \tag{8.82h}$$

$$\overset{9}{\nabla}_{[i} T^m_{\ jk]l} = 2S^m_{\ i[n} T^n_{\ jk]l} + 2S^n_{\ [ji} T^m_{\ k]nl}, \tag{8.82i}$$

$$\overset{10}{\nabla}_{[i} T^m_{\ jk]l} = 2S^m_{\ i[n} T^n_{\ jk]l} + 2S^n_{\ [ji} T^m_{\ k]nl} + 2S^n_{\ i[l} T^m_{\ kj]n}, \tag{8.82j}$$

$$\overset{11}{\nabla}_{[i} T^m_{\ jk]l} = 2S^m_{\ i[n} T^n_{\ jk]l}, \tag{8.82k}$$

$$\overset{12}{\nabla}_{[i} T^m_{\ jk]l} = 2S^m_{\ i[n} T^n_{\ jk]l} + 2S^n_{\ i[l} T^m_{\ kj]n}, \tag{8.82l}$$

$$\overset{13}{\nabla}_{[i} T^m_{\ jk]l} = 2S^m_{\ i[n} T^n_{\ jk]l}, \tag{8.82m}$$

$$\overset{14}{\nabla}_{[i} T^m_{\ jk]l} = 2S^m_{\ i[n} T^n_{\ jk]l} + 2S^n_{\ i[l} T^m_{\ kj]n}, \tag{8.82n}$$

$$\overset{15}{\nabla}_{[i} T^m_{\ jk]l} = 2S^m_{\ i[n} T^n_{\ jk]l} + 2S^n_{\ [ki} T^m_{\ j]nl}, \tag{8.82o}$$

$$\overset{16}{\nabla}_{[i} T^m_{\ jk]l} = 2S^n_{\ [ki} T^m_{\ j]nl} + 2S^m_{\ i[n} T^n_{\ jk]l} + 2S^n_{\ i[l} T^m_{\ kj]n}. \tag{8.82p}$$

In a space in which the linear connection is symmetric and the torsion tensor $S^i_{\ jk} = 0$, all of the foregoing 32 identities reduce to the classical Bianchi identity

$$\nabla_{[i} R^m_{\ jk]l} = 0. \tag{8.83}$$

8.15. Summary

This chapter has dealt with two principal questions:

(a) Does

$$\overset{s}{\nabla}_i \overset{t}{\nabla}_j \tilde{U}^{kl\cdots m}_{pq\cdots r} = \overset{s}{\nabla}_j \overset{t}{\nabla}_i \tilde{U}^{kl\cdots m}_{pq\cdots r}?$$

(b) If an akinetor field is defined by the condition

$$\overset{s}{\nabla}_i \tilde{U}^{kl\cdots m}_{pq\cdots r} = 0$$

and $(\tilde{U}^{kl\cdots m}_{pq\cdots r})_A$ is given at point A, can the akinetor field be defined uniquely by the equation

$$(\tilde{U}^{kl\cdots m}_{pq\cdots r})_B = (\tilde{U}^{kl\cdots m}_{pq\cdots r})_A + \int_A^B \frac{\partial \tilde{U}^{kl\cdots m}_{pq\cdots r}}{\partial x^i}\, dx^i?$$

The remarkable fact is that the answer to both of these apparently unrelated questions can be expressed in terms of two quadrivalent tensors

Table 8.01. *Properties of second covariant derivatives*

Equation	Condition
$\overset{s}{\nabla}_i \overset{t}{\nabla}_j \tilde{U}^{kl\cdots m}_{pq\cdots r} \neq \overset{s}{\nabla}_j \overset{t}{\nabla}_i \tilde{U}^{kl\cdots m}_{pq\cdots r}$	In general
$\nabla_i \nabla_j \tilde{U}^{kl\cdots m}_{pq\cdots r} = \nabla_j \nabla_i \tilde{U}^{kl\cdots m}_{pq\cdots r}$	$S^i_{jk}=0, \quad R^i_{jkl}=0, \quad R_{ij}=0$
$\nabla_i \nabla_j \tilde{S} = \nabla_j \nabla_i \tilde{S}$	$S^i_{jk}=0, \quad R_{ij}=0$
$\overset{s}{\nabla}_i \overset{t}{\nabla}_j U^{kl\cdots m}_{pq\cdots r} \neq \overset{s}{\nabla}_j \overset{t}{\nabla}_i U^{kl\cdots m}_{pq\cdots r}$	In general
$\nabla_i \nabla_j U^{kl\cdots m}_{pq\cdots r} = \nabla_j \nabla_i U^{kl\cdots m}_{pq\cdots r}$	$S^i_{jk}=0, \quad R^i_{jkl}=0$
$\nabla_i \nabla_j \phi = \nabla_j \nabla_i \phi$	$S^i_{jk}=0$

R^i_{kjl} and T^i_{kjl} and the bivalent tensors R_{kj} and T_{kj}, which can be obtained from the quadrivalent tensors by contraction on the last subscript.

The theorems of the chapter are summarized in Tables 8.01 and 8.02. We see from Table 8.01 that in general the order of covariant derivatives does matter. The difference is expressed as a linear function of the torsion tensor and both the first and second Riemann–Christoffel tensors. The order of second covariant derivatives can be interchanged for all akinetors if $S^i_{jk}=0$, $R^i_{jkl}=0$, and $R_{ij}=0$. For nilvalent akinetors, only the torsion tensor and the Ricci tensor R_{ij} must vanish. For tensors the conditions are less stringent. In general, we must require $S^i_{jk}=0$ and $R^i_{jkl}=0$. For scalars the only restriction is that $S^i_{jk}=0$.

Table 8.02 summarizes the conditions under which a unique akinetor field for which a covariant derivative of the akinetor is zero is defined by integration. If the first kind of covariant derivative of the akinetor is zero, the integrability condition is $R^i_{kjl}=0$ and $R_{kj}=0$. But if instead another kind of covariant derivative of the akinetor is set equal to zero, the condition shifts to being expressed in terms of the second kind of Riemann–Christoffel tensor or possibly both kinds. For nilvalent akinetors, the only integrability condition is the vanishing of either R_{ij} or T_{ij} depending on which of the covariant derivatives of S is equated to zero. The integrability conditions for tensors are similar but do not involve the Ricci tensors.

Finally, some of the properties of the Riemann–Christoffel tensors and of their covariant derivatives are studied.

Table 8.02. *Integrability conditions*

Field satisfying condition $\overset{\boxed{s}}{\nabla}_i \tilde{U}^{jk\cdots l}_{pq\cdots r}=0$	Uniquely determined if
$s=1$	$R^i_{kjl}=0,\quad R_{kj}=0$
$1<s<2^m$	$R^i_{kjl}=0,\quad T^i_{kjl}=0,\quad R_{kj}=0$
$2^{m+1}<s<2^{m+1}-1$	$R^i_{kjl}=0,\quad T^i_{kjl}=0,\quad T_{kj}=0$
$s=2^{m+1}$	$T^i_{kjl}=0,\quad T_{kj}=0$
$\overset{\boxed{1}}{\nabla}_k \tilde{S}=0$	$R_{ij}=0$
$\overset{\boxed{2}}{\nabla}_k \tilde{S}=0$	$T_{ij}=0$
$\overset{\boxed{s}}{\nabla}_i \tilde{U}^{jk\cdots l}_{pq\cdots r}=0$	
$s=1$	$R^i_{klm}=0$
$1<s<2^m$	$R^i_{klm}=0,\quad T^i_{klm}=0$
$s=2^m$	$T^i_{klm}=0$
$\nabla_i\phi=0$	Always
If $S^i_{jk}=0$	
$\nabla_i \tilde{U}^{jk\cdots l}_{pq\cdots r}=0$	$R^i_{kjl}=0,\quad R_{kj}=0$
$\nabla_k \tilde{S}=0$	$R_{ij}=0$
$\nabla_i U^{jk\cdots l}_{pq\cdots r}=0$	$R^i_{kjl}=0$
$\nabla_i\phi=0$	Always

Problems

8.01 From Equations (7.20) and (7.39), prove Equation (8.10).

8.02 From Equations (7.28) and (7.37), prove Equation (8.14).

8.03 From Equations (7.28) and (7.37), prove Equation (8.15).

8.04 For a linear connection in 2-space with the merates

$$\Gamma^1_{jk}=\begin{bmatrix}0 & 1\\ 0 & 0\end{bmatrix},\qquad \Gamma^2_{jk}=\begin{bmatrix}1 & 0\\ 1 & 0\end{bmatrix},$$

(a) find $\overset{1}{\nabla}_{[i}\overset{1}{\nabla}_{j]} v^k$ and
(b) find $\overset{2}{\nabla}_{[i}\overset{2}{\nabla}_{j]} v^k$.

8.05 For the linear connection of Probem 8.04,
(a) find $\overset{1}{\nabla}_{[i}\overset{1}{\nabla}_{j]} u_k$ and
(b) find $\overset{2}{\nabla}_{[i}\overset{2}{\nabla}_{j]} u_k$.

8.06 Derive an expression for $\overset{1}{\nabla}_{[i}\overset{1}{\nabla}_{j]} U^k_{l\cdot}$.

8.07 Derive an expression for $\overset{2}{\nabla}_{[i}\overset{2}{\nabla}_{j]} U^k_{l\cdot}$.

8.08 Derive an expression for $\overset{3}{\nabla}_{[i}\overset{3}{\nabla}_{j]} U_{kl}$.

8.09 From Equations (7.79) and (8.21) derive Equation (8.22).

8.10 From Equations (7.79) and (8.21) derive Equation (8.24).

8.11 For the linear connection of Problem 8.04,
(a) find $S^{\;i}_{jk}$,
(b) evaluate $R^{\;i}_{jkl}$,
(c) evaluate $T^{\;i}_{jkl}$,
(d) find R_{ij}, and
(e) find T_{ij}.

8.12 Evaluate $\overset{s}{\nabla}_{[i}\overset{t}{\nabla}_{j]}\tilde{S}$ for the linear connection of Problem 8.04.

8.13
(a) A linear connection $\Gamma^{\;i}_{jk}$ is known to be constant. What conditions must its merates satisfy if $R^{\;i}_{jkl}=0$?
(b) What conditions must the merates of a constant linear connection satisfy if $T^{\;i}_{jkl}=0$?

8.14 Derive an expression for $\overset{1}{\nabla}_{[i}\overset{1}{\nabla}_{j]}\tilde{v}^k$.

8.15 Derive an expression for $\overset{2}{\nabla}_{[i}\overset{2}{\nabla}_{j]}\tilde{v}^k$.

8.16 For the linear connection of Problem 8.04, evaluate $\overset{1}{\nabla}_{[i}\overset{1}{\nabla}_{j]}\tilde{v}^k$.

8.17 Suppose that

$$\frac{\partial\phi}{\partial x^i}=(0,1,-1)$$

for all values of x^i and that $\phi=0$ at the point $x^i=(1,1,1)$. By Theorem XXXV, can you find an expression for ϕ at any other point? Find this expression if it exists.

8.18 A linear connection is given by the expression

$$\Gamma^1_{jk} = \begin{bmatrix} 0 & 1 \\ 0 & 0 \end{bmatrix}, \qquad \Gamma^2_{jk} = \begin{bmatrix} 0 & 0 \\ -1 & 0 \end{bmatrix}.$$

(a) Write the differential equations $\overset{1}{\nabla}_j v^k = 0$ for this linear connection.
(b) If $v^k = (1,0)$ at point $x^i = 0$, find v^k at any other point if possible or prove that the field cannot be defined uniquely.

8.19 A linear connection is given by the matrices

$$\Gamma^1_{jk} = \begin{bmatrix} 1 & 1 \\ 0 & 0 \end{bmatrix}, \qquad \Gamma^2_{jk} = \begin{bmatrix} 1 & 1 \\ 0 & 0 \end{bmatrix}.$$

(a) A vector field is defined by the condition $\overset{1}{\nabla}_j v^k = 0$ and $v^k = (1,0)$ at $x^i = 0$. Find the vector field v^k at any other point if possible or prove that the field cannot be defined uniquely.
(b) Repeat if $\overset{2}{\nabla}_j v^k = 0$.

8.20
(a) A nilvalent akinetor field $\tilde{S}$ is generated by the condition $\overset{1}{\nabla}_k \tilde{S} = 0$ and $\tilde{S} = 1$ at $x^i = 0$. Either find a definition of $\tilde{S}$ at any point or prove that it cannot be found if the linear connection is

$$\Gamma^1_{jk} = \begin{bmatrix} 0 & 1 \\ 0 & 0 \end{bmatrix} \quad \text{and} \quad \Gamma^2_{jk} = \begin{bmatrix} 1 & 0 \\ 1 & 0 \end{bmatrix}.$$

(b) Repeat if instead $\overset{2}{\nabla}_k \tilde{S} = 0$.
(c) Sketch $\tilde{S}$ as a function of x^1 and x^2 for (a) and (b).

8.21 For the linear connection of Problem 8.19 and an akinetor field of weight -1, $\tilde{v}^k$, such that $\tilde{v}^k = (0,1)$ at $x^i = 0$,
(a) find the akinetor field at any other point or show that it cannot be uniquely defined if $\overset{1}{\nabla}_j \tilde{v}^k = 0$, and
(b) repeat if $\overset{4}{\nabla}_j \tilde{v}^k 0$.

8.22
(a) In 5-space, how many linearly independent merates does $R^i_{\ jkl}$ have that do not necessarily vanish?
(b) List them in the simplest possible way.

8.23 Derive Equation (8.83a).

8.24 Derive Equation (8.83e).

8.25 Derive Equation (8.84a).

8.26 Derive Equation (8.84c).

IV

Space structure

9

Non-Riemannian spaces

The Klein classification of spaces was discussed in Chapter 6. We now consider some spaces where structure is obtained by introducing the linear connection Γ^i_{jk}. This connection acts as a link to tie together nearby points in the space.

The study of linearly connected spaces was instigated principally by the development of unified field theories[1] in physics. Starting with Einstein's theory of 1916, he and others tried a great number of artificial spaces in an attempt to obtain a unified formulation of gravitation and electrodynamics. Practical applications of this new knowledge to a wide variety of other problems may well be of great importance when the theory becomes more generally known.

In the present chapter, we shall consider (a) spaces with a linear connection, (b) Weyl spaces, and (c) spaces with projective connection. These are all nonmetric spaces. The introduction of a metric will be deferred to Chapter 10, and Euclidean space will be considered in Chapter 11.

9.01. The absolute differential

In Chapter 7, we introduced the covariant derivative and the linear connection. Let us now reexamine these ideas in somewhat greater detail with the idea of using them in a very general space.

Closely associated with the covariant derivative is the *absolute* differential.[2] Given a tensor field $T^{kl\cdots m}_{pq\cdots r}$ in a region $\mathcal{R}$, the absolute differential $\delta T^{kl\cdots m}_{pq\cdots r}$ may be defined by the following four postulates:

(A) The absolute differential $\delta T^{kl\cdots m}_{pq\cdots r}$ transforms in the same way as the tensor $T^{kl\cdots m}_{pq\cdots r}$.

(B) The absolute differential of a sum of tensors is the sum of their absolute differentials:

$$\delta(T^{kl\cdots m}_{pq\cdots r}+U^{kl\cdots m}_{pq\cdots r})=\delta T^{kl\cdots m}_{pq\cdots r}+\delta U^{kl\cdots m}_{pq\cdots r}. \tag{9.01}$$

(C) The absolute differential of a product (contracted or uncontracted)* of tensors follows Leibniz's rule:

$$\delta(T^{kl\cdots m}_{pq\cdots r}U^{ab\cdots c}_{ef\cdots g})=T^{kl\cdots m}_{pq\cdots r}\delta U^{ab\cdots c}_{ef\cdots g}+U^{ab\cdots c}_{ef\cdots g}\delta T^{kl\cdots m}_{pq\cdots r}. \tag{9.02}$$

(D) If the tensor is univalent, the absolute differentials are

$$\underset{1}{\delta}v^i=dv^i+\Gamma^i_{jk}v^k\,dx^j, \tag{9.03a}$$

$$\underset{2}{\delta}v^i=dv^i+\Gamma^i_{kj}v^k\,dx^j, \tag{9.03b}$$

$$\underset{1}{\delta}u_i=du_i-\Gamma^k_{ji}u_k\,dx^j, \tag{9.04a}$$

$$\underset{2}{\delta}u_i=du_i-\Gamma^k_{ij}u_k\,dx^j. \tag{9.04b}$$

If the holor Γ^i_{jk} is not required to be symmetric in its subscripts, there are two equally important absolute differentials of a vector.

The absolute derivative of a univalent contravariant vector along a curve with parameter u is obtained from Equation (9.03),

$$\left(\frac{\delta v^i}{du}\right)_1=\frac{dv^i}{du}+\Gamma^i_{jk}v^k\frac{dx^j}{du}, \tag{9.05a}$$

$$\left(\frac{\delta v^i}{du}\right)_2=\frac{dv^i}{du}+\Gamma^i_{kj}v^k\frac{dx^j}{du}. \tag{9.05b}$$

To obtain the absolute differential of a nilvalent tensor S, we take

$$S=u_iv^i.$$

Then, according to postulates C and D,

$$\underset{1}{\delta}S=\underset{1}{\delta}(u_iv^i)=u_i\underset{1}{\delta}v^i+v^i\underset{1}{\delta}u_i.$$

Substituting Equations (9.03a) and (9.04a),

$$\begin{aligned}\underset{1}{\delta}S&=u_i(dv^i+\Gamma^i_{jk}v^k\,dx^j)+v^i(du_i-\Gamma^k_{ji}u_k\,dx^j)\\&=u_i\,dv^i+v^i\,du_i+\Gamma^i_{jk}u_iv^k\,dx^j-\Gamma^k_{ji}u_kv^i\,dx^j.\end{aligned}$$

* A still more general formulation can be obtained by limiting (c) to *uncontracted* products. The Γ's then become different in the transformation formulas for covariant and contravariant tensors. See J. A. Schouten and V. Hlavaty, "Zur Theorie der allgemeinen Übertragung," *Math. Zs.*, **30**, 1929, p. 414; J. L. Synge and A. Schild, *Tensor calculus* (University of Toronto Press, 1949), Chapter 8.

Since dummy indices can be renamed, the last two terms cancel, leaving

$$\underset{1}{\delta} S = d(u_i v^i) = dS. \tag{9.06a}$$

Similarly,

$$\begin{aligned}\underset{2}{\delta} S = \underset{2}{\delta}(u_i v^i) &= u_i \underset{2}{\delta} v^i + v^i \underset{2}{\delta} u_i \\ &= u_i(dv^i + \Gamma^i_{kj} v^k\, dx^j) + v^i(du_i - \Gamma^k_{ij} u_k\, dx^j) \\ &= u_i\, dv^i + v^i\, du_i + \Gamma^i_{kj} u_i v^k\, dx^j - \Gamma^k_{ij} u_k v^i\, dx^j.\end{aligned}$$

Canceling the last two terms,

$$\underset{2}{\delta} S = d(u_i v^i) = dS. \tag{9.06b}$$

Combining Equations (9.06a) and (9.06b),

$$\underset{1}{\delta} S = \underset{2}{\delta} S = d(u_i v^i) = dS. \tag{9.07}$$

Thus, *the absolute differential of a nilvalent tensor is identical with the ordinary differential.*

The foregoing definition of the absolute differential can easily be shown to uniquely determine the transformation equation of the linear connection. According to Equation (9.03),

$$\underset{1}{\delta} v^i = dv^i + \Gamma^i_{jk} v^k\, dx^j$$

or

$$\overset{1}{\nabla}_j v^i = \frac{\partial v^i}{\partial x^j} + \Gamma^i_{jk} v^k \tag{9.08a}$$

and

$$\underset{2}{\delta} v^i = dv^i + \Gamma^i_{kj} v^k\, dx^j$$

or

$$\overset{2}{\nabla}_j v^i = \frac{\partial v^i}{\partial x^j} + \Gamma^i_{kj} v^k. \tag{9.08b}$$

Here $\overset{1}{\nabla}_j v^i$, $\overset{2}{\nabla}_j v^i$, and v^k are tensors but $\partial v^i/\partial x^j$ and Γ^i_{jk} are not tensors. The transformation equations of the tensor quantities are

$$\overset{1}{\nabla}_j v^i = \frac{\partial x^i}{\partial x^{i'}} \frac{\partial x^{j'}}{\partial x^j} \overset{1}{\nabla}_{j'} v^{i'}, \tag{9.09a}$$

$$\overset{2}{\nabla}_j v^i = \frac{\partial x^i}{\partial x^{i'}} \frac{\partial x^{j'}}{\partial x^j} \overset{2}{\nabla}_{j'} v^{i'}, \tag{9.09b}$$

$$v^i = \frac{\partial x^i}{\partial x^{i'}} v^{i'}. \tag{9.10}$$

Differentiation of Equation (9.10) with respect to x^j gives the nontensor transformation equation,

$$\frac{\partial v^i}{\partial x^j} = \frac{\partial x^i}{\partial x^{i'}} \frac{\partial x^{j'}}{\partial x^j} \frac{\partial v^{i'}}{\partial x^{j'}} + \frac{\partial^2 x^i}{\partial x^{i'} \partial x^{j'}} \frac{\partial x^{j'}}{\partial x^j} v^{i'}. \tag{9.11}$$

Substituting Equations (9.09a) and (9.10) and (9.11) into (9.08a) gives

$$\frac{\partial x^i}{\partial x^{i'}} \frac{\partial x^{j'}}{\partial x^j} \overset{1}{\nabla}_{j'} v^{i'} = \frac{\partial x^i}{\partial x^{i'}} \frac{\partial x^{j'}}{\partial x^j} \frac{\partial v^{i'}}{\partial x^{j'}} + \frac{\partial^2 x^i}{\partial x^{i'} \partial x^{j'}} \frac{\partial x^{j'}}{\partial x^j} v^{i'} + \Gamma^i_{jk} \frac{\partial x^k}{\partial x^{k'}} v^{k'}.$$

On rearranging terms and changing the names of dummy indices,

$$\frac{\partial x^i}{\partial x^{i'}} \frac{\partial x^{j'}}{\partial x^j} \left(\overset{1}{\nabla}_{j'} v^{i'} - \frac{\partial v^{i'}}{\partial x^{j'}} \right) = \left(\frac{\partial^2 x^i}{\partial x^{k'} \partial x^{j'}} \frac{\partial x^{j'}}{\partial x^j} + \Gamma^i_{jk} \frac{\partial x^k}{\partial x^{k'}} \right) v^{k'}. \tag{9.12}$$

But if we assume that the same definition of the covariant derivative holds in all coordinate systems, then by Equation (9.08a)

$$\overset{1}{\nabla}_{j'} v^{i'} = \frac{\partial v^{i'}}{\partial x^{j'}} + \Gamma^{i'}_{j'k'} v^{k'}$$

or

$$\overset{1}{\nabla}_{j'} v^{i'} - \frac{\partial v^{i'}}{\partial x^{j'}} = \Gamma^{i'}_{j'k'} v^{k'}. \tag{9.13}$$

Substituting Equation (9.13) into (9.12) gives

$$\frac{\partial x^i}{\partial x^{i'}} \frac{\partial x^{j'}}{\partial x^j} \left(\Gamma^{i'}_{j'k'} v^{k'} \right) = \left(\frac{\partial^2 x^i}{\partial x^{k'} \partial x^{j'}} \frac{\partial x^{j'}}{\partial x^j} + \Gamma^i_{jk} \frac{\partial x^k}{\partial x^{k'}} \right) v^{k'}.$$

Factoring out $v^{k'}$,

$$\left(\frac{\partial x^i}{\partial x^{i'}} \frac{\partial x^{j'}}{\partial x^j} \Gamma^{i'}_{j'k'} - \Gamma^i_{jk} \frac{\partial x^k}{\partial x^{k'}} - \frac{\partial^2 x^i}{\partial x^{k'} \partial x^{j'}} \frac{\partial x^{j'}}{\partial x^j} \right) v^{k'} = 0. \tag{9.14}$$

Since Equation (9.14) is true for any arbitrary vector $v^{k'}$, the quantity in parentheses, which multiplies $v^{k'}$, must be zero. Therefore,

$$\frac{\partial x^i}{\partial x^{i'}} \frac{\partial x^{j'}}{\partial x^j} \Gamma^{i'}_{j'k'} = \Gamma^i_{jk} \frac{\partial x^k}{\partial x^{k'}} + \frac{\partial^2 x^i}{\partial x^{k'} \partial x^{j'}} \frac{\partial x^{j'}}{\partial x^j}. \tag{9.15}$$

Now multiply the equation by the inverses of coefficients $\partial x^i / \partial x^{i'}$ and $\partial x^{j'} / \partial x^j$, which are defined by the equations

$$\frac{\partial x^{m'}}{\partial x^i} \frac{\partial x^i}{\partial x^{i'}} = \delta^{k'}_{i'}, \qquad \frac{\partial x^j}{\partial x^{i'}} \frac{\partial x^{j'}}{\partial x^j} = \delta^{j'}_{n'}$$

and contract on indices i and j. Then

$$\frac{\partial x^{m'}}{\partial x^i}\frac{\partial x^j}{\partial x^{n'}}\frac{\partial x^i}{\partial x^{i'}}\frac{\partial x^{j'}}{\partial x^j}\Gamma^{i'}_{j'k'} = \frac{\partial x^{m'}}{\partial x^i}\frac{\partial x^j}{\partial x^{n'}}\frac{\partial x^k}{\partial x^{k'}}\Gamma^i_{jk} + \frac{\partial^2 x^i}{\partial x^{k'}\partial x^{j'}}\frac{\partial x^{j'}}{\partial x^j}\frac{\partial x^{m'}}{\partial x^i}\frac{\partial x^j}{\partial x^{n'}} \tag{9.16}$$

or, in terms of Kronecker deltas,

$$\delta^{m'}_{i'}\delta^{j'}_{n'}\Gamma^{i'}_{j'k'} = \frac{\partial x^{m'}}{\partial x^i}\frac{\partial x^j}{\partial x^{n'}}\frac{\partial x^k}{\partial x^{k'}}\Gamma^i_{jk} + \frac{\partial^2 x^i}{\partial x^{k'}\partial x^{j'}}\frac{\partial x^{m'}}{\partial x^i}\delta^{j'}_{n'}$$

or

$$\Gamma^{m'}_{n'k'} = \frac{\partial x^{m'}}{\partial x^i}\frac{\partial x^j}{\partial x^{n'}}\frac{\partial x^k}{\partial x^{k'}}\Gamma^i_{jk} + \frac{\partial^2 x^i}{\partial x^{k'}\partial x^{n'}}\frac{\partial x^{m'}}{\partial x^i}. \tag{9.17}$$

which is equivalent to the transformation with which we started in Chapter 7. Thus, we can start with the transformation equation of Γ^i_{jk} and derive the expression for the covariant derivative, or we can postulate the definition of the covariant derivative and from it derive the transformation equation of Γ^i_{jk}.

The transformation equation of the linear connection can also be derived by postulating any of the definitions of the covariant derivative and from it derive the transformation equation of Γ^i_{jk}.

The transformation equation of the linear connection can also be derived by postulating any of the definitions of the covariant derivative of a tensor.

Other forms of the transformation equation of the linear connection are sometimes useful:

$$\Gamma^{i'}_{j'k'} = \frac{\partial x^{i'}}{\partial x^i}\frac{\partial x^j}{\partial x^{j'}}\frac{\partial x^k}{\partial x^{k'}}\Gamma^i_{jk} - \frac{\partial^2 x^{i'}}{\partial x^j\partial x^k}\frac{\partial x^j}{\partial x^{j'}}\frac{\partial x^k}{\partial x^{k'}}, \tag{9.17a}$$

$$\Gamma^i_{jk} = \frac{\partial x^i}{\partial x^{i'}}\frac{\partial x^{j'}}{\partial x^j}\frac{\partial x^{k'}}{\partial x^k}\Gamma^{i'}_{j'k'} + \frac{\partial^2 x^{i'}}{\partial x^j\partial x^k}\frac{\partial x^i}{\partial x^{i'}}, \tag{9.17b}$$

$$\Gamma^i_{jk} = \frac{\partial x^i}{\partial x^{i'}}\frac{\partial x^{j'}}{\partial x^j}\frac{\partial x^{k'}}{\partial x^k}\Gamma^{i'}_{j'k'} - \frac{\partial^2 x^i}{\partial x^{j'}\partial x^{k'}}\frac{\partial x^{j'}}{\partial x^j}\frac{\partial x^{k'}}{\partial x^k}. \tag{9.17c}$$

9.02. Homeomorphic displacements*

Consider a tensor field $T^{kl\cdots m}_{pq\cdots r}$ in a region $\mathcal{R}$, the tensor being of any valence.[3] In general, if one moves from point P to a neighboring point

* From ὅμοιος *same* and μορφή *form*.

Q in $\mathcal{R}$, $T^{kl\cdots m}_{pq\cdots r}$ will change. It is possible, however, that a direction can be found at P such that the absolute differential $\delta T^{kl\cdots m}_{pq\cdots r}$ is zero for the displacement PQ.[4] At Q, there is perhaps a direction in which a further displacement causes no change in $\delta T^{kl\cdots m}_{pq\cdots r}$. In this way, one may be able to map a curve C at any point of which

$$\delta T^{kl\cdots m}_{pq\cdots r} = 0. \tag{9.18}$$

Evidently, the criterion is the vanishing of the absolute differential, not the ordinary differential, since only in this way will the result hold for all coordinate systems.*

Definition XIII: An infinitesimal displacement of a tensor $T^{kl\cdots m}_{pq\cdots r}$ for which

$$\delta T^{kl\cdots m}_{pq\cdots r} = 0$$

is called a homeomorphic displacement.

If $T^{kl\cdots m}_{pq\cdots r}$ is a nilvalent tensor S, then according to Equation (9.07),

$$\delta S = dS.$$

Thus, a homeomorphic displacement of a nilvalent tensor gives

$$dS = 0,$$

and S is a constant for a homeomorphic displacement.

If the tensor is a univalent contravariant vector, the absolute differentials are

$$\underset{1}{\delta} v^i = dv^i + \Gamma^i_{jk} v^k \, dx^j \tag{9.03a}$$

or

$$\underset{2}{\delta} v^i = dv^i + \Gamma^i_{kj} v^k \, dx^j. \tag{9.03b}$$

Thus, for an asymmetric linear connection, two homeomorphic displacements of a univalent contravariant tensor are determined by the equations

* Usually called *parallel displacement*. We object on two counts to this use of the word *parallel*:

(1) In an X_n, parallelism is meaningless. The space must have at least the structure of affine space for the concept of parallelism to have any significance. To circumvent this difficulty, geometers attempt to associate with each point of the X_n a local affine space. The result tends to be both clumsy and ambiguous.

(2) The idea of parallelism applies to vectors. But it is convenient to apply homeomorphic displacements to tensors of all valences. What is the meaning of "the scalar at Q is parallel to the scalar at P" or "the trivalent tensor at Q is parallel to the trivalent tensor at P"?

$$dv^i + \Gamma^i_{jk} v^k \, dx^j = 0 \tag{9.19a}$$

or

$$dv^i + \Gamma^i_{kj} v^k \, dx^j = 0. \tag{9.19b}$$

Similarly, there are two homeomorphic displacements of a covariant tensor u_k that satisfy the equations

$$du_i - \Gamma^k_{ji} u_k \, dx^j = 0 \tag{9.20a}$$

or

$$du_i - \Gamma^k_{ij} u_k \, dx^j = 0. \tag{9.20b}$$

In particular, if $\Gamma^i_{jk} = 0$, then $dv^i = 0$ and $du_i = 0$ and along a homeomorphic path v^i = constant and u_i = constant. In general, $\Gamma^i_{jk} \neq 0$ and the tensors v^k and u_i vary along the homeomorphic paths.

If the tensor is bivalent, w^{ij}, four homeomorphic paths are defined by the equations

$$dw^{ij} + \Gamma^i_{kl} w^{lj} \, dx^k + \Gamma^j_{km} w^{mi} \, dx^k = 0, \tag{9.21a}$$

or

$$dw^{ij} + \Gamma^i_{lk} w^{lj} \, dx^k + \Gamma^j_{km} w^{mi} \, dx^k = 0, \tag{9.21b}$$

or

$$dw^{ij} + \Gamma^i_{kl} w^{lj} \, dx^k + \Gamma^j_{mk} w^{mi} \, dx^k = 0, \tag{9.21c}$$

or

$$dw^{ij} + \Gamma^i_{lk} w^{lj} \, dx^k + \Gamma^j_{mk} w^{mi} \, dx^k = 0. \tag{9.21d}$$

Each of these four homeomorphic displacements causes a change in the merates of the tensor w^{ij} except in the special case $\Gamma^i_{jk} = 0$, which is associated with w^{ij} = constant along the single homeomorphic path.

The case of a vector v^i and a curve C along which the absolute differential is zero is particularly important. A parameter u can be introduced along the curve $x^j(u)$,

$$\left(\frac{\delta v^i}{du}\right)_1 = \frac{dv^i}{du} + \Gamma^i_{jk} v^k \frac{dx^j}{du} = 0, \tag{9.22a}$$

or

$$\left(\frac{\delta v^i}{du}\right)_2 = \frac{dv^i}{du} + \Gamma^i_{kj} v^k \frac{dx^j}{du} = 0. \tag{9.22b}$$

Definition XIV: The curves C_1 and C_2 along which the two absolute derivatives of a vector v^i are zero,

$$\left(\frac{\delta v^i}{du}\right)_1 = 0 \quad \text{or} \quad \left(\frac{\delta v^i}{du}\right)_2 = 0,$$

are called *paths*.* If the linear connection is not symmetric, two distinct paths are defined at every point.

In one interesting special case, the two paths become identical even for an asymmetric linear connection. If

$$v^k = \frac{dx^k}{du}, \tag{9.23}$$

then both Equations (9.22a) and (9.22b) give the same path defined by

$$\frac{\delta v^k}{du} = \frac{dv^k}{du} + \Gamma^i_{jk} v^j v^k = 0. \tag{9.24}$$

The *geometry of paths*[2] is a comparatively recent development in differential geometry and one that promises to be of great importance.

9.03. The infinitesimal pentagon

Consider an arbitrary point 0 in region $\mathcal{R}$ and two infinitesimal increments $(dx^i)_1$ and $(dx^i)_2$ defined at point 0 (Figure 9.01). Two homeomorphic displacements[3] of a contravariant vector v^k are defined by the equations

$$\underset{1}{\delta} v^i = dv^i + \Gamma^i_{jk} v^k \, dx^j = 0 \tag{9.25a}$$

and

$$\underset{2}{\delta} v^i = dv^i + \Gamma^i_{kj} v^k \, dx^j = 0. \tag{9.25b}$$

Suppose that we now consider a particular vector v^i. Let

$$(v^i)_{0A} = (dx^i)_1 = \overrightarrow{0A}.$$

If we displace this vector homeomorphically to point B, the change in v^i is by Equation (9.25),

$$(dv^i)_1 = -(\Gamma^i_{jk})_0 (dx^k)_1 (dx^j)_2 = -da^i \tag{9.26a}$$

* Sometimes called a *geodesic*. It seems preferable to use the word *path* in a nonmetric space, however, reserving *geodesic* for metric spaces, where a stationary *distance* has significance. See Synge and Schild,[2] Eq. (8.117).

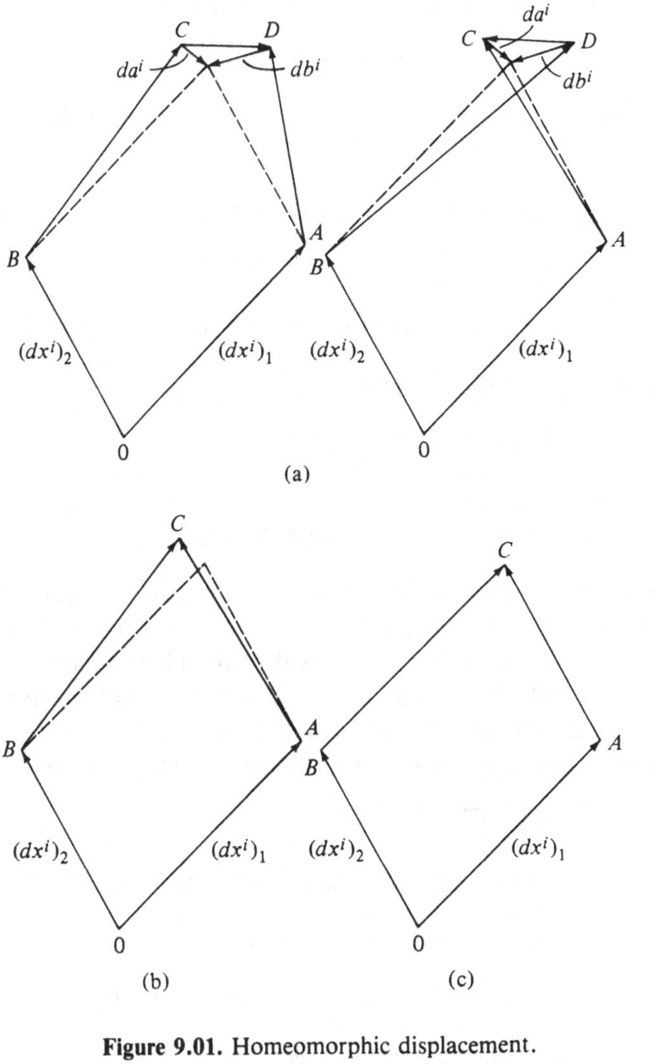

Figure 9.01. Homeomorphic displacement.

or

$$(dv^i)_2 = -(\Gamma^i_{kj})_0 (dx^k)_1 (dx^j)_2 = -db^i. \tag{9.26b}$$

The vector field v^i at point B is either

$$\overrightarrow{BC} = (v^i)_{B1} = (v^i)_{0A} + (dv^i)_1 = (v^i)_{0A} - da^i \tag{9.27a}$$

or

$$\overrightarrow{BD} = (v^i)_{B2} = (v^i)_{0A} + (dv^i)_2 = (v^i)_{0A} - db^i. \tag{9.27b}$$

Now consider another particular vector v^i. Let

$$(v^i)_{0B} = (dx^i)_2 = \overrightarrow{0B}.$$

If we displace this vector to point A, the change in v^i is, by Equation (9.25),

$$(dv^i)_3 = -(\Gamma^i_{jk})_0(dx^k)_2(dx^j)_1 = -(\Gamma^i_{kj})_0(dx^k)_1(dx^j)_2 = -db^i \quad (9.28a)$$

or

$$(dv^i)_4 = -(\Gamma^i_{kj})_0(dx^k)_2(dx^j)_1 = -(\Gamma^i_{jk})_0(dx^k)_1(dx^j)_2 = -da^i. \quad (9.28b)$$

The vector field v^i at point A is either

$$\overrightarrow{AD} = (v^i)_{A1} = (v^i)_{0A} + (dv^i)_3 = (v^i)_{0A} - db^i \quad (9.29a)$$

or

$$\overrightarrow{AC} = (v^i)_{A2} = (v^i)_{0A} + (dv^i)_4 = (v^i)_{0A} - da^i. \quad (9.29b)$$

The polygons formed by this process are shown in Figure 9.01. If the condition $\underset{1}{\delta} v^i = 0$ is employed, then a pentagon is formed as shown in Figure 9.01a. In general, the vectors $\overrightarrow{0A}, \overrightarrow{0B}$ and the vectors obtained by moving $\overrightarrow{0A}$ and $\overrightarrow{0B}$ to B and A, respectively (keeping their first differentials zero), do not quite form a closed quadrilateral. There is a fifth side so a pentagon is generated. The fifth side can readily be expressed as

$$\begin{aligned}
\overrightarrow{CD} &= \overrightarrow{0A} - \overrightarrow{0B} + \overrightarrow{AD} - \overrightarrow{BC} \\
&= (dx^i)_1 - (dx^i)_2 + (dx^i)_2 - (\Gamma^i_{jk})_0(dx^k)_2(dx^j)_1 \\
&\quad - (dx^i)_1 + (\Gamma^i_{jk})_0(dx^k)_1(dx^j)_2 \\
&= (\Gamma^i_{jk})_0(dx^k)_2(dx^j)_1 - (\Gamma^i_{jk})_0(dx^k)_1(dx^j)_2 \\
&= (\Gamma^i_{jk})_0(dx^k)_2(dx^j)_1 - (\Gamma^i_{kj})_0(dx^k)_2(dx^j)_1 \\
&= ((\Gamma^i_{jk})_0 - (\Gamma^i_{kj})_0)(dx^j)_1(dx^k)_2 \\
&= 2(S^i_{jk})_0(dx^j)_1(dx^k)_2.
\end{aligned} \quad (9.30)$$

On the other hand, if we employ the second kind of homeomorphic displacement $\underset{2}{\delta} v^i = 0$, then the pentagon becomes that shown in Figure 9.01b. Note that one of the two pentagons of Figure 9.01a or 9.01b must be a folded pentagon. It could be the pentagon associated with the first or second kind of homeomorphic displacement that is folded. But the fifth side is always plus or minus the same vector. We can write for Figure 9.01a

$$\overrightarrow{0A}+\overrightarrow{AD}=\overrightarrow{0B}+\overrightarrow{BC}+\overrightarrow{CD}, \tag{9.31a}$$

or, for Figure 9.01b,

$$\overrightarrow{0A}+\overrightarrow{AC}=\overrightarrow{0B}+\overrightarrow{BD}+\overrightarrow{DC}. \tag{9.31b}$$

Whenever the torsion tensor $S^i_{\;jk}\neq 0$, there are two pentagons, one of which is a folded pentagon.

However, for a symmetric linear connection, the fifth side disappears since $S^i_{\;jk}=0$. The polygon formed by the original vectors and the vectors obtained by homeomorphic displacement is now a closed quadrilateral (Figure 9.01c). There is only one polygon, and it is closed. But this polygon is not generally a parallelogram.

In the still more special case in which $\Gamma^i_{\;jk}=0$, the polygon becomes a parallelogram and is also closed (Figure 9.01).

9.04. Linearly connected spaces

As in Chapter 6, we begin with a set of points x^i, which constitute an arithmetic space X_n. A general coordinate transformation is allowed:

$$x^{i'}=x^{i'}(x^i), \tag{9.32}$$

where $x^{i'}$ is any single-valued analytic function with unique inverse.

A structure is imposed on this arithmetic space by introducing an arbitrary linear connection $\Gamma^i_{\;jk}$, which is a geometric object with transformation equation

$$\Gamma^i_{\;jk}=\frac{\partial x^i}{\partial x^{i'}}\frac{\partial x^{j'}}{\partial x^j}\frac{\partial x^{k'}}{\partial x^k}\Gamma^{i'}_{\;j'k'}+\frac{\partial^2 x^{i'}}{\partial x^j\,\partial x^k}\frac{\partial x^i}{\partial x^{i'}}. \tag{7.07}$$

The space is now said to be a linearly connected space.

Consider a tensor field $T^{kl\cdots m}_{pq\cdots r}$ of arbitrary valence in a linearly connected space. At each point on a path C,

$$T^{kl\cdots m}_{pq\cdots r}=0.$$

For the special case of a univalent tensor v^i defined as

$$v^i=\frac{dx^i}{du} \tag{9.33}$$

(tangent vector), a path is determined according to Equation (9.05):

$$\frac{\delta v^i}{du}=\frac{dv^i}{du}+\Gamma'^i_{jk}v^k\frac{dx^j}{du}=0$$

or

$$\frac{d^2x^i}{(du)^2}+\Gamma'^i_{jk}\frac{dx^j}{du}\frac{dx^k}{du}=0. \tag{9.34}$$

This differential equation allows us to determine the equations of paths C for the special case of a tangent vector that remains tangent along C.

A particularly simple result is obtained if

$$\Gamma'^i_{jk}=0.$$

Then the differential equation for the paths reduces to

$$\frac{d^2x^i}{(du)^2}=0,$$

whose general solution is

$$x^i=A^i+B^iu$$

or, in 3-space,

$$x^1=A^1+B^1u,$$

$$x^2=A^2+B^2u,$$

$$x^3=A^3+B^3u.$$

Thus, the paths are straight lines determined by the constants $\{A^i\}$ and $\{B^i\}$.

As a somewhat more complicated example, consider a 2-space with spherical coordinates

$$x^1=a, \qquad x^2=\theta, \qquad x^3=\psi.$$

Suppose that the only nonzero merates of Γ'^i_{jk} are

$$\Gamma^2_{33}=-\sin\theta\cos\theta,$$

$$\Gamma^3_{23}=\Gamma^3_{32}=\cot\theta.$$

Thus, for $i=2$, Equation (9.34) becomes

$$\frac{d^2\theta}{(du)^2}-\sin\theta\cos\theta\left(\frac{d\psi}{du}\right)^2=0, \tag{9.35}$$

and for $i=3$,

$$\frac{d^2x^3}{(du)^2}+\Gamma^3_{23}\frac{dx^2}{du}\frac{dx^3}{du}+\Gamma^3_{32}\frac{dx^3}{du}\frac{dx^2}{du}=0$$

or

$$\frac{d^2\psi}{(du)^2}+2\cot\theta\frac{d\theta}{du}\frac{d\psi}{du}=0. \tag{9.36}$$

Consider a great circle on the sphere: Does it satisfy the above differential equations for a path or geodesic? Evidently, we can always orient our spherical coordinate system so that the specified great circle is $\theta=\frac{1}{2}\pi$, with $\psi=0$ at $u=0$. Equation (9.35) is then satisfied; and Equation (9.36) becomes

$$d^2\psi/du^2=0,$$

whose general solution is

$$\psi=A+Bu.$$

If we let

$$u=a\psi,$$

the unique solution of the differential equation is

$$\psi=u/a.$$

Thus any great circle satisfies Equation (9.34) and is a path.

Now take a small circle $\theta=$ constant. Equation (9.35) reduces to

$$\sin\theta\cos\theta\frac{d\psi}{du}=0, \tag{9.35}$$

whereas Equation (9.36) becomes

$$d^2\psi/(ds)^2=0. \tag{9.36}$$

The general solution of Equation (9.36a) is

$$\psi=C+Du.$$

But Equation (9.35a) requires that either

$$\psi=\text{constant} \quad \text{or} \quad \sin\theta\cos\theta=0.$$

Discarding a fixed point ($\psi=$ constant) as a solution, we obtain $\theta=\frac{1}{2}\pi$ as the only possibility. Therefore, one concludes that for this linear connection the only paths in the 2-space are great circles.

9.05. Equations of paths

The differential equation that determines a path[4] C is

$$\frac{d^2x^i}{(du)^2}+\Gamma^i_{jk}\frac{dx^j}{du}\frac{dx^k}{du}=0. \tag{9.34}$$

This is a nonlinear equation, and its general solution cannot be written in a simple form. As boundary conditions, let us take the coordinates $x^i = p^i$ and the derivatives $dx^i/du=\xi^i$ at the fixed point a at which we set $u=0$. The unique solution of Equation (9.34) for these boundary conditions may be expressed as a power series in u:

$$x^i(u)=p^i+\xi^i u-\frac{1}{2!}(\Gamma^i_{jk})_a\xi^j\xi^k(u)^2-\frac{1}{3!}(\Gamma^i_{jkl})_a\xi^j\xi^k\xi^l(u)^3$$
$$-\frac{1}{4!}(\Gamma^i_{jklm})_a\xi^j\xi^k\xi^l\xi^m(u)^4-\cdots. \tag{9.37}$$

Here the higher Γ's are obtained from the recurrence formula

$$\Gamma^i_{jk\cdots pq}=\frac{1}{M}\Sigma\left[\frac{\partial\Gamma^i_{jk\cdots p}}{\partial x^q}-\Gamma^i_{ak\cdots p}\Gamma^a_{jq}-\Gamma^i_{ja\cdots p}\Gamma^a_{kq}-\cdots-\Gamma^i_{jk\cdots a}\Gamma^a_{pq}\right], \tag{9.38}$$

and the summation is for all cyclic permutations of the subscripts of $\Gamma^i_{jk\cdots pq}$. The constant M represents the number of subscripts. For example,

$$\Gamma^i_{jkl}=\frac{1}{3}\Sigma\left[\frac{\partial\Gamma^i_{jk}}{\partial x^l}-\Gamma^i_{pk}\Gamma^p_{jl}-\Gamma^i_{jp}\Gamma^p_{kl}\right], \tag{9.38a}$$

$$\Gamma^i_{jklm}=\frac{1}{4}\Sigma\left[\frac{\partial\Gamma^i_{jkl}}{\partial x^m}-\Gamma^i_{pkl}\Gamma^p_{jm}-\Gamma^i_{jpl}\Gamma^p_{km}-\Gamma^i_{jkp}\Gamma^p_{lm}\right]. \tag{9.38b}$$

To prove that Equation (9.37) is the solution of Equation (9.34), we merely perform a substitution. From Equation (9.37),

$$x^i=p^i+\xi^i u-\frac{1}{2!}(\Gamma^i_{jk})_a\xi^j\xi^k(u)^2-\frac{1}{3!}(\Gamma^i_{jkl})_a\xi^j\xi^k\xi^l(u)^3-\cdots,$$

$$\frac{dx^i}{du}=\xi^i-(\Gamma^i_{jk})_a\xi^j\xi^k u-\tfrac{1}{2}(\Gamma^i_{jkl})_a\xi^j\xi^k\xi^l(u)^2-\cdots,$$

$$\frac{d^2x^i}{(du)^2}=-(\Gamma^i_{jk})_a\xi^j\xi^k-(\Gamma^i_{jkl})_a\xi^j\xi^k\xi^l u-\tfrac{1}{2}(\Gamma^i_{jklm})_a\xi^j\xi^k\xi^l\xi^m(u)^2-\cdots.$$

Substitution into Equation (9.34) gives

$$-\Gamma^i_{jk}\xi^j\xi^k - \Gamma^i_{jkl}\xi^j\xi^k\xi^l u - \tfrac{1}{2}\Gamma^i_{jklm}\xi^j\xi^k\xi^l\xi^m(u)^2 - \cdots$$
$$+\Gamma^i_{jk}(\xi^j - \Gamma^j_{mn}\xi^m\xi^n u - \tfrac{1}{2}\Gamma^j_{mnp}\xi^m\xi^n\xi^p(u)^2 - \cdots)$$
$$\times(\xi^k - \Gamma^k_{mn}\xi^m\xi^n u - \tfrac{1}{2}\Gamma^k_{mnp}\xi^m\xi^n\xi^p(u)^2 - \cdots) = 0.$$

The coefficients of the various powers of u are now evaluated. Evidently the zero-power terms give zero, as they should. The coefficient of the first power of u is

$$-\Gamma^i_{jk}\xi^j\xi^k\xi^l - \Gamma^i_{jk}\Gamma^k_{mn}\xi^j\xi^m\xi^n - \Gamma^i_{jk}\Gamma^j_{mn}\xi^k\xi^m\xi^n.$$

Substitution of Equation (9.38a) leads to

$$\tfrac{1}{3}(\Gamma^i_{pk}\Gamma^p_{jl} + \Gamma^i_{jp}\Gamma^p_{kl} + \Gamma^i_{pl}\Gamma^p_{kj} + \Gamma^i_{jp}\Gamma^p_{lj} + \Gamma^i_{pj}\Gamma^p_{lk} + \Gamma^i_{lp}\Gamma^p_{jk})\xi^j\xi^k\xi^l$$
$$-\Gamma^i_{jk}\Gamma^k_{mn}\xi^j\xi^m\xi^n - \Gamma^i_{jk}\Gamma^j_{mn}\xi^k\xi^m\xi^n = 0.$$

That this coefficient is equal to zero is evident by proper interchange of dummy indices. The Γ's need not be symmetric. Similarly, coefficients of higher powers of u are proved to be zero.

Thus, Equation (9.37) is a solution of Equation (9.34). Our solution satisfies the differential equation and the boundary conditions. By the theory of ordinary differential equations, therefore, Equation (9.37) is the unique solution of the problem. The path is unique:

Theorem XLVI. *Given an arbitrary point P in region $\mathcal{R}$ of an arithmetic n-space with linear connection, and given x^i and dx^i/du at P, then there is one and only one path in $\mathcal{R}$ that satisfies these conditions.*

Alternatively, one may specify coordinates at two points instead of specifying two conditions at P:

Theorem XLVII. *Given arbitrary points P and Q in region $\mathcal{R}$ of an arithmetic n-space with linear connection, there is one and only one path joining P and Q.*

Let us return to the spherical 2-space (Section 9.04) and see how it is handled by the method of the present section. The linear connection is specified by

$$\Gamma^2_{33} = -\sin x^2 \cos x^2,$$
$$\Gamma^3_{23} = \Gamma^3_{32} = \cot x^2.$$

According to Equation (9.38a),

$$\Gamma^i_{jk} = \frac{1}{3}\left(\frac{\partial \Gamma^i_{jk}}{\partial x^l} + \frac{\partial \Gamma^i_{kl}}{\partial x^j} + \frac{\partial \Gamma^i_{lj}}{\partial x^k} - \Gamma^i_{pk}\Gamma^p_{jl} - \Gamma^i_{pl}\Gamma^p_{kj}\right.$$
$$\left. - \Gamma^i_{pj}\Gamma^p_{lk} - \Gamma^i_{jp}\Gamma^p_{kl} - \Gamma^i_{kp}\Gamma^p_{lj} - \Gamma^i_{lp}\Gamma^p_{jk}\right),$$

where $p = 2, 3$. Substitution gives

$$\Gamma^2_{233} = \Gamma^2_{323} = \Gamma^2_{332} = \tfrac{1}{3}(1 - 2\cos^2\theta),$$
$$\Gamma^3_{223} = \Gamma^3_{232} = \Gamma^3_{322} = -2\csc^2\theta,$$

the other Γ's being zero.

Without loss of generality, we can take point P at the top of the sphere and start the path along $\psi = 0$. Then, at $u = 0$,

$$p^2 = 0, \qquad p^3 = 0, \qquad \xi^2 = 1/a, \qquad \xi^3 = 0.$$

There is no reason why the path should continue along $\psi = 0$ unless Equation (9.34) forces it to do so. Except for this equation, the path might follow a small circle or it might go off on any peculiar curve.

According to Equation (9.37), however,

$$x^2(u) = p^2 + \xi^2 u - \frac{1}{2!}\Gamma^2_{jk}\xi^j\xi^k(u)^2 - \frac{1}{3!}\Gamma^2_{jk}\xi^j\xi^k\xi^l(u)^3 - \cdots$$

or

$$\theta(u) = 0 + u/a - \tfrac{1}{2}(\Gamma^2_{22}(\xi^2)^2 + \Gamma^2_{23}\xi^2\xi^3 + \Gamma^2_{32}\xi^3\xi^2 + \Gamma^2_{33}(\xi^3)^2)(u)^2$$
$$- \tfrac{1}{6}\Gamma^2_{222}(\xi^2)^3(u)^3 - \cdots.$$

But the terms in the bracket are zero, as in Γ^2_{222}, so the one and only path has

$$\theta(u) = u/a. \tag{9.39}$$

Similarly,

$$x^3(u) = \psi(u) = p^3 + \xi^3 u - \tfrac{1}{2}(\Gamma^3_{22}(\xi^2)^2 + 0)(u)^2 - \tfrac{1}{6}\Gamma^3_{222}(\xi^2)^2(u)^3 - \cdots$$

or

$$\psi(u) = 0. \tag{9.40}$$

In other words, the path must continue along the great circle $\psi = 0$. Thus, we have proved the well-known result that the only geodesics on a sphere are the great circles (if the linear connection on the sphere is the one assumed in Section 9.04).

9.06. Parameter transformation

The reader is now familiar with the differential equation for paths in an n-space:

$$\frac{d^2x^i}{(du)^2}+\Gamma^i_{jk}\frac{dx^j}{du}\frac{dx^k}{du}=0. \tag{9.34}$$

Here u is a parameter measured along the path C. One can think of a measuring rod of arbitrary length applied to the curve to establish a uniform scale along it. This procedure, however, does not make a metric space (see Chapter 10).

Now suppose that the parameter u is to be changed to another parameter v by arbitrary functional transformation[5]:

$$v=f(u). \tag{9.41}$$

Here f is a single-valued analytic function. Thus,

$$\frac{dx^i}{du}=\frac{dx^i}{dv}\frac{dv}{du},$$

$$\frac{d^2x^i}{(du)^2}=\frac{d^2x^i}{(dv)^2}\left(\frac{dv}{du}\right)^2+\frac{dx^i}{dv}\frac{d^2v}{du^2}.$$

Substitution into Equation (9.34) gives

$$\left(\frac{d^2x^i}{(dv)^2}+\Gamma^i_{jk}\frac{dx^j}{dv}\frac{dx^k}{dv}\right)\left(\frac{dv}{du}\right)^2=-\frac{dx^i}{dv}\frac{d^2v}{du^2}$$

or

$$\left(\frac{dx^i}{dv}\right)^{-1}\left(\frac{d^2x^i}{(dv)^2}+\Gamma^i_{jk}\frac{dx^j}{dv}\frac{dx^k}{dv}\right)=-\frac{d^2t/(du)^2}{(dv/du)^2}. \tag{9.42}$$

Equation (9.42) is a more general form of Equation (9.34), allowing logarithmic or other nonuniform scales along C.

Note that the right side of Equation (9.42) is fixed for a given functional transformation [Equation (9.41)]. Thus, Equation (9.42) applies to any coordinate x^i, and we can write it equally well as

$$\frac{dx^l}{dv}\left(\frac{d^2x^i}{(dv)^2}+\Gamma^i_{jk}\frac{dx^j}{dv}\frac{dx^k}{dv}\right)=\frac{dx^i}{dv}\left(\frac{d^2x^l}{(dv)^2}+\Gamma^l_{mp}\frac{dx^m}{dv}\frac{dx^p}{dv}\right), \tag{9.42a}$$

where $i, j, k, l, m, p=1, 2, \dots, n$.

On the other hand, if the differential equation for a path is to retain its original form [Equation (9.34)], the functional transformation is strictly limited. If

$$\frac{d^2x^i}{(dv)^2}+\Gamma^i_{jk}\frac{dx^j}{dv}\frac{dx^k}{dv}=0,$$

the right side of Equation (9.42) must be zero, or

$$d^2v/(du)^2=0,$$

whose general solution is

$$v=au+b. \tag{9.43}$$

Therefore, the original form of the differential equation continues to apply only if Equation (9.41) is limited to linear transformations. In other words, if Equation (9.34) is to represent a path, only linear transformations of the parameter u are allowable.[6]

9.07. Weyl space

We now consider a particular space of symmetric connection introduced by Hermann Weyl[7] in 1918.

Definition XV: A Weyl space is a space of symmetric connection in which tensors γ_{ij} and ϕ_k exist such that

$$\nabla_k\gamma_{ij}+\phi_k\gamma_{ij}=0, \qquad \gamma_{ij}=\gamma_{ji}, \quad |\gamma_{ij}|\neq 0. \tag{9.44}$$

The existence of the tensor γ_{ij} allows the raising and lowering of indices:

$$\Gamma^i_{jk}\gamma_{il}=\Gamma_{jkl}, \qquad \Gamma_{jkl}\gamma^{il}=\Gamma^i_{jk}, \tag{9.45}$$

where

$$\gamma_{ij}\gamma^{jk}=\delta^k_i.$$

Also, from Equation (7.37),

$$\nabla_k\gamma_{ij}=\frac{\partial\gamma_{ij}}{\partial x^k}-\Gamma^l_{ki}\gamma_{jl}-\Gamma^h_{kj}\gamma_{hi}.$$

Thus, for a symmetric connection,

$$\frac{\partial\gamma_{ij}}{\partial x^k}-\Gamma^l_{ki}\gamma_{jl}-\Gamma^h_{jk}\gamma_{hi}+\phi_k\gamma_{ij}=0.$$

By Equation (9.45),

$$\Gamma_{kij}+\Gamma_{jki}=\frac{\partial\gamma_{ij}}{\partial x^k}+\phi_k\gamma_{ij}. \tag{9.46}$$

and by cyclic interchange of indices,

$$\Gamma_{ijk}+\Gamma_{kij}=\frac{\partial \gamma_{jk}}{\partial x^i}+\phi_i\gamma_{jk},$$

$$\Gamma_{jki}+\Gamma_{ijk}=\frac{\partial \gamma_{ki}}{\partial x^j}+\phi_j\gamma_{ki}.$$

Addition of the latter pair of equations and subtraction of Equation (9.46) gives

$$\Gamma_{ijk}=\frac{1}{2}\left(\frac{\partial \gamma_{jk}}{\partial x^i}+\frac{\partial \gamma_{ki}}{\partial x^j}-\frac{\partial \gamma_{ij}}{\partial x^k}\right)+\frac{1}{2}(\phi_i\gamma_{jk}+\phi_j\gamma_{ki}-\phi_k\gamma_{ij}). \tag{9.47}$$

Equation (9.47), when multiplied by γ^{kl}, yields

$$\Gamma^l_{ij}=\frac{\gamma^{kl}}{2}\left(\frac{\partial \gamma_{jk}}{\partial x^i}+\frac{\partial \gamma_{ki}}{\partial x^j}-\frac{\partial \gamma_{ij}}{\partial x^k}\right)+\frac{1}{2}(\gamma_{jk}\gamma^{kl}\phi_i+\gamma_{ki}\gamma^{kl}\phi_j-\gamma_{ij}\gamma^{kl}\phi_k)$$

or

$$\Gamma^i_{jk}=\frac{\gamma^{il}}{2}\left(\frac{\partial \gamma_{kl}}{\partial x^j}+\frac{\partial \gamma_{lj}}{\partial x^k}-\frac{\partial \gamma_{jk}}{\partial x^l}\right)+\frac{1}{2}(\delta^i_k\phi_j+\delta^i_j\phi_k-\gamma_{jk}\gamma^{li}\phi_l). \tag{9.48}$$

Readers who are familiar with Riemannian geometry will recognize the similarity of the above to the Christoffel symbols, which are usually expressed in terms of the metric tensor g_{ij}. What Weyl has done is to derive analogous equations for the more general case of an arbitrary tensor γ_{ij}. The Weyl space is not a metric space.

Given tensors γ_{ij} and ϕ_k, arbitrary except for the restrictions of Equation (9.44), the linear connection is uniquely determined by Equation (9.48). A given Γ^i_{jk}, however, does not ordinarily determine a unique γ_{ij} and ϕ_k. With Γ^i_{jk} chosen, there still remains a certain flexibility in the choice of γ_{ij} and ϕ_k. Suppose that the latter satisfy Equation (9.48) at a given point P, and we ask if another pair of tensors, say $\bar{\gamma}_{ij}$ and $\bar{\phi}_k$, also satisfy this equation for the same Γ^i_{jk}. Let

$$\bar{\gamma}_{ij}=\lambda\gamma_{ij}, \qquad \bar{\phi}_k=\phi_k-\frac{1}{\lambda}\frac{\partial\lambda}{\partial x^k}, \tag{9.49}$$

where λ is a nilvalent function of position and is called the gauge.

Then

$$\nabla_k\bar{\gamma}_{ij}+\bar{\phi}_k\bar{\gamma}_{ij}=\nabla_k(\lambda\gamma_{ij})+\left(\phi_k-\frac{1}{\lambda}\frac{\partial\lambda}{\partial x^k}\right)\lambda\gamma_{ij}$$

$$=\lambda(\nabla_k\gamma_{ij}+\phi_k\gamma_{ij})+\gamma_{ij}\left(\nabla_k\lambda-\frac{\partial\lambda}{\partial x^k}\right).$$

The first bracket is zero according to Equation (9.44); the second is zero by the definition of a covariant derivative of a nilvalent tensor. Therefore,

$$\nabla_k \bar{\gamma}_{ij} + \bar{\phi}_k \bar{\gamma}_{ij} = 0, \tag{9.44a}$$

which means that the barred tensors can be used interchangeably with the unbarred. In other words, there are infinitely many pairs of tensors for a given Γ^i_{jk}.

The replacement of a pair of tensors γ_{ij}, ϕ_k by an equivalent pair $\bar{\gamma}_{ij}, \bar{\phi}_k$ is called a gauge transformation. Weyl space is concerned with invariance under gauge transformation, or gauge invariance.[8]

9.08. Projectively connected space

In Chapter 6, we considered projective space in the sense of Klein. A feature of such a space was projection along straight lines. We now obtain a generalization of projective space where the straight lines of Chapter 6 are replaced by paths. The new space is called projectively connected space,[9] or the projective space of paths.

With a linearly connected space, the differential equation for the paths is

$$\frac{d^2x^i}{(du)^2} + \Gamma^i_{jk} \frac{dx^j}{du} \frac{dx^k}{du} = 0. \tag{9.34}$$

Under what circumstances can the same paths be obtained with a different connection Υ^i_{jk}? That is,

$$\frac{d^2x^i}{(du)^2} + \Upsilon^i_{jk} \frac{dx^j}{du} \frac{dx^k}{du} = 0. \tag{9.34a}$$

Assume that a path $x^i = x^i(v)$ can be found such that Equation (8.21) is satisfied with either Γ^i_{jk} or Υ^i_{jk}:

$$\frac{dx^i}{dv}\left(\frac{d^2x^i}{(dv)^2} + \Gamma^i_{jk} \frac{dx^j}{dv} \frac{dx^k}{dv}\right) = \frac{dx^i}{dv}\left(\frac{d^2x^l}{(dv)^2} + \Gamma^l_{mn} \frac{dx^m}{dv} \frac{dx^n}{dv}\right),$$

$$\frac{dx^l}{dv}\left(\frac{d^2x^i}{(dv)^2} + \Upsilon^i_{jk} \frac{dx^j}{dv} \frac{dx^k}{dv}\right) = \frac{dx^i}{dv}\left(\frac{d^2x^l}{(dv)^2} + \Upsilon^l_{mn} \frac{dx^m}{dv} \frac{dx^n}{dv}\right).$$

Elimination of the second derivative yields

$$\frac{dx^l}{dv}(\Gamma^i_{jk} - \Upsilon^i_{jk}) \frac{dx^j}{dv} \frac{dx^k}{dv} = \frac{dx^i}{dv}(\Gamma^l_{mn} - \Upsilon^l_{mn}) \frac{dx^m}{dv} \frac{dx^n}{dv}. \tag{9.50}$$

Let

$$\Gamma^i_{jk} - \Upsilon^i_{jk} = \Phi^i_{jk}. \tag{9.51}$$

Then Equation (9.50) becomes

$$(\Phi^i_{jk}\delta^l_p - \Phi^l_{jk}\delta^i_p)\frac{dx^j}{dv}\frac{dx^k}{dv}\frac{dx^p}{dv} = 0. \tag{9.52}$$

Since dx^i/dt is arbitrary,

$$\Phi^i_{jk}\delta^l_p = \Phi^l_{jk}\delta^i_p \tag{9.53}$$

or, introducing additional terms with cyclic interchange of indices,

$$\Phi^i_{jk}\delta^l_p + \Phi^i_{kp}\delta^l_j + \Phi^i_{pj}\delta^l_k = \Phi^l_{jk}\delta^i_p + \Phi^l_{kp}\delta^i_j + \Phi^l_{pj}\delta^i_k.$$

Suppose we now contract by letting $l = p$. Then

$$\Phi^i_{jk} + \Phi^i_{kp}\delta^p_j + \Phi^i_{pj}\delta^p_k = \Phi^p_{jk}\delta^i_p + \Phi_k\delta^i_j + \Phi_j\delta^i_k,$$

where

$$\Phi_k = \frac{\Phi^p_{kp}}{n+1}. \tag{9.54}$$

If the connections are symmetric, four terms combine, giving

$$\Gamma^i_{jk} - \Upsilon^i_{jk} = \Phi_k\delta^i_j + \Phi_j\delta^i_k. \tag{9.55}$$

Equation (9.55) is therefore the necessary condition that identical paths be obtained with Γ^i_{jk} and with Υ^i_{jk}. One can easily show that this condition is also sufficient:

Theorem XLVIII. *The necessary and sufficient condition that Equations* (9.34) *and* (9.34a) *shall represent the same set of paths is that functions* Φ_i *exist that satisfy Equation* (9.55).

The projective connection Π^i_{jk} is now obtained. Suppose that $i = j$ in Equation (9.55). Then

$$\Gamma^i_{ik} - \Upsilon^i_{ik} = \Phi_k\delta^i_i + \Phi_i\delta^i_k = (n+1)\Phi_k.$$

Substitution into Equation (9.55) gives

$$\Gamma^i_{jk} - \frac{1}{n+1}(\delta^i_j\Gamma^l_{lk} + \delta^i_k\Gamma^l_{lj}) = \Upsilon^i_{jk} - \frac{1}{n+1}(\delta^i_j\Upsilon^l_{lk} + \delta^i_k\Upsilon^l_{lj}).$$

The quantity that is invariant under projective transformation of the linear connection is therefore

$$\Pi^i_{jk} = \Gamma^i_{jk} - \frac{1}{n+1}(\delta^i_j \Gamma^l_{lk} + \delta^i_k \Gamma^l_{lj}). \tag{9.56}$$

The differential equation of the paths may now be written

$$\frac{d^2x^i}{(dp)^2} + \Pi^i_{jk}\frac{dx^j}{dp}\frac{dx^k}{dp} = 0, \tag{9.57}$$

where the affine parameter p is independent of projective transformation of the linear connection.

9.09. Projective invariants

The projective connection Π^i_{jk} is not a tensor. In fact, as can be determined from Equation (9.56), it transforms as

$$\begin{aligned}\Pi^i_{jk} = {} & \frac{\partial x^i}{\partial x^{i'}}\frac{\partial x^{j'}}{\partial x^j}\frac{\partial x^{k'}}{\partial x^k}\Pi^{i'}_{j'k'} + \frac{\partial x^i}{\partial x^{i'}}\frac{\partial^2 x^{i'}}{\partial x^j \partial x^k} \\ & - \frac{1}{n+1}\left(\frac{\partial x^i}{\partial x^j}\frac{\partial(\ln J)}{\partial x^k} + \frac{\partial x^i}{\partial x^k}\frac{\partial(\ln J)}{\partial x^j}\right).\end{aligned} \tag{9.58}$$

The first two terms correspond to the terms in the transformation equation of the geometric object Γ^i_{jk} [Equation (7.07)]. But the additional terms in Equation (9.58) eliminate the possibility of expressing covariant differentiation in terms of Π^i_{jk}.

One remedy for this difficulty would be to restrict our allowable coordinate transformations to those with $j = 1$. Another possibility is to retain our general functional transformation but to introduce an additional dimension in the space, so instead of $i, j, k = 1, 2, \ldots, n$, we have $i, j, k = 0, 1, 2, \ldots, n$. Such a step is common in classical projective geometry when homogeneous coordinates are employed.

Consider the transformation

$$x^0 = x^{0'} + \ln J, \qquad x^i = x^i(x^{1'}, x^{2'}, \ldots, x^{n'}).$$

We now define $\Pi^\alpha_{\beta\gamma}$ in the new $(n+1)$-space as equal to Γ^i_{jk} if zero is not involved, but

$$\Pi^\alpha_{0\beta} = \Pi^\alpha_{\beta 0} = -\frac{1}{n+1}\delta^\alpha_\beta,$$

$$\Pi^0_{jk} = \Pi^0_{kj} = \frac{n+1}{n-1}\left[\frac{\partial \Pi^l_{jk}}{\partial x^l} - \Pi^i_{lj}\Pi^l_{ik}\right]$$

if one of the indices is zero. The transformation equation then becomes

Table 9.01. *Designation of spaces*

X_n	= arithmetic space
L_n	= linearly connected space
	= an X_n with linear connection $\Gamma^{\,i}_{jk}$
A_n	= an X_n with *symmetric* linear connection
W_n	= Weyl space
V_n	= Riemannian space
R_n	= Euclidean space
E_n	= affine space

$$\Pi^{\alpha}_{\beta\gamma} = \frac{\partial x^{\alpha}}{\partial x^{\alpha'}} \frac{\partial x^{\beta'}}{\partial x^{\beta}} \frac{\partial x^{\gamma'}}{\partial x^{\gamma}} \Pi^{\alpha'}_{\beta'\gamma'} + \frac{\partial x^{\alpha}}{\partial x^{\alpha'}} \frac{\partial^2 x^{\alpha'}}{\partial x^{\beta}\, \partial x^{\gamma}}. \qquad (9.59)$$

In this way, the projective geometry of paths in an n-space becomes equivalent to the geometry of a linearly connected $(n+1)$-space.

9.10. Summary

The chapter has applied the linear connection $\Gamma^{\,i}_{jk}$ to three kinds of spaces:
(a) spaces L_n with asymmetric linear connection,
(b) Weyl spaces W_n, and
(c) spaces with projective connection.
A list of spaces is given in Table 9.01.

Problems

9.01 From the definition of the second kind of covariant derivative of a contravariant vector $\overset{2}{\nabla}_j v^i$ [Equation (9.08b)], derive an expression for the transformation equation of $\Gamma^{\,i}_{jk}$.

9.02 Repeat Problem 9.01 starting with the definition

$$\overset{1}{\nabla}_j u_i = \frac{\partial u_i}{\partial x^j} - \Gamma^{k}_{ji} u_k.$$

9.03 Repeat Problem 9.01 starting with the definition

$$\overset{2}{\nabla}_j u_i = \frac{\partial u_i}{\partial x^j} - \Gamma^{k}_{ij} u_k.$$

9.04 From Equation (9.17), derive Equation (9.17a).

9.05 From Equation (9.17), derive Equation (9.17b).

9.06 From Equation (9.17), derive Equation (9.17c).

9.07 A thin homogeneous metal disk conducts electricity from a circular terminal of radius a to a concentric terminal of radius b. The portion of the disk between the two terminals is regarded as a region $\mathcal{R}$ of a 2-space. Describe the paths determined by Equation (9.18), where tensor $T^{kl\cdots m}_{pq\cdots r}$ is taken as the electrical potential ϕ.

9.08 In a 2-space with rectangular coordinates (x, y), a scalar field S is found to be

$$S = Ae^{-\pi y/a}\sin(\pi x/a),$$

where A and a are constants.[10] Obtain the equation $y = f(x)$ of paths such that $\delta S = 0$.

9.09 If the linear connection $\Gamma^i_{jk} = 0$, and a bivalent tensor field is defined by the equation

$$T^{ij} = \begin{bmatrix} A^{11} & A^{12} \\ A^{21} & A^{22} \end{bmatrix} e^{-(x^1+x^2)},$$

where A^{ij} are constants, find the equation of paths $x^2 = f(x^1)$ such that $\delta T^{ij} = 0$.

9.10 A linear connection is given by the equations

$$\Gamma^1_{jk} = \begin{bmatrix} 1 & 0 \\ 0 & 0 \end{bmatrix}, \qquad \Gamma^2_{jk} = \begin{bmatrix} 0 & 1 \\ 0 & 0 \end{bmatrix}.$$

(a) Find the differential equation satisfied by v^i if $\underset{1}{\delta} v^i = 0$.
(b) Solve this differential equation if at $x^i = 0$, $v^i = (1, 1)$.
(c) Find the differential equation satisfied by v^i if $\underset{2}{\delta} v^i = 0$.
(d) Solve the differential equations of (c) if at $x^i = 0$, $v^i = (1, 1)$.
(e) Calculate the merates of the vector fields generated by the conditions $\underset{1}{\delta} v^i = 0$ and $\underset{2}{\delta} v^i = 0$ at points $(0,0)$, $(0,1)$, $(0,2)$, $(1,0)$, and $(2,0)$ and plot these pairs of vectors.

9.11 Consider an infinitesimal pentagon in which

$$(dx^k)_1 = (dx, 0), \qquad (dx^k)_2 = (0, dy)$$

and

$$\Gamma^1_{jk} = \begin{bmatrix} \frac{1}{2} & 0 \\ 0 & 0 \end{bmatrix}, \qquad \Gamma^2_{jk} = \begin{bmatrix} 0 & \frac{1}{2} \\ 0 & 0 \end{bmatrix}.$$

Draw the two pentagons calculating each vector carefully and evaluate the vectors $\overrightarrow{C_1C_3}$ and $\overrightarrow{C_2C_4}$.

9.12 Successive differentiation of Equation (9.34) gives

$$\frac{d^3x^i}{(du)^3} + 3\Gamma^i_{jkl}\frac{dx^j}{du}\frac{dx^k}{du}\frac{dx^l}{du} = 0,$$

$$\frac{d^4x^i}{(du)^4} + 4\Gamma^i_{jklm}\frac{dx^j}{du}\frac{dx^k}{du}\frac{dx^l}{du}\frac{dx^m}{du} = 0, \quad \text{etc.}$$

where the Γ's are obtained from Equation (9.38). Show that the first of these equations is true.

9.13 In transforming the parameter u to v in Section 9.06, we obtained the equation

$$\left[\frac{d^2x^i}{(dv)^2} + \Gamma^i_{jk}\frac{dx^j}{dv}\frac{dx^k}{dv}\right]\left(\frac{dv}{du}\right)^2 = -\frac{dx^i}{dv}\frac{d^2v}{du^2},$$

where dx^i/dv is a univalent holor. We have learned that division by a univalent holor is generally meaningless. Yet such a division was performed in obtaining Equation (9.42). How can this step be justified?

9.14 Show that the transformation equation for the projective connection can be reduced to the form of Equation (9.59) by changing from n-space to $(n+1)$-space.

10

Riemannian space

Previous chapters have dealt with spaces in which distances in different directions cannot be compared. We now take the momentous step of introducing a metric, which allows distances to be measured in any direction. As soon as space is provided with a metric, a new set of concepts springs into being: length, area, volume, angle. Also, gradient, divergence, and curl acquire definite physical meanings and the scalar and vector Laplacians can be defined.

The idea of a general metric dates from Riemann's celebrated *Habilitationsschrift*[1] of 1854. In this dissertation, Riemann suggests that the concept of distance be introduced into arithmetic space by means of a quadratic differential form

$$(ds)^2 = g_{ij}\, dx^i\, dx^j. \qquad (10.01)$$

The resulting space is called a *Riemannian space.*

The metric coefficients g_{ij} in Riemannian space can be determined in three ways:

(a) Any arbitrary mathematical functions can be employed as the metric coefficients.

(b) The metric coefficients can be fitted to experimental data. This procedure has been employed, for example, in the study of color discrimination data.[2]

(c) In the special case of Euclidean space, the metric coefficients are obtained by transformation of coordinates from the Pythagorean equation

$$(ds)^2 = (dx)^2 + (dy)^2 + (dz)^2.$$

This derivation will be given in Section 11.02.

A still more general metric was developed by Finsler[3] in 1918; and considerable research has been done on this and similar extensions of the Riemannian metric. In this chapter, however, we shall limit ourselves to Riemannian space and in Chapter 11 to its degenerate form, Euclidean space.

10.01. The metric tensor

In Euclidean 3-space and rectangular coordinates (x, y, z), an infinitesimal distance ds may be expressed as

$$(ds)^2 = (dx)^2 + (dy)^2 + (dz)^2.$$

With spherical coordinates (r, θ, ψ), the same distance is written

$$(ds)^2 = (dr)^2 + (r)^2(d\theta)^2 + (r \sin\theta)^2(d\psi)^2.$$

In parabolic coordinates (μ, ν, ψ),

$$(ds)^2 = [(\mu)^2 + (\nu)^2][(d\mu)^2 + (d\nu)^2] + (\mu\nu)^2(d\psi)^2.$$

With other coordinate systems in Euclidean 3-space, we have similar differential forms. In each case,[4] $(ds)^2$ is written as a sum of products of dx^i and dx^j, generally with coefficients that are functions of the coordinates.

Riemann suggested a generalization of this idea by writing

$$(ds)^2 = g_{ij}\, dx^i\, dx^j, \tag{10.01}$$

where $i, j = 1, 2, \ldots, n$ and $g_{ij} = g_{ji}$. This is a generalization since it now applies in n-space and with skew coordinate systems ($i \neq j$ as well as $i = j$). If we require that $(ds)^2$ be an invariant, then, according to Equation (10.01), g_{ij} is a symmetric tensor, which is called the *metric tensor.* Its merates are arbitrary functions of the coordinates.

If a transformation of coordinates exists such that g_{ij} reduces to the unit matrix, then the metric is said to be Euclidean and we have a Euclidean space (Chapter 11). In a Riemannian 3-space, the matrix is symmetric and has the form

$$g_{ij} = \begin{bmatrix} g_{11} & g_{12} & g_{13} \\ g_{12} & g_{22} & g_{23} \\ g_{13} & g_{23} & g_{33} \end{bmatrix},$$

but if the coordinate system is an orthogonal one, the matrix becomes a diagonal matrix:

$$g_{ij} = \begin{bmatrix} g_{11} & 0 & 0 \\ 0 & g_{22} & 0 \\ 0 & 0 & g_{33} \end{bmatrix}.$$

With the familiar spherical coordinates of Euclidean 3-space, for instance,[4]

$$g_{ij} = \begin{bmatrix} 1 & 0 & 0 \\ 0 & (x^1)^2 & 0 \\ 0 & 0 & (x^1 \sin x^2)^2 \end{bmatrix}.$$

In the present chapter, however, we shall deal with the general case of Riemannian n-space. Corresponding to g_{ij}, it is convenient to have the contravariant tensor g^{jk}, defined by the relation

$$g_{ij} g^{jk} = \delta_i^k. \tag{10.02}$$

Since both g's are tensors, they transform as

$$g_{i'j'} = \frac{\partial x^i}{\partial x^{i'}} \frac{\partial x^j}{\partial x^{j'}} g_{ij}, \qquad g^{j'k'} = \frac{\partial x^{j'}}{\partial x^j} \frac{\partial x^{k'}}{\partial x^k} g^{jk}.$$

The determinant of the metric tensor g_{ij} is written g:

$$g = \begin{vmatrix} g_{11} & g_{12} & \cdots & g_{1n} \\ g_{21} & g_{22} & \cdots & g_{2n} \\ \cdot & \cdot & \cdots & \cdot \\ g_{n1} & g_{n2} & \cdots & g_{nn} \end{vmatrix}. \tag{10.03}$$

Similarly,

$$(g)^{-1} = \begin{vmatrix} g^{11} & g^{12} & \cdots & g^{1n} \\ g^{21} & g^{22} & \cdots & g^{2n} \\ \cdot & \cdot & \cdots & \cdot \\ g^{n1} & g^{n2} & \cdots & g^{nn} \end{vmatrix}, \tag{10.04}$$

where

$$(g)^{-1} = 1/g. \tag{10.05}$$

The transformation equation of g can be obtained directly from the known transformation equation of g_{ij}. Since the determinant of the product of two matrices is equal to the product of the two determinants,

$$g' = \left| \frac{\partial x^i}{\partial x^{i'}} \right| \left| \frac{\partial x^j}{\partial x^{j'}} \right| g$$

or

$$g' = (\Delta)^2 g. \tag{10.06}$$

Therefore, g transforms as an akinetor of weight $+2$. Likewise,

$$(g')^{-1} = (\Delta)^{-2}(g)^{-1}, \tag{10.07}$$

so $(g)^{-1}$ transforms as an akinetor of weight -2. We shall find these facts useful in transforming general akinetors into tensors in Riemannian space. It is convenient to remember that

$$(g')^{1/2} = \Delta(g)^{1/2},$$
$$((g')^{-1})^{1/2} = \Delta^{-1}((g)^{-1})^{1/2}.$$

As an example, take parabolic coordinates[5] in Euclidean 3-space. Here $x^1 = \text{constant}$ and $x^2 = \text{constant}$ represent paraboloids of revolution, and x^3 is the angle measured about the axis of revolution. Equations relating the parabolic coordinates x^1, x^2, x^3 to rectangular coordinates $x^{1'}, x^{2'}, x^{3'}$ are

$$x^{1'} = x^1 x^2 \cos x^3,$$
$$x^{2'} = x^1 x^2 \sin x^3,$$
$$x^{3'} = \tfrac{1}{2}[(x^1)^2 - (x^2)^2].$$

It will be shown in Chapter 11 that the metric coefficients can be derived from these relations in Euclidean space using Equations (11.05) so

$$g_{ij} = \begin{bmatrix} (x^1)^2 + (x^2)^2 & 0 & 0 \\ 0 & (x^1)^2 + (x^2)^2 & 0 \\ 0 & 0 & (x^1 x^2)^2 \end{bmatrix}$$

and

$$(g)^{1/2} = x^1 x^2 [(x^1)^2 + (x^2)^2].$$

An infinitesimal distance anywhere in the space is written

$$(ds)^2 = [(x^1)^2 + (x^2)^2][(dx^1)^2 + (dx^2)^2] + (x^1 x^2)^2 (dx^3)^2.$$

What is the distance from $x^3 = 0$ to $x^3 = \tfrac{1}{2}\pi$ on the intersection of two paraboloids $x^1 = a$ and $x^2 = b$? Evidently,

$$s = \int ds = \int (g_{ij}\, dx^i\, dx^j)^{1/2}$$

or

$$s = \int [g_{11}(dx^1)^2 + g_{22}(dx^2)^2 + g_{33}(dx^3)^2]^{1/2}$$

$$= ab \int_0^{\pi/2} dx^3 = \tfrac{1}{2}\pi ab.$$

Alternatively, we may restrict ourselves to a paraboloidal surface $x^1 = a$. Points in this 2-space are represented by $x^1 = (x^2, x^3)$, and

$$g_{22} = (a)^2 + (x^2)^2, \qquad g_{33} = (ax^2)^2,$$

$$(ds)^2 = [(a)^2 + (x^2)^2](dx^2)^2 + (ax^2)^2(dx^3)^2.$$

Since there is obviously no way in which rectangular coordinates can be applied in this 2-space, it is not a Euclidean space, but it is a Riemannian space.

10.02. Modifications

Chapter 3 showed how holors can be modified in various ways by multiplication by arbitrary γ's. The modifications were of three kinds:
(a) change of index positions,
(b) change of valence, and
(c) change of weight.

A convenient way of making these changes in Riemannian space is to identify the arbitrary γ's with the metric tensors g_{ij} and g^{jk}. Let

$$\gamma_{ij} = g_{ij}, \qquad \gamma^{jk} = g^{jk}. \tag{10.08}$$

(a) Raising and lowering of indices is then easily accomplished. Thus, to obtain a covariant tensor from a contravariant tensor v^j, we write

$$g_{ij} v^j = v_i. \tag{10.09}$$

Similarly,

$$g^{jk} v_k = v^j, \tag{10.10}$$

where the same base letter may be used without ambiguity. Thus, in Riemannian space, indices can be raised or lowered at will, and from a contravariant tensor, we can obtain a related covariant tensor, or vice versa. It should be emphasized, however, that these related tensors are by no means identical.

In accordance with this idea,

$$dx_i = g_{ij}\, dx^j, \qquad dx^j = g^{jk}\, dx_k. \tag{10.11}$$

An elementary distance can therefore be written in several ways:

$$(ds)^2 = g_{ij}\, dx^i\, dx^j = dx_i\, dx^j = g^{ij}\, dx_i\, dx_j. \tag{10.12}$$

The use of the metric tensors allows considerable freedom in the writing of tensor equations. To obtain a nilvalent tensor S, for instance, one may write

$$g_{ij} u^i v^j = S. \tag{10.13}$$

Since g_{ij} is a tensor, S must be a tensor if $u^i v^j$ is a tensor. Other examples are

$$g_{ij} w^{ij} = S, \qquad g_{ij} w^j_k = w_{ik}, \qquad g_{ij} w^{jk} = w^k_i,$$
$$g_{ij} w^{ijk} = v^k, \qquad g^{ij} U_{ijkl} = U_{kl}.$$

(b) Not only can we associate tensors with different arrangements of indices but we can also associate tensors of different valence. Suppose, for example, that we have a univalent tensor v^i and wish to associate with it a nilvalent tensor v. Let us define the magnitude of the vector v^i as

$$v = [g_{ij} v^i v^j]^{1/2}. \tag{10.14}$$

The justification in using the word *magnitude* for this Riemannian invariant is that Equation (10.14) reduces to the usual magnitude of a vector if we take the special case of a Euclidean 3-space with rectangular coordinates. Then $g_{11} = g_{22} = g_{33} = 1$ and $g_{ij} = 0$ for $i \neq j$. Thus, Equation (10.14) reduces to the familiar

$$v = [(v^1)^2 + (v^2)^2 + (v^3)^2]^{1/2}.$$

For a covariant tensor v_i, similarly,

$$v = [g^{ij} v_i v_j]^{1/2}. \tag{10.15}$$

Also, from Equation (10.12),

$$v = [g_{ij} v^i v^j]^{1/2} = [g^{ij} v_i v_j]^{1/2} = [v_i v^i]^{1/2}. \tag{10.16}$$

Another important invariant in Riemannian n-space can be obtained from two arbitrary vectors u^i and v^j and their magnitudes:

$$\frac{g_{ij} u^i v^j}{[g_{ij} u^i u^j g_{ij} v^i v^j]^{1/2}} = \frac{g_{ij} u^i v^j}{uv}.$$

Evidently, this combination is a nilvalent tensor. It may be called

$$\cos\theta = \frac{g_{ij} u^i v^j}{uv}, \tag{10.17}$$

since it agrees with the ordinary concept of angle in Euclidean 2-space. Equation (10.17), however, is not limited to this particular interpretation but may be used in a Riemannian space with any number of dimensions.

Note that the concept of orthogonality, which has no meaning in previous spaces, now has significance. Two vectors u^i and v^j are orthogonal, according to Equation (10.17), if

$$g_{ij}u^iv^j=0. \tag{10.17a}$$

In particular, if the coordinate system is orthogonal, the vectors $u^i=(dx^1,0,0)$ and $v^j=(0,dx^2,0)$ are orthogonal, so

$$g_{ij}u^iv^j=g_{12}\,dx^1\,dx^2=0.$$

Therefore,

$$g_{12}=0.$$

Likewise, the vectors $u^1=(dx^1,0,0)$ and $w^j=(0,0,dx^3)$ are orthogonal, so

$$g_{ij}u^iw^j=g_{13}\,dx^1\,dx^3=0.$$

Therefore,

$$g_{13}=0.$$

Also, v^i is orthogonal to w^j,

$$g_{ij}v^iw^j=g_{23}\,dx^2\,dx^3=0.$$

Therefore,

$$g_{23}=0.$$

Therefore, in any orthogonal coordinate system in Riemannian space, the metric tensor takes on the diagonal form

$$g_{ij}=\begin{bmatrix} g_{11} & 0 & 0 \\ 0 & g_{22} & 0 \\ 0 & 0 & g_{33} \end{bmatrix}. \tag{10.18}$$

The concept of the component of a vector is ordinarily reserved for Euclidean space. However, it can readily be introduced into Riemannian space. Suppose that there are unit vectors $\mathbf{a}_i$ in each coordinate direction. In 3-space, a vector can be written as a linear combination of these unit vectors,

$$\mathbf{v}=\mathbf{a}_1(v)_1+\mathbf{a}_2(v)_2+\mathbf{a}_3(v)_3. \tag{10.19}$$

The coefficients $(v)_i$ are called the *components* of the vector v^i. To find the relation between components and contravariant merates, consider the definition of the length of the vector. The length in an orthogonal coordinate system will be defined as the square root of the sum of the squares of the components of the vector,

$$v=[((v)_1)^2+((v)_2)^2+((v)_3)^2]^{1/2}. \tag{10.20}$$

Equating this expression to Equation (10.14) in the special case of orthogonal coordinates,

$$v=[g_{11}(v^1)^2+g_{22}(v^2)^2+g_{33}(v^3)^2]^{1/2}. \tag{10.21}$$

Equating corresponding terms of Equations (10.20) and (10.21),

$$(v)_1=\sqrt{g_{11}}\,v^1, \qquad (v)_2=\sqrt{g_{22}}\,v^2, \qquad (v)_3=\sqrt{g_{33}}\,v^3. \tag{10.22}$$

This definition of components can be employed for all coordinate systems even though it was derived in the special case of an orthogonal coordinate system. In an oblique system, Equation (10.22) may be retained if we remember to write the length as

$$v=\left[((v)_1)^2+((v)_2)^2+((v)_3)^2+2\frac{g_{12}}{g_{11}g_{22}}(v)_1(v)_2\right.$$
$$\left.+2\frac{g_{13}}{g_{11}g_{33}}(v)_1(v)_3+2\frac{g_{23}}{g_{22}g_{33}}(v)_2(v)_3\right]^{1/2}. \tag{10.23}$$

(c) The Riemannian metric also allows us to change the weight of an akinetor. In particular, an akinetor of weight w can be changed into a tensor. For example, an akinetor of weight -1 transforms as

$$\tilde{v}^{i'}=\Delta^{-1}\frac{\partial x^{i'}}{\partial x^i}\tilde{v}^i.$$

Multiplication of both sides of the equation by $(g^{1/2})'$, which is an akinetor of weight $+1$, gives

$$(g^{1/2})'\tilde{v}^{i'}=(g^{1/2})'\Delta^{-1}\frac{\partial x^{i'}}{\partial x^i}\tilde{v}^i$$
$$=g^{1/2}\Delta\Delta^{-1}\frac{\partial x^{i'}}{\partial x^i}\tilde{v}^{i'}$$

or

$$[(g^{1/2})'\tilde{v}^{i'}]=\frac{\partial x^{i'}}{\partial x^i}[g^{1/2}\tilde{v}^i].$$

Thus, the bracketed quantity transforms as a univalent tensor, which may be written v^i:

$$v^i = g^{1/2}\tilde{v}^i. \tag{10.24}$$

Evidently, the same process can be used to associate a tensor with an arbitrary akinetor of any weight w.

10.03. The linear connection

In Section 9.01, a general definition was given for the absolute differential in terms of the linear connection Γ^i_{jk}. The latter holor is not a tensor but transforms as

$$\Gamma^{i'}_{j'k'} = \frac{\partial x^{i'}}{\partial x^i}\frac{\partial x^j}{\partial x^{j'}}\frac{\partial x^k}{\partial x^{k'}}\Gamma^i_{jk} + \frac{\partial^2 x^i}{\partial x^{j'}\partial x^{k'}}\frac{\partial x^{i'}}{\partial x^i}. \tag{7.07}$$

For arithmetic space, there is no other restriction on the merates of Γ^i_{jk}. Thus, the linear connection is a holor of great flexibility, allowing a great variety of linearly connected spaces.[6]

This flexibility is reduced if we confine our attention to metric spaces. Consider, for example, the homeomorphic displacement of a vector v^i in a Riemannian n-space with a symmetric connection. Since the displacement is homeomorphic, $\delta v^i = 0$ and the covariant derivative is

$$\nabla_k v^i = \frac{\partial v^i}{\partial x^k} + \Gamma^i_{kl}v^l = 0$$

or

$$\frac{\partial v^i}{\partial x^k} = -\Gamma^i_{kl}v^l. \tag{10.25}$$

Magnitude of v^i, which had no significance in X_n, is expressed as

$$v = [g_{ij}v^i v^j]^{1/2} \tag{10.14}$$

in a Riemannian space V_n. Let us impose the requirement that the magnitude of the vector must not change in a homeomorphic displacement. Then

$$\frac{\partial v}{\partial x^k} = 0. \tag{10.26}$$

Thus, in a V_n, both Equations (10.25) and (10.26) must hold. One might expect that the additional restriction would reduce the freedom of choice of Γ^i_{jk}; and such turns out to be the case.

Substitution of Equation (10.14) into (10.26) gives

$$\tfrac{1}{2}(g_{ij}v^iv^j)^{-1/2}\left(\frac{\partial g_{ij}}{\partial x^k}v^iv^j+g_{ij}v^i\frac{\partial v^j}{\partial x^k}+g_{ij}v^j\frac{\partial v^i}{\partial x^k}\right)=0$$

or

$$\frac{\partial g_{ij}}{\partial x^k}v^iv^j+g_{ij}v^i\frac{\partial v^j}{\partial x^k}+g_{ij}v^j\frac{\partial v^i}{\partial x^k}=0. \quad (10.27)$$

Substitution of Equation (10.25) into (10.27) yields

$$\frac{\partial g_{ij}}{\partial x^k}=g_{il}\Gamma^l_{kj}+g_{lj}\Gamma^l_{ki}, \quad (10.28)$$

and cyclic permutation of indices gives

$$\frac{\partial g_{jk}}{\partial x^i}=g_{jl}\Gamma^l_{ik}+g_{lk}\Gamma^l_{ij}, \quad (10.29)$$

$$\frac{\partial g_{ki}}{\partial x^j}=g_{kl}\Gamma^l_{ji}+g_{li}\Gamma^l_{jk}. \quad (10.30)$$

Evidently, considerable cancellation occurs if we add the first and third equations and subtract the second:

$$\begin{aligned}\frac{\partial g_{ij}}{\partial x^k}+\frac{\partial g_{ik}}{\partial x^j}-\frac{\partial g_{jk}}{\partial x^i}&=g_{il}\Gamma^l_{kj}+g_{mj}\Gamma^m_{ki}+g_{kl}\Gamma^l_{ji}\\&\quad+g_{li}\Gamma^l_{jk}-g_{jl}\Gamma^l_{ik}-g_{lk}\Gamma^l_{ij}\\&=2g_{il}\Gamma^l_{kj},\end{aligned}$$

remembering the symmetry of both g_{il} and Γ^l_{kj}. Therefore,

$$\Gamma^i_{jk}=\frac{g^{il}}{2}\left(\frac{\partial g_{lj}}{\partial x^k}+\frac{\partial g_{lk}}{\partial x^j}-\frac{\partial g_{jk}}{\partial x^l}\right). \quad (10.31)$$

This quantity was developed by Christoffel[7] some time before the advent of tensor analysis. It is called the *Christoffel symbol of the second kind.* Equation (10.31) shows that in Riemannian space we no longer have free choice of Γ^i_{jk}: This geometric object is now determined by the metric tensor.

In the special case in which the coordinate surfaces are orthogonal and the metric tensor g_{ij} is a diagonal matrix [Equation (10.18)], the linear connection can be written much more simply. Equation (10.31) becomes

$$\Gamma^1_{jk} = \frac{1}{2g_{11}} \begin{bmatrix} \dfrac{\partial g_{11}}{\partial x^1} & \dfrac{\partial g_{11}}{\partial x^2} & \dfrac{\partial g_{11}}{\partial x^3} \\ \dfrac{\partial g_{11}}{\partial x^2} & -\dfrac{\partial g_{22}}{\partial x^1} & 0 \\ \dfrac{\partial g_{11}}{\partial x^3} & 0 & -\dfrac{\partial g_{33}}{\partial x^1} \end{bmatrix},$$

$$\Gamma^2_{jk} = \frac{1}{2g_{22}} \begin{bmatrix} -\dfrac{\partial g_{11}}{\partial x^2} & \dfrac{\partial g_{22}}{\partial x^1} & 0 \\ \dfrac{\partial g_{22}}{\partial x^1} & \dfrac{\partial g_{22}}{\partial x^2} & \dfrac{\partial g_{22}}{\partial x^3} \\ 0 & \dfrac{\partial g_{22}}{\partial x^3} & -\dfrac{\partial g_{33}}{\partial x^2} \end{bmatrix}, \tag{10.32}$$

$$\Gamma^3_{jk} = \frac{1}{2g_{33}} \begin{bmatrix} -\dfrac{\partial g_{11}}{\partial x^3} & 0 & \dfrac{\partial g_{33}}{\partial x^1} \\ 0 & -\dfrac{\partial g_{22}}{\partial x^3} & \dfrac{\partial g_{33}}{\partial x^2} \\ \dfrac{\partial g_{33}}{\partial x^1} & \dfrac{\partial g_{33}}{\partial x^2} & \dfrac{\partial g_{33}}{\partial x^3} \end{bmatrix}.$$

In an orthogonal cylindrical coordinate system,[4]

$$g_{11} = g_{11}(x^1, x^2), \qquad g_{22} = g_{22}(x^1, x^2), \qquad g_{33} = 1,$$

so Equation (10.32) becomes

$$\Gamma^1_{jk} = \frac{1}{2g_{11}} \begin{bmatrix} \dfrac{\partial g_{11}}{\partial x^1} & \dfrac{\partial g_{11}}{\partial x^2} & 0 \\ \dfrac{\partial g_{11}}{\partial x^2} & -\dfrac{\partial g_{22}}{\partial x^1} & 0 \\ 0 & 0 & 0 \end{bmatrix},$$

$$\Gamma^2_{jk} = \frac{1}{2g_{22}} \begin{bmatrix} -\dfrac{\partial g_{11}}{\partial x^2} & \dfrac{\partial g_{22}}{\partial x^1} & 0 \\ \dfrac{\partial g_{22}}{\partial x^1} & \dfrac{\partial g_{22}}{\partial x^2} & 0 \\ 0 & 0 & 0 \end{bmatrix}, \tag{10.32a}$$

$$\Gamma^3_{jk} = \begin{bmatrix} 0 & 0 & 0 \\ 0 & 0 & 0 \\ 0 & 0 & 0 \end{bmatrix}.$$

In the special case of circular cylinder coordinates,

$$g_{11} = 1, \qquad g_{22} = (x^1)^2 = r^2, \qquad g_{33} = 1,$$

and Equation (10.27a) reduces to

$$\Gamma^1_{jk} = \begin{bmatrix} 0 & 0 & 0 \\ 0 & -r & 0 \\ 0 & 0 & 0 \end{bmatrix},$$

$$\Gamma^2_{jk} = \begin{bmatrix} 0 & 1/r & 0 \\ 1/r & 0 & 0 \\ 0 & 0 & 0 \end{bmatrix}, \qquad (10.32b)$$

$$\Gamma^3_{jk} = \begin{bmatrix} 0 & 0 & 0 \\ 0 & 0 & 0 \\ 0 & 0 & 0 \end{bmatrix}.$$

In orthogonal rotational coordinates,[4]

$$g_{11} = g_{11}(x^1, x^2), \qquad g_{22} = g_{22}(x^1, x^2), \qquad g_{33} = g_{33}(x^1, x^2),$$

so Equation (10.32) can be written

$$\Gamma^1_{jk} = \frac{1}{2g_{11}} \begin{bmatrix} \dfrac{\partial g_{11}}{\partial x^1} & \dfrac{\partial g_{11}}{\partial x^2} & 0 \\ \dfrac{\partial g_{11}}{\partial x^2} & -\dfrac{\partial g_{22}}{\partial x^1} & 0 \\ 0 & 0 & -\dfrac{\partial g_{33}}{\partial x^1} \end{bmatrix},$$

$$\Gamma^2_{jk} = \frac{1}{2g_{22}} \begin{bmatrix} -\dfrac{\partial g_{11}}{\partial x^2} & \dfrac{\partial g_{22}}{\partial x^1} & 0 \\ \dfrac{\partial g_{22}}{\partial x^1} & \dfrac{\partial g_{22}}{\partial x^2} & 0 \\ 0 & 0 & -\dfrac{\partial g_{33}}{\partial x^2} \end{bmatrix}, \qquad (10.32c)$$

$$\Gamma^3_{jk} = \frac{1}{2g_{33}} \begin{bmatrix} 0 & 0 & \dfrac{\partial g_{33}}{\partial x^1} \\ 0 & 0 & \dfrac{\partial g_{33}}{\partial x^2} \\ \dfrac{\partial g_{33}}{\partial x^1} & \dfrac{\partial g_{33}}{\partial x^2} & 0 \end{bmatrix}.$$

For example, in spherical coordinates,

$$g_{11} = 1, \qquad g_{22} = (x^1)^2 = r^2, \qquad g_{33} = (x^1)^2 \sin^2 x^2 = r^2 \sin\theta,$$

and Equation (10.27c) becomes

$$\Gamma^1_{jk} = \begin{bmatrix} 0 & 0 & 0 \\ 0 & -r & 0 \\ 0 & 0 & -r\sin^2\theta \end{bmatrix},$$

$$\Gamma^2_{jk} = \begin{bmatrix} 0 & 1/r & 0 \\ 1/r & 0 & 0 \\ 0 & 0 & -\sin\theta\cos\theta \end{bmatrix}, \tag{10.32d}$$

$$\Gamma^3_{jk} = \begin{bmatrix} 0 & 0 & 1/r \\ 0 & 0 & \cot\theta \\ 1/r & \cot\theta & 0 \end{bmatrix}.$$

A very simple equation can be derived for the contracted linear connection Γ^j_{ij} in terms of the metric coefficients. From the definition of a determinant (Section 1.07) and the definition of the inverse metric coefficients [Equation (10.02)], it can readily be shown that

$$\frac{\partial g}{\partial x^i} = g g^{kj} \frac{\partial g_{kj}}{\partial x^i}, \tag{10.33}$$

permuting indices [Equation (10.31)] can be rewritten as

$$\Gamma^l_{ij} = \frac{g^{lm}}{2}\left(\frac{\partial g_{mi}}{\partial x^j} + \frac{\partial g_{mj}}{\partial x^i} - \frac{\partial g_{ij}}{\partial x^m}\right) \tag{10.31a}$$

and

$$\Gamma^l_{ki} = \frac{g^{lm}}{2}\left(\frac{\partial g_{mk}}{\partial x^i} + \frac{\partial g_{mi}}{\partial x^k} - \frac{\partial g_{ki}}{\partial x^m}\right). \tag{10.31b}$$

Now multiply Equation (10.31a) by g_{lk} and (10.31b) by g_{lj}. Then

$$g_{lk}\Gamma^l_{ij} = \frac{g_{lk}g^{lm}}{2}\left(\frac{\partial g_{mi}}{\partial x^j} + \frac{\partial g_{mj}}{\partial x^i} - \frac{\partial g_{ij}}{\partial x^m}\right)$$

$$= \frac{\delta^m_k}{2}\left(\frac{\partial g_{mi}}{\partial x^j} + \frac{\partial g_{mj}}{\partial x^i} - \frac{\partial g_{ij}}{\partial x^m}\right)$$

$$= \frac{1}{2}\left(\frac{\partial g_{ki}}{\partial x^j} + \frac{\partial g_{kj}}{\partial x^i} - \frac{\partial g_{ij}}{\partial x^k}\right) \tag{10.34a}$$

and

$$g_{lj}\,\Gamma^l_{ki} = \frac{1}{2}\left(\frac{\partial g_{jk}}{\partial x^i} + \frac{\partial g_{ji}}{\partial x^k} - \frac{\partial g_{ki}}{\partial x^j}\right). \tag{10.34b}$$

Addition of Equations (10.34a) and (10.34b) gives

$$\frac{\partial g_{kj}}{\partial x^i} = g_{lk}\Gamma^l_{ij} + g_{lj}\Gamma^l_{ki}. \tag{10.35}$$

Substitution of Equation (10.35) into (10.33) gives

$$\begin{aligned}\frac{\partial g}{\partial x^i} &= gg^{kj}(g_{lk}\Gamma^l_{ij} + g_{lj}\Gamma^l_{ki}) \\ &= g(\delta^j_l\Gamma^l_{ij} + \delta^k_l\Gamma^l_{ki}) \\ &= 2g\delta^j_l\Gamma^l_{ij} \\ &= 2g\Gamma^j_{ij}.\end{aligned}$$

Therefore, we can express Γ^j_{ij} in a very simple form in terms of the metric tensor, as

$$\Gamma^j_{ij} = \frac{1}{2g}\frac{\partial g}{\partial x^i} = \frac{\partial \ln\sqrt{g}}{\partial x^i}. \tag{10.36}$$

In an orthogonal coordinate system,

$$g = g_{11}g_{22}g_{33},$$

so, from Equation (10.36),

$$\begin{aligned}\Gamma^j_{1j} &= \frac{1}{2g_{11}}\frac{\partial g_{11}}{\partial x^1} + \frac{1}{2g_{22}}\frac{\partial g_{22}}{\partial x^1} + \frac{1}{2g_{33}}\frac{\partial g_{33}}{\partial x^1}, \\ \Gamma^j_{2j} &= \frac{1}{2g_{11}}\frac{\partial g_{11}}{\partial x^2} + \frac{1}{2g_{22}}\frac{\partial g_{22}}{\partial x^2} + \frac{1}{2g_{33}}\frac{\partial g_{33}}{\partial x^2}, \\ \Gamma^j_{3j} &= \frac{1}{2g_{11}}\frac{\partial g_{11}}{\partial x^3} + \frac{1}{2g_{22}}\frac{\partial g_{22}}{\partial x^3} + \frac{1}{2g_{33}}\frac{\partial g_{33}}{\partial x^3}.\end{aligned} \tag{10.37}$$

In an orthogonal cylindrical coordinate system,

$$g = g_{11}(x^1, x^2) g_{22}(x^1, x^2).$$

Thus, Equation (10.37) becomes

$$\Gamma^j_{1j} = \frac{1}{2g_{11}} \frac{\partial g_{11}}{\partial x^1} + \frac{1}{2g_{22}} \frac{\partial g_{22}}{\partial x^1},$$
$$\Gamma^j_{2j} = \frac{1}{2g_{11}} \frac{\partial g_{11}}{\partial x^2} + \frac{1}{2g_{22}} \frac{\partial g_{22}}{\partial x^2}, \qquad (10.37a)$$
$$\Gamma^j_{3j} = 0.$$

In circular cylinder coordinates, where

$$g_{11} = 1 \quad \text{and} \quad g_{22} = (x^1)^2 = r^2,$$

Equation (10.37a) becomes

$$\Gamma^j_{1j} = 1/r, \qquad \Gamma^j_{2j} = 0, \qquad \Gamma^j_{3j} = 0. \qquad (10.37b)$$

With orthogonal rotational coordinates where

$$g_{11} = g_{11}(x^1, x^2), \qquad g_{22} = g_{22}(x^1, x^2), \qquad g_{33} = g_{33}(x^1, x^2),$$

Equation (10.37) simplifies to

$$\Gamma^j_{1j} = \frac{1}{2g_{11}} \frac{\partial g_{11}}{\partial x^1} + \frac{1}{2g_{22}} \frac{\partial g_{22}}{\partial x^1} + \frac{1}{2g_{33}} \frac{\partial g_{33}}{\partial x^1},$$
$$\Gamma^j_{2j} = \frac{1}{2g_{11}} \frac{\partial g_{11}}{\partial x^2} + \frac{1}{2g_{22}} \frac{\partial g_{22}}{\partial x^2} + \frac{1}{2g_{33}} \frac{\partial g_{33}}{\partial x^2}, \qquad (10.37c)$$
$$\Gamma^j_{3j} = 0.$$

In spherical coordinates, where

$$g_{11} = 1, \qquad g_{22} = r^2, \qquad g_{33} = r^2 \sin^2 \theta,$$

Equation (10.37c) becomes

$$\Gamma^j_{1j} = 2/r, \qquad \Gamma^j_{2j} = \cot \theta, \qquad \Gamma^j_{3j} = 0. \qquad (10.37d)$$

10.04. Geodesics

The simplest geodesic in Euclidean space is a straight line. It is the path of a light ray in free space or of an unaccelerated space ship. A geodesic in Euclidean space has two characteristics:

(a) It is the shortest path between two given points.
(b) A tangent vector moved along a geodesic by homeomorphic displacement remains tangent to the curve.

In generalizing this idea to nonmetric spaces, we must relinquish (a), since no general ds is available. The closest approach to a geodesic in a nonmetric space is the path (Chapter 9), which is based on (b). In Riemannian space, however, a metric exists, and the distance between points A and B is

$$L=\int_A^B ds=\int_A^B\left(g_{ij}\frac{\partial x^i}{\partial u}\frac{\partial x^j}{\partial u}\right)^{1/2}du.$$

A geodesic is the curve whose length L has a stationary value with respect to small variations in the curve, the end points A and B being fixed. This optimum curve is evaluated by use of the calculus of variations,[8] and the differential equation for a geodesic is

$$\frac{d^2x^i}{(ds)^2}+\Gamma^i_{jk}\frac{dx^j}{ds}\frac{dx^k}{ds}=0, \tag{10.38}$$

which is just the equation found for paths in Section 9.05, except that the parameter is now s rather than u.

The principal difference in the two formulations is that Γ^i_{jk} for the general case of paths is arbitrary, while the calculus-of-variations approach requires a Γ^i_{jk} given by Equation (10.31). In other words, in Riemannian space, the linear connection is definitely fixed by the metric tensor. The relation [Equation (10.31)] is necessary so that homeopathic displacement of a vector will not change the magnitude. The same relation is necessary so that the path will also be the shortest distance.

On the surface of a circular cylinder, $x^1=r=a$. So $dx^1/ds=0$. Expansion of Equation (10.38) gives

$$\begin{aligned}&\frac{d^2x^2}{(ds)^2}+\Gamma^2_{22}\left(\frac{dx^2}{ds}\right)^2+2\Gamma^2_{23}\frac{dx^2}{ds}\frac{dx^3}{ds}+\Gamma^2_{33}\left(\frac{dx^3}{ds}\right)^2=0,\\&\frac{d^2x^3}{(ds)^2}+\Gamma^3_{22}\left(\frac{dx^2}{ds}\right)^2+2\Gamma^3_{23}\frac{dx^2}{ds}\frac{dx^3}{ds}+\Gamma^3_{33}\left(\frac{dx^3}{ds}\right)^2=0.\end{aligned} \tag{10.38a}$$

On the surface of this cylinder, the linear connection is [by Equation (10.32b)]

$$\Gamma^1_{jk}=\begin{bmatrix}0&0&0\\0&-a&0\\0&0&0\end{bmatrix},\quad \Gamma^2_{jk}=\begin{bmatrix}0&1/a&0\\1/a&0&0\\0&0&0\end{bmatrix},\quad \Gamma^3_{jk}=\begin{bmatrix}0&0&0\\0&0&0\\0&0&0\end{bmatrix}.$$

So Equation (10.38a) reduces to

$$\frac{d^2x^2}{(ds)^2}=0, \quad \frac{d^2x^3}{(ds)^2}=0, \quad \text{or} \quad \frac{d^2\psi}{(ds)^2}=0, \quad \frac{d^2z}{(ds)^2}=0. \tag{10.38b}$$

Integration of Equation (10.38b) gives

$$\psi=bs+c, \qquad z=es+f, \tag{10.39}$$

or if at $s=0$, $\psi=z=0$,

$$\psi=bs, \quad z=es, \quad \text{or} \quad z=\left(\frac{e}{b}\right)\psi. \tag{10.40}$$

If $e=0$, the geodesic is a horizontal circle. But if $b=0$, the geodesic becomes a vertical line. If neither b nor e vanishes, the geodesic is a helix on the cylinder (Figure 10.01). Note that if the surface of the cylinder is developed onto a plane, all of the geodesics on the circular cylinder become straight lines.

Now consider the geodesics on the surface of a circular cone. Since the cone can be defined in spherical coordinates as $x^2=\theta=\theta_0$, $0<\theta_0<\frac{1}{2}\pi$ and $d\theta=0$, from Equations (10.38) and (10.32d) (since we are confined to the surface of the cone the indices take on the values 1 and 3),

$$\begin{aligned} \frac{d^2r}{(ds)^2}-r\sin^2\theta_0\left(\frac{d\psi}{ds}\right)^2&=0, \\ \frac{d^2\psi}{(ds)^2}+\frac{2}{r}\frac{dr}{ds}\frac{d\psi}{ds}&=0. \end{aligned} \tag{10.38c}$$

One particular solution of these equations can be found quite easily. Suppose that $d\psi/ds=0$, $\psi=\psi_0$ and $d^2\psi/(ds)^2=0$. Then substitution into Equation (10.33b) gives

$$d^2r/(ds)^2=0.$$

Integrating twice we have the solution

$$r=a+bs, \qquad \psi=\psi_0. \tag{10.41}$$

Thus, one family of geodesics on the surface of the cone is the family of radial straight lines that passes through the apex of the cone.

It can be shown that any other curve on the cone that maps into a straight line when the cone is developed is also a geodesic (Figure 10.02).

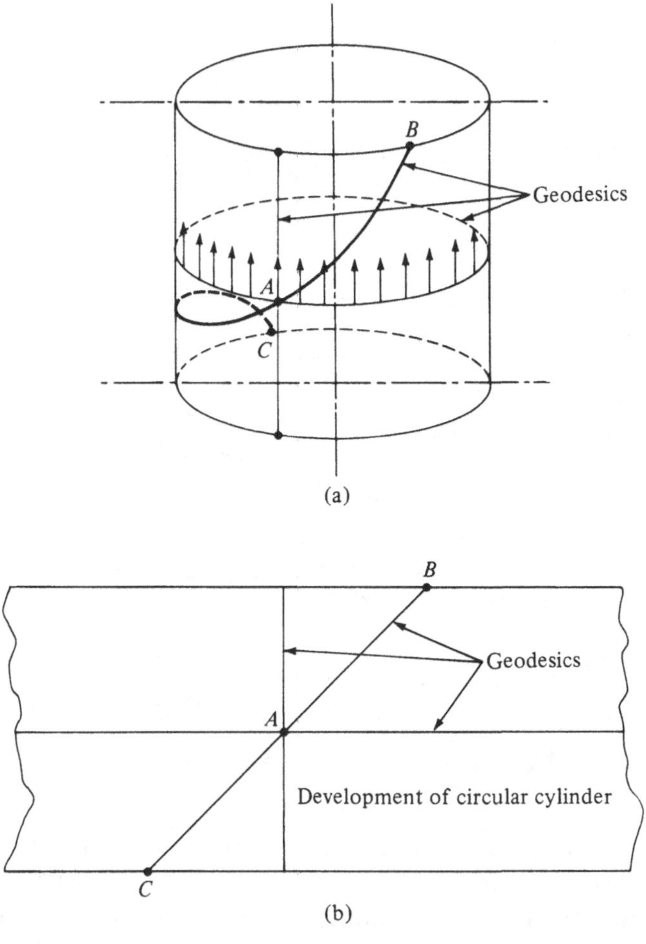

Figure 10.01. Geodesics on a circular cylinder.

Next consider a sphere of radius a. For this surface, $x^1 = r = a$ and $dr = 0$. From Equations (10.38) and (10.32d) (letting indices take on values 2 and 3), the differential equations of the geodesics on the sphere are

$$\frac{d^2\theta}{(ds)^2} - \sin\theta\cos\theta\left(\frac{d\psi}{ds}\right)^2 = 0,$$
$$\frac{d^2\psi}{(ds)^2} + 2\cot\theta\frac{d\theta}{ds}\frac{d\psi}{ds} = 0. \tag{10.38d}$$

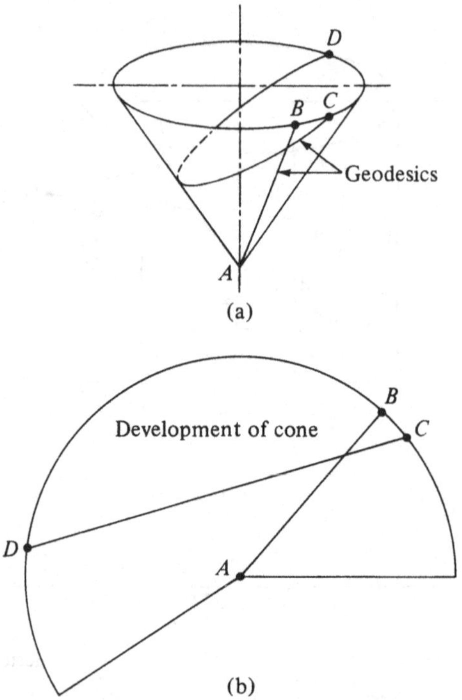

Figure 10.02. Geodesics on a circular cone.

Consider the equation of a great circle through the north pole, $\psi = \psi_0$. Then $d\psi/ds = 0$, $d^2\psi/ds^2 = 0$, and Equation (10.38c) reduces to $d^2\theta/ds^2 = 0$. Integrating, the solution of the differential equation is

$$\theta = a + bs, \qquad \psi = \psi_0. \tag{10.42}$$

Therefore, every great circle passing through the north pole is a geodesic. By rotation of coordinate axes, the result can be generalized to the statement that all great circles of a sphere are geodesics.

10.05. Curvature

Our previous investigation of the Riemann–Christoffel tensors (Chapter 8) is still valid in Riemannian space,[9] but here, since the linear connection is symmetric, the two Riemann–Christoffel tensors become identical. The single Riemann–Christoffel tensor is

$$R^i_{jkl} = \frac{\partial \Gamma^i_{kl}}{\partial x^j} - \frac{\partial \Gamma^i_{jl}}{\partial x^k} + \Gamma^i_{jp}\Gamma^p_{kl} - \Gamma^i_{kp}\Gamma^p_{jl}, \qquad (8.09a)$$

where the Γ's are now obtained from Equation (10.31).

A covariant curvature tensor is defined by taking the contracted product of the Riemann–Christoffel tensor and the metric coefficients,

$$R_{ijkl} = g_{im} R^m_{jkl}. \qquad (10.43)$$

The Ricci tensor of the third kind is defined by contracting Equation (8.09a) on indices i and j,

$$\mathcal{R}_{kl} = R^i_{ikl}, \qquad (10.44)$$

and the nilvalent curvature tensor is

$$R = g^{kl} \mathcal{R}_{kl}. \qquad (10.45)$$

The Gaussian curvature[10] κ is given by the equation

$$\kappa = [n(n-1)]^{-1} R. \qquad (10.46)$$

For example, consider a circular cylinder, $r = a$. On this surface, $x^2 = \psi$ and $x^3 = z$ vary. So the indices in the foregoing equations take on the values 2 and 3.

$$\Gamma^2_{jk} = \begin{bmatrix} 0 & 0 \\ 0 & 0 \end{bmatrix}, \qquad \Gamma^3_{jk} = \begin{bmatrix} 0 & 0 \\ 0 & 0 \end{bmatrix}.$$

Therefore, $R^i_{jkl} = 0$, $R_{ijkl} = 0$, $R_{kl} = 0$, and $\kappa = 0$. The Gaussian curvature of the circular cylinder is zero and it is a developable surface.

As another example, consider the circular cone, $\theta = \theta_0$. On this surface, $x^1 = r$ and $x^3 = \psi$ vary. So the indices take on the values 1 and 3. From Equation (10.32a),

$$\Gamma^1_{jk} = \begin{bmatrix} 0 & 0 \\ 0 & -r\sin^2\theta_0 \end{bmatrix}, \qquad \Gamma^3_{jk} = \begin{bmatrix} 0 & 1/r \\ 1/r & 0 \end{bmatrix}.$$

According to Section 8.13, there are only four distinct linearly independent merates of the Riemann–Christoffel tensor that can be evaluated by substitution of these Γ's into Equation (8.09a),

$$R^j_{13k} = \begin{bmatrix} 0 & 0 \\ 0 & 0 \end{bmatrix}.$$

Therefore, $R_{ijkl}=0$, $\mathcal{R}_{kl}=0$, and $\kappa=0$. The surface of a cone is a surface of zero Gaussian curvature and can be developed into a plane.

Now let us consider the surface of a sphere, $x^1=r=a$. On this surface, the angles $x^2=\theta$ and $x^3=\psi$ vary. So on this surface indices take on the values 2 and 3. From Equation (10.32d),

$$\Gamma^2_{jk} = \begin{bmatrix} 0 & 0 \\ 0 & -\sin\theta\cos\theta \end{bmatrix}, \qquad \Gamma^3_{jk} = \begin{bmatrix} 0 & \cot\theta \\ \cot\theta & 0 \end{bmatrix}.$$

The distinct merates of the Riemann-Christoffel tensor on the surface of the sphere are found by substituting these Γ's into Equation (8.09a),

$$R^k_{23l} = \begin{bmatrix} 0 & \sin^2\theta \\ -1 & 0 \end{bmatrix}.$$

The Ricci tensor can be evaluated from Equation (10.44),

$$\mathcal{R}_{jk} = R^2_{2jk} + R^3_{3jk}$$

$$= \begin{bmatrix} 1 & 0 \\ 0 & \sin^2\theta \end{bmatrix}.$$

The nilvalent curvature tensor R is found from Equation (10.45),

$$R = g^{kl}\mathcal{R}_{kl} = \frac{2}{a^2}.$$

Finally, the Gaussian curvature κ is found from Equation (10.46) to be

$$\kappa = \tfrac{1}{2}R = 1/a^2.$$

For all points on the surface of the sphere, the Gaussian curvature is the same and is equal to the reciprocal of the square of the radius of the sphere.

10.06. Homeomorphic transport of a vector

In Riemannian space, instead of two homeomorphic displacements, there is a single homeomorphic displacement defined by the equation

$$dv^i = -\Gamma^i_{jk} v^k \, dx^j. \tag{10.47}$$

If the vector v^i is given at a single point, and if the linear connection Γ'^i_{jk} is known, then Equation (10.47) enables us to define the vector at every other point of space. However, the value will generally depend on the path of integration.

In Cartesian coordinates, the $\Gamma'^i_{jk}=0$, so for any homeomorphic transport in a plane $x^1=0$,

$$dv^1=0, \qquad dv^2=0, \qquad dv^3=0.$$

Thus, the vector remains constant, and homeomorphic transport along any path in a plane reduces to parallel transport.

In an orthogonal cylindrical coordinate system,

$$\begin{aligned} dv^1 &= -\Gamma^1_{11}v^1\,dx^1-\Gamma^1_{12}v^2\,dx^1-\Gamma^1_{21}v^1\,dx^2-\Gamma^1_{22}v^2\,dx^2, \\ dv^2 &= -\Gamma^2_{11}v^1\,dx^1-\Gamma^2_{12}v^2\,dx^1-\Gamma^2_{21}v^1\,dx^2-\Gamma^2_{22}v^2\,dx^2, \\ dv^3 &= 0. \end{aligned} \tag{10.48}$$

Therefore, $v^3=$ constant for all homeomorphic transports in any orthogonal cylindrical coordinate system.

Suppose that we confine the path to the surface of a cylinder $x^1=$ constant, then $dx^1=0$ and Equation (10.48) becomes

$$\begin{aligned} dv^1/dx^2 &= -\Gamma^1_{21}v^1-\Gamma^1_{22}v^2, \\ dv^2/dx^2 &= -\Gamma^2_{21}v^1-\Gamma^2_{22}v^2, \\ dv^3 &= 0. \end{aligned} \tag{10.48a}$$

In particular, if the cylinder is circular, by Equation (10.32b) Equation (10.48) becomes on the surface $r=a$

$$dv^1/d\psi=av^2, \qquad dv^2/d\psi=-v^1/a. \tag{10.48b}$$

Differentiating the first equation with respect to ψ and substituting the second,

$$d^2v^1/d\psi^2=-v^1, \tag{10.49}$$

whose solution is

$$v^1=A\sin\psi+B\cos\psi. \tag{10.50}$$

Substitution into the first of Equations (10.48) gives

$$v^2=(A\cos\psi-B\sin\psi)/a, \tag{10.51}$$

and from the third equation of (10.48),

$$v^3=C. \tag{10.52}$$

The components of the vector obtained by homeomorphic transport along any path in the circular cylinder are found from Equation (10.22):

$$\begin{aligned} v_r &= \sqrt{g_{11}}\, v^1 = A \sin\psi + B\cos\psi, \\ v_\psi &= \sqrt{g_{22}}\, v^2 = A\cos\psi - B\sin\psi, \\ v_z &= \sqrt{g_{33}}\, v^3 = C. \end{aligned} \tag{10.53}$$

The length of the vector obtained by homeomorphic transport is a constant,

$$v = \sqrt{A^2 + B^2 + C^2}. \tag{10.54}$$

Suppose that at $\psi = 0$, $\mathbf{v}(0) = \mathbf{a}_r$. Then, by Equation (10.53), $A = 0$, $B = 1$, and $C = 0$, so the vector obtained by homeomorphic transport is

$$\mathbf{v}(\psi) = \mathbf{a}_r \cos\psi - \mathbf{a}_\psi \sin\psi,$$

which is parallel to $\mathbf{v}(0)$.

On the other hand, if at $\psi = 0$, $\mathbf{v}(0) = \mathbf{a}_\psi$, then $A = 1$, $B = 0$, $C = 0$, and

$$\mathbf{v}(\psi) = \mathbf{a}_r \sin\psi + \mathbf{a}_\psi \cos\psi,$$

which is also parallel to $\mathbf{v}(0)$. Consequently, we can conclude that any vector moved homeomorphically along any path in a circular cylinder moves parallel to itself (Figure 10.01).

In spherical coordinates, Equation (10.47) becomes

$$\begin{aligned} dv^1 &= rv^2\, d\theta + r\sin^2\theta v^3\, d\psi, \\ dv^2 &= -(v^2\, dr + v^1\, d\theta)/r + (\sin\theta\cos\theta) v^3\, d\psi, \\ dv^3 &= -(v^3\, dr + v^1\, d\psi)/r - \cot\theta (v^3\, d\theta + v^2\, d\psi). \end{aligned} \tag{10.55}$$

On the surface of a cone $\theta = \theta_0$, $dx^i = (dr, 0, d\psi)$, so Equation (10.55) becomes

$$\begin{aligned} dv^1 &= r\sin^2\theta_0 v^3\, d\psi, \\ dv^2 &= -v^2/r\, dr + \sin\theta_0 \cos\theta_0 v^3\, d\psi, \\ dv^3 &= -v^3/r\, dr - v^1/r\, d - \cot\theta_0 v^2\, d\psi. \end{aligned} \tag{10.55a}$$

Consider a radial path in the cone, $d\psi = 0$. Then Equation (10.55a) reduces to

$$dv^1 = 0, \qquad dv^2 = -\frac{dr}{r} v^2, \qquad dv^3 = -\frac{dr}{r} v^3. \tag{10.55b}$$

Integration gives

$$v^1 = A, \qquad v^2 = B/r, \qquad v^3 = C/r.$$

The components of the vector obtained by homeomorphic transport are

$$v_r = A, \qquad v_\theta = B, \qquad v_\psi = C \sin\theta_0.$$

The vectors obtained by homeomorphic transport along this radial line in the cone are all parallel.

Consider a circular path on the surface of a sphere of radius r. Let the circle be defined by $r = a$, $\theta = \theta_0$. Then along the circular path $dx^i = (0, 0, d\psi)$ and Equation (10.55) becomes

$$\begin{aligned} dv^1 &= a \sin^2\theta_0 v^3 \, d\psi, \\ dv^2 &= \sin\theta_0 \cos\theta_0 v^3 \, d\psi, \\ dv^3 &= -(v^1/a) \, d\psi - \cot\theta_0 v^2 \, d\psi. \end{aligned} \tag{10.55c}$$

The general solution of these equations is

$$\begin{aligned} v^1 &= a \sin^2\theta_0(-A\cos\psi + B\sin\psi) + aC\cot\theta_0, \\ v^2 &= \sin\theta_0\cos\theta_0(-A\cos\psi + B\sin\psi) + C, \\ v^3 &= A\sin\psi + B\cos\psi. \end{aligned}$$

The components of the vector obtained by homeomorphic transport are

$$\begin{aligned} v_r &= a\sin^2\theta_0(-A\cos\psi + B\sin\psi) + aD\cot\theta_0, \\ v_\theta &= a\sin\theta_0\cos\theta_0(-A\cos\psi + B\sin\psi) + aD, \\ v_\psi &= a\sin\theta_0(A\sin\psi + B\cos\psi). \end{aligned}$$

In the special case in which the path is a great circle of the sphere, $\theta_0 = \frac{1}{2}\pi$,

$$\begin{aligned} v_r &= -aA\cos\psi + aB\sin\psi, \\ v_\theta &= aD, \\ v_\psi &= aA\sin\psi + aB\cos\psi. \end{aligned}$$

Suppose that at $\psi = 0$ the vector is

$$\mathbf{v} = \mathbf{a}_r(v)_r + \mathbf{a}_\theta(v)_\theta + \mathbf{a}_\psi(v)_\psi,$$

then the vector obtained by homeomorphic transport along the great circle is

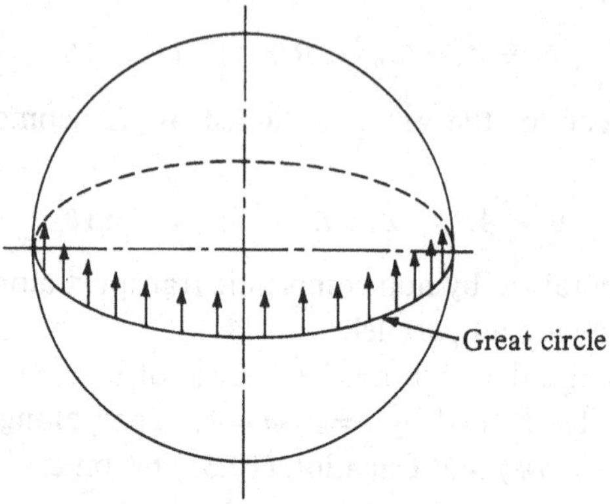

Figure 10.03. A great circle is a geodesic of a sphere.

$$\mathbf{v}_T=\mathbf{a}_r[(v)_r\cos\psi+(v)_\psi\sin\psi]+\mathbf{a}_\theta(v)_\theta$$
$$+\mathbf{a}_\psi[-(v)_r\sin\psi+(v)_\psi\cos\psi].$$

This vector $\mathbf{v}_T$ is always parallel to the given vector $\mathbf{v}$ (Figure 10.03). Thus, we can conclude that homeomorphic transport of any vector along a great circle of a sphere is parallel transport. However, if $\theta_0\neq\frac{1}{2}\pi$, homeomorphic transport along a small circle is not generally parallel transport.

10.07. Volume and area

In Chapter 5, pseudovolume was defined in arithmetic 3-space as

$$d\tilde{\mathcal{V}}=\delta^{123}_{ijk}\,du^i\,dv^j\,dw^k, \tag{5.21}$$

where du^i, dv^j, and dw^k are three arbitrary infinitesimal vectors. Pseudovolume transformed as an akinetor of weight -1. Since $g^{1/2}$ is an akinetor of weight $+1$, we can take $d\mathcal{V}$ as a nilvalent tensor in Riemannian space by multiplying Equation (5.21) by $g^{1/2}$. Therefore, the element of volume in Riemannian 3-space is

$$d\mathcal{V}=g^{1/2}\delta^{123}_{ijk}\,du^i\,dv^j\,dw^k, \tag{10.56}$$

the corresponding hypervolume in Riemannian n-space is

$$d\mathcal{V}=g^{1/2}\delta^{12\cdots n}_{ij\ldots p}\,du^i\,dv^j\cdots dz^p. \tag{10.56a}$$

Equation (10.56) reduces to the ordinary expression for volume in Euclidean 3-space with rectangular coordinate systems.

Pseudoarea (Chapter 5) may be defined in arithmetic 3-space as

$$d\tilde{\mathfrak{A}}_i = \delta^{123}_{ijk}\, dv^j\, dw^k, \tag{5.19}$$

which is a covariant akinetor of weight -1. Multiplication by $g^{1/2}$ gives a tensor area,

$$g^{1/2}\, d\tilde{\mathfrak{A}}_i = d\mathfrak{A}_i = g^{1/2}\delta^{123}_{ijk}\, dv^j\, dw^k. \tag{10.57}$$

The magnitude of this area is

$$d\mathfrak{A} = [g^{ij}\, d\mathfrak{A}_i\, d\mathfrak{A}_j]^{1/2}. \tag{10.58}$$

Substitution of Equation (10.57) into (10.58) leads to

$$d\mathfrak{A} = g^{1/2}[g^{ij}\delta^{123}_{ikl}\, dv^k\, dw^l\, \delta^{123}_{jmn}\, dv^m\, dw^n]^{1/2}. \tag{10.59}$$

Another way of expressing area in arithmetic 3-space is as a contravariant antisymmetric tensor,

$$d\mathfrak{A}^{jk} = dv^{[j}\, dw^{k]}. \tag{10.60}$$

Evidently, Equation (10.60) applies without change in Riemannian space.

As an example of area, consider a Riemannian 2-space (Figure 10.04) consisting of a paraboloid of revolution $x^1 = a = \text{constant}$ embedded in a Euclidean 3-space. What is the area of the paraboloidal cup extending from $x^2 = 0$ to $x^2 = b$?

Let us take infinitesimal vectors along the coordinate axes in the Riemannian 2-space:

$$dv^j = (0, dx^2, 0), \qquad dw^k = (0, 0, dx^3).$$

Then, according to Equation (10.57),

$$\begin{aligned} d\mathfrak{A}_1 &= g^{1/2}[\delta^{123}_{123}\, dv^2\, dw^3 + \delta^{123}_{132}\, dv^3\, dw^2] \\ &= ax^2[(a)^2 + (x^2)^2]\, dx^2\, dx^3. \end{aligned}$$

The total area is obtained by integration:

$$\begin{aligned} \mathfrak{A}_1 &= a\int_0^{2\pi} dx^3 \int_0^b x^2[(a)^2 + (x^2)^2]\, dx^2 \\ &= 2\pi a\int_0^b x^2[(a)^2 + (x^2)^2]\, dx^2 \\ &= \pi ab^2[(a)^2 + \tfrac{1}{2}(b)^2]. \end{aligned}$$

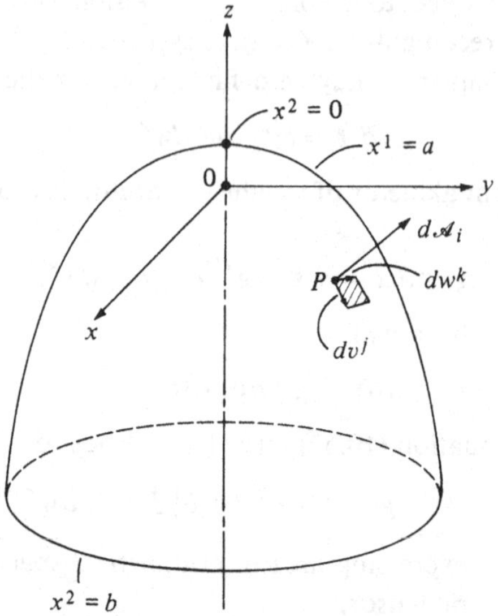

Figure 10.04. A paraboloid of revolution, $x^1 =$ constant. The surface area is $\mathcal{A}_1 = \pi ab^2[(a)^2 + \frac{1}{2}(b)^2]$.

In this simple case, the area could be found by elementary geometry: Though, even here, Equation (10.57) introduces a helpful systematization. For more complicated cases, the equations of this section are of real importance.

10.08. Scalar and vector products

In Riemannian n-space, a scalar that is the product of two contravariant vectors u^i and v^j has been defined by the equation

$$S = g_{ij} u^i v^j. \tag{10.13}$$

This quantity has been called the scalar product of two vectors. According to Equation (10.17), the scalar product S is equal to the product of the magnitudes of the two vectors times the cosine of the angle between them. In terms of components, the scalar product can be written

$$S = \sum_{i=1}^{n} \sum_{j=1}^{n} g_{ij} \left(\frac{(u)_i}{(g_{ii})^{1/2}} \frac{(v)_j}{(g_{jj})^{1/2}} \right). \tag{10.61}$$

In Riemannian 3-space, Equation (10.61) becomes

$$S=(u)_1(v)_1+(u)_2(v)_2+(u)_3(v)_3+\frac{g_{12}}{(g_{11}g_{22})^{1/2}}[(u)_1(v)_2+(u)_2(v)_1]$$
$$+\frac{g_{13}}{(g_{11}g_{33})^{1/2}}[(u)_1(v)_3+(u)_3(v)_1]+\frac{g_{23}}{(g_{22}g_{33})^{1/2}}[(u)_2(v)_3+(u)_3(v)_2]. \tag{10.61a}$$

If the vectors u^i and v^j are expressed in terms of unit vectors $\mathbf{a}_i$ in each of the coordinate directions times corresponding components,

$$\begin{aligned}\mathbf{u}&=\mathbf{a}_1(u)_1+\mathbf{a}_2(u)_2+\mathbf{a}_3(u)_3+\cdots+\mathbf{a}_n(u)_n,\\ \mathbf{v}&=\mathbf{a}_1(v)_1+\mathbf{a}_2(v)_2+\mathbf{a}_3(v)_3+\cdots+\mathbf{a}_n(v)_n.\end{aligned} \tag{10.62}$$

In classical vector analysis,[11] it is customary to write the scalar product of two vectors **u** and **v**

$$S=\mathbf{u}\cdot\mathbf{v}. \tag{10.63}$$

This notation can be extended without change to Riemannian n-space[12] if a suitable multiplication table is defined for the unit vectors. Substitution of Equation (10.62) into (10.63) gives

$$\begin{aligned}S&=[\mathbf{a}_1(u)_1+\mathbf{a}_2(u)_2+\mathbf{a}_3(u)_3+\cdots+\mathbf{a}_n(u)_n]\\ &\quad\cdot[\mathbf{a}_1(v)_1+\mathbf{a}_2(v)_2+\mathbf{a}_3(v)_3+\cdots+\mathbf{a}_n(v)_n]\\ &=(\mathbf{a}_1\cdot\mathbf{a}_1)(u)_1(v)_1+(\mathbf{a}_2\cdot\mathbf{a}_2)(u)_2(v)_2+(\mathbf{a}_3\cdot\mathbf{a}_3)(u)_3(v)_3+\cdots\\ &\quad+(\mathbf{a}_n\cdot\mathbf{a}_n)(u)_n(v)_n+\cdots+(\mathbf{a}_i\cdot\mathbf{a}_j)(u)_i(v)_j+(\mathbf{a}_j\cdot\mathbf{a}_i)(u)_j(v)_i+\cdots.\end{aligned} \tag{10.64}$$

Equating coefficients in Equations (10.61) and (10.64),

$\cdot$	$\mathbf{a}_1$	$\mathbf{a}_2$	$\mathbf{a}_3$	$\cdots$	$\mathbf{a}_n$
$\mathbf{a}_1$	1	$\frac{g_{12}}{(g_{11}g_{22})^{1/2}}$	$\frac{g_{13}}{(g_{11}g_{33})^{1/2}}$	$\cdots$	$\frac{g_{1n}}{(g_{11}g_{nn})^{1/2}}$
$\mathbf{a}_2$	$\frac{g_{12}}{(g_{11}g_{22})^{1/2}}$	1	$\frac{g_{23}}{(g_{22}g_{33})^{1/2}}$	$\cdots$	$\frac{g_{2n}}{(g_{22}g_{nn})^{1/2}}$
$\mathbf{a}_3$	$\frac{g_{13}}{(g_{11}g_{33})^{1/2}}$	$\frac{g_{23}}{(g_{22}g_{33})^{1/2}}$	1	$\cdots$	$\frac{g_{3n}}{(g_{33}g_{nn})^{1/2}}$
.	.	.	.	$\cdots$	.
$\mathbf{a}_n$	$\frac{g_{1n}}{(g_{11}g_{nn})^{1/2}}$	$\frac{g_{2n}}{(g_{22}g_{nn})^{1/2}}$	$\frac{g_{3n}}{(g_{33}g_{nn})^{1/2}}$	$\cdots$	1

(10.65)

Thus, in any coordinate system in Riemannian n-space, the scalar product is commutative

$$\mathbf{u}\cdot\mathbf{v}=\mathbf{v}\cdot\mathbf{u}. \tag{10.66}$$

In any orthogonal coordinate system in Riemannian n-space, the scalar multiplication table for the unit vectors takes on the simple form

$\cdot$	$\mathbf{a}_1$	$\mathbf{a}_2$	$\mathbf{a}_3$	$\cdots$	$\mathbf{a}_n$
$\mathbf{a}_1$	1	0	0	$\cdots$	0
$\mathbf{a}_2$	0	1	0	$\cdots$	0
$\mathbf{a}_3$	0	0	1	$\cdots$	0
.	.	.	.	$\cdots$	.
$\mathbf{a}_n$	0	0	0	$\cdots$	1

(10.65a)

In order to define a contravariant vector w^k that is a product of the two contravariant vectors u^i and v^j, coefficients γ_{ij}^k (Chapter 3) must be defined in the equation

$$w^k=\gamma_{ij}^k u^i v^j. \tag{10.67}$$

This general equation is valid in n-dimensional arithmetic space. Since the three vectors all transform as tensors, the coefficient γ_{ij}^k must also transform as a tensor. The question is how γ_{ij}^k can be defined in Riemannian space in terms of previously introduced tensors and akinetors. The definition that reduces to the vector product of classical vector analysis[11] is

$$\gamma_{ij}^k=(g)^{1/2}g^{kl}\delta_{lij}^{123}. \tag{10.68}$$

Because of the inclusion of the trivalent alternator, which is defined only in 3-space, the vector product, unlike the scalar product, is restricted to Riemannian 3-space.

Expansion of Equation (10.68) gives the three matrices

$$\gamma_{ij}^1=g^{1/2}\begin{bmatrix}0 & g^{13} & -g^{12}\\ -g^{13} & 0 & g^{11}\\ g^{12} & -g^{11} & 0\end{bmatrix},\qquad \gamma_{ij}^2=g^{1/2}\begin{bmatrix}0 & g^{23} & -g^{22}\\ -g^{23} & 0 & g^{12}\\ g^{22} & -g^{12} & 0\end{bmatrix},$$

$$\gamma_{ij}^3=g^{1/2}\begin{bmatrix}0 & g^{33} & -g^{23}\\ -g^{33} & 0 & g^{13}\\ g^{23} & -g^{13} & 0\end{bmatrix}. \tag{10.69}$$

Substitution of Equation (10.68) into (10.67) gives the defining equation of the vector product

$$w^k=g^{1/2}g^{kl}\delta_{lij}^{123}u^i v^j. \tag{10.70}$$

Utilizing Equation (10.69), Equation (10.70) becomes

$$
\begin{aligned}
w^1 &= g^{1/2}[g^{11}(u^2v^3-u^3v^2)+g^{12}(u^3v^1-u^1v^3)+g^{13}(u^1v^2-u^2v^1)],\\
w^2 &= g^{1/2}[g^{21}(u^2v^3-u^3v^2)+g^{22}(u^3v^1-u^1v^3)\\
&\qquad +g^{23}(u^1v^2-u^2v^1)],\\
w^3 &= g^{1/2}[g^{31}(u^2v^3-u^3v^2)+g^{32}(u^3v^1-u^1v^3)+g^{33}(u^1v^2-u^2v^1)].
\end{aligned} \tag{10.70a}
$$

Instead, Equation (10.70a) can be expressed in terms of components as

$$
\begin{aligned}
(w)_1 = (gg_{11})^{1/2}\Big\{&\frac{g^{11}}{(g_{22}g_{33})^{1/2}}[(u)_2(v)_3-(u)_3(v)_2]\\
&+\frac{g^{12}}{(g_{11}g_{33})^{1/2}}[(u)_3(v)_1-(u)_1(v)_3]\\
&+\frac{g^{13}}{(g_{11}g_{22})^{1/2}}[(u)_1(v)_2-(u)_2(v)_1]\Big\},\\
(w)_2 = (gg_{22})^{1/2}\Big\{&\frac{g^{21}}{(g_{22}g_{33})^{1/2}}[(u)_2(v)_3-(u)_3(v)_2]\\
&+\frac{g^{22}}{(g_{11}g_{22})^{1/2}}[(u)_3(v)_1-(u)_1(v)_3]\\
&+\frac{g^{23}}{(g_{11}g_{22})^{1/2}}[(u)_1(v)_2-(u)_2(v)_1]\Big\},\\
(w)_3 = (gg_{33})^{1/2}\Big\{&\frac{g^{31}}{(g_{22}g_{33})^{1/2}}[(u)_2(v)_3-(u)_3(v)_2]\\
&+\frac{g^{32}}{(g_{11}g_{33})^{1/2}}[(u)_3(v)_1-(u)_1(v)_3]\\
&+\frac{g^{33}}{(g_{11}g_{22})^{1/2}}[(u)_1(v)_2-(u)_2(v)_1]\Big\}.
\end{aligned} \tag{10.70b}
$$

The notation of classical vector analysis[11] can readily be extended to Riemannian 3-space: The vector product is written

$$\mathbf{w}=\mathbf{u}\times\mathbf{v}. \tag{10.71}$$

Expansion of Equation (10.71) in terms of unit vectors and components gives

$$
\begin{aligned}
&\mathbf{a}_1(w)_1+\mathbf{a}_2(w)_2+\mathbf{a}_3(w)_3\\
&\quad=[\mathbf{a}_1(u)_1+\mathbf{a}_2(u)_2+\mathbf{a}_3(u)_3]\times[\mathbf{a}_1(v)_1+\mathbf{a}_2(v)_2+\mathbf{a}_3(v)_3]\\
&\quad=(\mathbf{a}_1\times\mathbf{a}_1)(u)_1(v)_1+(\mathbf{a}_2\times\mathbf{a}_2)(u)_2(v)_2+(\mathbf{a}_3\times\mathbf{a}_3)(u)_3(v)_3\\
&\qquad+(\mathbf{a}_1\times\mathbf{a}_2)(u)_1(v)_2+(\mathbf{a}_2\times\mathbf{a}_1)(u)_2(v)_1+(\mathbf{a}_1\times\mathbf{a}_3)(u)_1(v)_3\\
&\qquad+(\mathbf{a}_3\times\mathbf{a}_1)(u)_3(v)_1+(\mathbf{a}_2\times\mathbf{a}_3)(u)_2(v)_3+(\mathbf{a}_3\times\mathbf{a}_2)(u)_3(v)_2.
\end{aligned} \tag{10.72}
$$

$\curvearrowleft\times$	$\mathbf{a}_1$	$\mathbf{a}_2$	$\mathbf{a}_3$
$\mathbf{a}_1$	0	$\left(\frac{g}{g_{11}g_{22}}\right)^{1/2}[\mathbf{a}_1(g_{11})^{1/2}g^{13} + \mathbf{a}_2(g_{22})^{1/2}g^{23} + \mathbf{a}_3(g_{33})^{1/2}g^{33}]$	$-\left(\frac{g}{g_{11}g_{33}}\right)^{1/2}[\mathbf{a}_1(g_{11})^{1/2}g^{12} + \mathbf{a}_2(g_{22})^{1/2}g^{22} + \mathbf{a}_3(g_{33})^{1/2}g^{32}]$
$\mathbf{a}_2$	$-\left(\frac{g}{g_{11}g_{22}}\right)^{1/2}[\mathbf{a}_1(g_{11})^{1/2}g^{13} + \mathbf{a}_2(g_{22})^{1/2}g^{23} + \mathbf{a}_3(g_{33})^{1/2}g^{33}]$	0	$\left(\frac{g}{g_{22}g_{33}}\right)^{1/2}[\mathbf{a}_1(g_{11})^{1/2}g^{11} + \mathbf{a}_2(g_{22})^{1/2}g^{21} + \mathbf{a}_3(g_{33})^{1/2}g^{31}]$
$\mathbf{a}_3$	$\left(\frac{g}{g_{11}g_{33}}\right)^{1/2}[\mathbf{a}_1(g_{11})^{1/2}g^{12} + \mathbf{a}_2(g_{22})^{1/2}g^{22} + \mathbf{a}_3(g_{33})^{1/2}g^{32}]$	$-\left(\frac{g}{g_{22}g_{33}}\right)^{1/2}[\mathbf{a}_1(g_{11})^{1/2}g^{11} + \mathbf{a}_2(g_{22})^{1/2}g^{21} + \mathbf{a}_3(g_{33})^{1/2}g^{31}]$	0

(10.73)

Equating coefficients of corresponding terms in Equation (10.70b) and Equation (10.72), the multiplication table (10.73) for the vector product of the unit vectors is obtained.

Because of the antisymmetry of the vector product multiplication table, we can conclude that in all coordinate systems in Riemannian n-space,

$$\mathbf{u} \times \mathbf{v} = -\mathbf{v} \times \mathbf{u}. \tag{10.74}$$

In the special case of orthogonal coordinates in Riemannian 3-space, the multiplication table for the vector product of the unit vectors simplifies drastically:

$\curvearrowleft\times$	$\mathbf{a}_1$	$\mathbf{a}_2$	$\mathbf{a}_3$
$\mathbf{a}_1$	0	$\mathbf{a}_3$	$-\mathbf{a}_2$
$\mathbf{a}_2$	$-\mathbf{a}_3$	0	$\mathbf{a}_1$
$\mathbf{a}_3$	$\mathbf{a}_2$	$-\mathbf{a}_1$	0

(10.73a)

and the vector product can be expressed compactly in determinant form as

$$\mathbf{w} = \mathbf{u} \times \mathbf{v} = \begin{vmatrix} \mathbf{a}_1 & \mathbf{a}_2 & \mathbf{a}_3 \\ (u)_1 & (u)_2 & (u)_3 \\ (v)_1 & (v)_2 & (v)_3 \end{vmatrix}. \tag{10.70c}$$

The vector $\mathbf{w}$ is perpendicular to the plane of $\mathbf{u}$ and $\mathbf{v}$, and its magnitude is the area of the parallelogram formed by the two vectors.

10.09. Gradient, divergence, and curl

In Chapter 5, akinetors resembling gradient, divergence, and curl were introduced into arithmetic space. For Riemannian space, these akinetors assume a more familiar - and more useful - form.

A covariant tensor was introduced in arithmetic n-space that may be called the gradient of a scalar,[13]

$$\mathrm{grad}_i\,\phi = \frac{\partial \phi}{\partial x^i}. \tag{5.25}$$

In Riemannian space, the gradient also has contravariant merates, which are defined by the equation

$$\mathrm{grad}^i\,\phi = g^{ij}\frac{\partial \phi}{\partial x^j}. \tag{10.75}$$

In terms of components, a gradient in Riemannian n-space may be defined as

$$\operatorname{grad}\phi=\mathbf{a}_1(g_{11})^{1/2}g^{1j}\frac{\partial\phi}{\partial x^j}+\mathbf{a}_2(g_{22})^{1/2}g^{2j}\frac{\partial\phi}{\partial x^j}+\cdots+\mathbf{a}_n(g_{nn})^{1/2}g^{nj}\frac{\partial\phi}{\partial x^j}. \tag{10.76}$$

Thus, in 3-dimensional Riemannian space, Equation (10.76) may be expanded into

$$\begin{aligned}\operatorname{grad}\phi=\mathbf{a}_1(g_{11})^{1/2}&\left[g^{11}\frac{\partial\phi}{\partial x^1}+g^{12}\frac{\partial\phi}{\partial x^2}+g^{13}\frac{\partial\phi}{\partial x^3}\right]\\ +\mathbf{a}_2(g_{22})^{1/2}&\left[g^{21}\frac{\partial\phi}{\partial x^1}+g^{22}\frac{\partial\phi}{\partial x^2}+g^{23}\frac{\partial\phi}{\partial x^3}\right]\\ +\mathbf{a}_3(g_{33})^{1/2}&\left[g^{31}\frac{\partial\phi}{\partial x^1}+g^{32}\frac{\partial\phi}{\partial x^2}+g^{33}\frac{\partial\phi}{\partial x^3}\right].\end{aligned} \tag{10.76a}$$

If the coordinate system in the Riemannian 3-space is orthogonal, Equation (10.76a) becomes

$$\operatorname{grad}\phi=\mathbf{a}_1\frac{1}{(g_{11})^{1/2}}\frac{\partial\phi}{\partial x^1}+\mathbf{a}_2\frac{1}{(g_{22})^{1/2}}\frac{\partial\phi}{\partial x^2}+\mathbf{a}_3\frac{1}{(g_{33})^{1/2}}\frac{\partial\phi}{\partial x^3}. \tag{10.76b}$$

Now consider the special case of Euclidean 3-space. If rectangular coordinates are employed, the gradient becomes

$$\operatorname{grad}\phi=\mathbf{a}_x\frac{\partial\phi}{\partial x}+\mathbf{a}_y\frac{\partial\phi}{\partial y}+\mathbf{a}_z\frac{\partial\phi}{\partial z}. \tag{10.76c}$$

For circular cylinder coordinates,

$$\operatorname{grad}\phi=\mathbf{a}_r\frac{\partial\phi}{\partial r}+\mathbf{a}_\psi\frac{1}{r}\frac{\partial\phi}{\partial\psi}+\mathbf{a}_z\frac{\partial\phi}{\partial z}, \tag{10.76d}$$

and in spherical coordinates,

$$\operatorname{grad}\phi=\mathbf{a}_r\frac{\partial\phi}{\partial r}+\mathbf{a}_\theta\frac{1}{r}\frac{\partial\phi}{\partial\theta}+\mathbf{a}_\psi\frac{1}{r\sin\theta}\frac{\partial\phi}{\partial\psi}. \tag{10.76e}$$

In Section 5.05, the generalized divergence was defined in arithmetic n-space as a nilvalent akinetor of weight +1,

$$\tilde{S}=\frac{\partial\tilde{v}^i}{\partial x^i}, \tag{5.37}$$

where $\tilde{v}^i$ is required to be a contravariant akinetor of weight +1. In Riemannian n-space, a related scalar can be introduced that applies to a contravariant vector v^i. The trick is to remember that $g^{1/2}$ is an akinetor of weight +1. Therefore, the product of $g^{1/2}$ and a contravariant vector v^i is a contravariant akinetor of weight +1,

$$\tilde{v}^i = g^{1/2} v^i.$$

Similarly, a scalar S can be formed from $\tilde{S}$ by dividing by $g^{1/2}$,

$$S = \tilde{S}/g^{1/2}.$$

This scalar is called the divergence[14] of the contravariant vector v^i,

$$S = \operatorname{div} \mathbf{v} = \frac{1}{g^{1/2}} \frac{\partial}{\partial x^i} (g^{1/2} v^i). \tag{10.77}$$

According to Equation (10.22), the contravariant merates of the vector v^i can be expressed in terms of components $(v)_i$ by the equation

$$v^i = (v)_i / (g_{ii})^{1/2}, \tag{10.22a}$$

where summation on the index i does not occur. Substitution of Equation (10.22a) into (10.77) gives an expression for the divergence of a vector in Riemannian n-space:

$$\operatorname{div} \mathbf{v} = \frac{1}{g^{1/2}} \left[\frac{\partial}{\partial x^1} \left(\frac{g}{g_{11}} \right)^{1/2} (v)_1 + \frac{\partial}{\partial x^2} \left(\frac{g}{g_{22}} \right)^{1/2} (v)_2 + \frac{\partial}{\partial x^3} \left(\frac{g}{g_{33}} \right)^{1/2} (v)_3 \right.$$
$$\left. + \cdots + \frac{\partial}{\partial x^n} \left(\frac{g}{g_{nn}} \right)^{1/2} (v)_n \right]. \tag{10.77a}$$

In Riemannian 3-space, Equation (10.77a) becomes, for an orthogonal coordinate system,

$$\operatorname{div} \mathbf{v} = \frac{1}{g^{1/2}} \left[\frac{\partial}{\partial x^1} ((g_{22} g_{33})^{1/2} (v)_1) + \frac{\partial}{\partial x^2} ((g_{11} g_{33})^{1/2} (v)_2) \right.$$
$$\left. + \frac{\partial}{\partial x^3} ((g_{11} g_{22})^{1/2} (v)_3) \right]. \tag{10.77b}$$

In Euclidean 3-space, Equation (10.77b) can be simplified further. In rectangular coordinates,

$$\operatorname{div} \mathbf{v} = \frac{\partial v_x}{\partial x} + \frac{\partial v_y}{\partial y} + \frac{\partial v_z}{\partial z}. \tag{10.77c}$$

In circular cylinder coordinates, there is one extra term and one functional coefficient,

$$\operatorname{div}\mathbf{v}=\frac{\partial v_r}{\partial r}+\frac{v_r}{r}+\frac{1}{r}\frac{\partial v_\psi}{\partial \psi}+\frac{\partial v_z}{\partial z}. \tag{10.77d}$$

In spherical coordinates, there are two extra terms and more complicated functional coefficients,

$$\operatorname{div}\mathbf{v}=\frac{\partial v_r}{\partial r}+\frac{2}{r}v_r+\frac{1}{r}\frac{\partial v_\theta}{\partial \theta}+\frac{\cot\theta}{r}v_\theta+\frac{1}{r\sin\theta}\frac{\partial v_\psi}{\partial \psi}. \tag{10.77e}$$

In arithmetic space, a bivalent alternating tensor was defined in terms of a covariant vector v_i as

$$w_{ij}=\frac{\partial v_i}{\partial x^j}-\frac{\partial v_j}{\partial x^i}. \tag{5.38a}$$

This may be called the generalized curl. In Riemannian space, this equation can also be expressed in terms of the contravariant merates of v_i since

$$v_i=g_{im}v^m \quad\text{and}\quad v_j=g_{jn}v^n,$$

$$w_{ij}=\frac{\partial}{\partial x^j}(g_{im}v^m)-\frac{\partial}{\partial x^i}(g_{jn}v^n). \tag{10.78}$$

The tensor w_{ij} is an alternating tensor. The merates on the principal diagonal are identically equal to zero, and the merates above the principal diagonal are the negatives of the merates below the principal diagonal. Therefore, the number N of independent merates of w_{ij} is

$$N=\tfrac{1}{2}(n^2-n)=\tfrac{1}{2}[n(n-1)]. \tag{10.79}$$

The number of merates of a contravariant vector is n. If $N=n$, it is possible to associate a contravariant vector with the generalized curl. So let us equate

$$n=\tfrac{1}{2}[n(n-1)]$$

and solve for n. The only solution of this equation is $n=3$. Thus, we conclude: The generalized curl w_{ij} of Equation (5.38a) is a concept valid in arithmetic n-space for any covariant vector. In Riemannian n-space, the generalized curl w_{ij} can also be extended to contravariant vectors as shown in Equation (10.78). But it is only in Riemannian 3-space that a curl vector can be introduced.

The pseudocurl was defined in arithmetic 3-space in Section 5.08 as

$$\tilde{W}^i = \delta^{ijk}_{123} \frac{\partial v_k}{\partial x^j}. \tag{5.46}$$

Since the alternator δ^{ijk}_{123} transforms as a trivalent akinetor of weight $+1$, $\tilde{W}^i$ is a contravariant akinetor of weight $+1$. A related contravariant vector W^i can be introduced in Riemannian 3-space by dividing Equation (5.46) by $g^{1/2}$,

$$W^i = \frac{1}{g^{1/2}} \tilde{W}^i = \frac{1}{g^{12}} \delta^{ijk}_{123} \frac{\partial v_k}{\partial x^j}. \tag{10.80}$$

This equation defines[15] the contravariant curl vector W^i for a covariant vector v_k. In Riemannian 3-space, Equation (10.80) can be extended to the contravariant merates v^l of the vector v_k,

$$W^i = \frac{1}{g^{1/2}} \delta^{ijk}_{123} \frac{\partial}{\partial x^j} (g_{kl} v^l). \tag{10.80a}$$

The components of W^i and of v can be found by multiplying and dividing by $(g_{ii})^{1/2}$, so in terms of unit vectors and components, curl **v** can be defined in Riemannian 3-space by the equation

$$\begin{aligned}
\operatorname{curl} \mathbf{v} = \mathbf{a}_1 \left(\frac{g_{11}}{g}\right)^{1/2} & \left[\frac{\partial}{\partial x^2}\left(g_{31}\frac{(v)_1}{(g_{11})^{1/2}} + g_{32}\frac{(v)_2}{(g_{22})^{1/2}} + g_{33}\frac{(v)_3}{(g_{33})^{1/2}}\right)\right. \\
& \left. - \frac{\partial}{\partial x^3}\left(g_{21}\frac{(v)_1}{(g_{11})^{1/2}} + g_{22}\frac{(v)_2}{(g_{22})^{1/2}} + g_{23}\frac{(v)_3}{(g_{33})^{1/2}}\right)\right] \\
+ \mathbf{a}_2 \left(\frac{g_{22}}{g}\right)^{1/2} & \left[\frac{\partial}{\partial x^3}\left(g_{11}\frac{(v)_1}{(g_{11})^{1/2}} + g_{12}\frac{(v)_2}{(g_{22})^{1/2}} + g_{13}\frac{(v)_3}{(g_{33})^{1/2}}\right)\right. \\
& \left. - \frac{\partial}{\partial x^1}\left(g_{31}\frac{(v)_1}{(g_{11})^{1/2}} + g_{32}\frac{(v)_2}{(g_{22})^{1/2}} + g_{33}\frac{(v)_3}{(g_{33})^{1/2}}\right)\right] \\
+ \mathbf{a}_3 \left(\frac{g_{33}}{g}\right)^{1/2} & \left[\frac{\partial}{\partial x^1}\left(g_{21}\frac{(v)_1}{(g_{11})^{1/2}} + g_{22}\frac{(v)_2}{(g_{22})^{1/2}} + g_{23}\frac{(v)_3}{(g_{33})^{1/2}}\right)\right. \\
& \left. - \frac{\partial}{\partial x^2}\left(g_{11}\frac{(v)_1}{(g_{11})^{1/2}} + g_{12}\frac{(v)_2}{(g_{22})^{1/2}} + g_{13}\frac{(v)_3}{(g_{33})^{1/2}}\right)\right].
\end{aligned} \tag{10.80b}$$

In an orthogonal coordinate system in Riemannian 3-space, Equation (10.80b) simplifies to

$$\operatorname{curl}\mathbf{v} = \mathbf{a}_1\left(\frac{g_{11}}{g}\right)^{1/2}\left[\frac{\partial}{\partial x^2}((g_{33})^{1/2}(v)_3) - \frac{\partial}{\partial x^3}((g_{22})^{1/2}(v)_2)\right]$$
$$+\mathbf{a}_2\left(\frac{g_{22}}{g}\right)^{1/2}\left[\frac{\partial}{\partial x^3}((g_{11})^{1/2}(v)_1) - \frac{\partial}{\partial x^1}((g_{33})^{1/2}(v)_3)\right]$$
$$+\mathbf{a}_3\left(\frac{g_{33}}{g}\right)^{1/2}\left[\frac{\partial}{\partial x^1}((g_{22})^{1/2}(v)_2) - \frac{\partial}{\partial x^2}((g_{11})^{1/2}(v)_1)\right]. \quad (10.80c)$$

A convenient way of remembering the expression for curl **v** that holds in any orthogonal coordinate system in Riemannian 3-space is to write

$$\operatorname{curl}\mathbf{v} = \frac{1}{g^{1/2}}\begin{vmatrix} (g_{11})^{1/2}\mathbf{a}_1 & (g_{22})^{1/2}\mathbf{a}_2 & (g_{33})^{1/2}\mathbf{a}_3 \\ \partial/\partial x^1 & \partial/\partial x^2 & \partial/\partial x^3 \\ (g_{11})^{1/2}(v)_1 & (g_{22})^{1/2}(v)_2 & (g_{33})^{1/2}(v)_3 \end{vmatrix}. \quad (10.80d)$$

In the special case of a Euclidean 3-space in rectangular coordinates,

$$\operatorname{curl}\mathbf{v} = \begin{vmatrix} \mathbf{a}_x & \mathbf{a}_y & \mathbf{a}_z \\ \partial/\partial x & \partial/\partial y & \partial/\partial z \\ v_x & v_y & v_z \end{vmatrix}. \quad (10.80e)$$

Similarly, in circular cylinder coordinates,

$$\operatorname{curl}\mathbf{v} = \frac{1}{r}\begin{vmatrix} \mathbf{a}_r & r\mathbf{a}_\psi & \mathbf{a}_z \\ \partial/\partial r & \partial/\partial \psi & \partial/\partial z \\ v_r & rv_\psi & v_z \end{vmatrix}. \quad (10.80f)$$

Likewise, in spherical coordinates,

$$\operatorname{curl}\mathbf{v} = \frac{1}{(r)^2\sin\theta}\begin{vmatrix} \mathbf{a}_r & r\mathbf{a}_\theta & r\sin\theta\,\mathbf{a}_\psi \\ \partial/\partial r & \partial/\partial \theta & \partial/\partial \psi \\ v_r & rv_\theta & r\sin\theta\, v_\psi \end{vmatrix}. \quad (10.80g)$$

10.10. Theorems of Gauss and Stokes

The generalized theorems of Gauss and Stokes obtained in Chapter 5 for arithmetic space are easily applied to Riemannian space. The generalized Gauss theorem (divergence theorem) relates a volume integral and a surface integral:

$$\oint_S \tilde{v}^i\, d\tilde{\mathfrak{a}}_i = \int_{\mathfrak{V}} \frac{\partial \tilde{v}^i}{\partial x^i}\, d\tilde{\mathfrak{V}}. \quad (5.45)$$

Now let $\tilde{v}^i = g^{1/2} v^i$ in Riemannian 3-space. Then

$$\oint_S g^{1/2} v^i \, d\tilde{\mathcal{A}}_i = \int_{\mathcal{V}} \frac{\partial}{\partial x^i} (g^{1/2} v^i) \, d\tilde{\mathcal{V}}.$$

But, from Section 10.07,

$$d\mathcal{A}_i = g^{1/2} \, d\tilde{\mathcal{A}}_i, \qquad d\mathcal{V} = g^{1/2} \, d\tilde{\mathcal{V}},$$

so the divergence theorem in Riemannian 3-space is

$$\oint_S v^i \, d\mathcal{A}_i = \int_{\mathcal{V}} \operatorname{div} \mathbf{v} \, d\mathcal{V}, \tag{10.81}$$

exactly as in the familiar Euclidean space. Equation (10.81) may be taken as the basis of a useful integral definition of divergence[14]:

$$\operatorname{div} \mathbf{v} = \lim_{\Delta\mathcal{V} \to 0} \frac{\oint v^i \, d\mathcal{A}_i}{\Delta \mathcal{V}}. \tag{10.82}$$

The generalized Stokes theorem for arithmetic 3-space was written

$$\int_S \tilde{W}^i \, d\tilde{\mathcal{A}}_i = \oint_C v_i \, dx^i, \tag{5.48a}$$

where $\tilde{W}^i$ is the pseudocurl,

$$\tilde{W}^i = \tfrac{1}{2} \delta^{ijk}_{123} \frac{\partial v_k}{\partial x^j}. \tag{5.38a}$$

But, from Equations (10.57) and (10.80) for Riemannian 3-space,

$$\tilde{W}^i = g^{1/2} W^i = g^{1/2} \operatorname{curl}^i \mathbf{v}, \qquad d\tilde{\mathcal{A}}_i = d\mathcal{A}_i / g^{1/2}.$$

Thus,

$$\int_S \operatorname{curl}^i \mathbf{v} \, d\mathcal{A}_i = \oint_C v_j \, dx^j. \tag{10.83}$$

Equation (10.83) corresponds to the usual statements of Stokes's theorem in textbooks on vector analysis. It may be employed also as an integral definition[15] of curl:

$$\operatorname{curl}^i \mathbf{v} = \lim_{\Delta\mathcal{A}_i \to 0} \frac{\oint v_j \, dx^j}{\Delta \mathcal{A}_i}. \tag{10.84}$$

10.11. The scalar and vector Laplacians

Another important concept in vector analysis is the scalar Laplacian.[16] In Riemannian n-space, the scalar Laplacian may be defined as

$$\nabla^2 \phi = \operatorname{div} \operatorname{grad} \phi. \tag{10.85}$$

Substitution of Equation (10.76) into (10.77a) gives

$$\nabla^2\phi = \frac{1}{g^{1/2}}\frac{\partial}{\partial x^i}\left(g^{1/2}g^{ij}\frac{\partial\phi}{\partial x^j}\right). \tag{10.86}$$

In an orthogonal coordinate system in a Riemannian n-space, the scalar Laplacian can be written

$$\nabla^2\phi = \frac{1}{g^{1/2}}\left[\frac{\partial}{\partial x^1}\left(\frac{g^{1/2}}{g_{11}}\frac{\partial\phi}{\partial x^1}\right)+\frac{\partial}{\partial x^2}\left(\frac{g^{1/2}}{g_{22}}\frac{\partial\phi}{\partial x^2}\right)+\frac{\partial}{\partial x^3}\left(\frac{g^{1/2}}{g_{33}}\frac{\partial\phi}{\partial x^3}\right)+\cdots\right.$$
$$\left.+\frac{\partial}{\partial x^n}\left(\frac{g^{1/2}}{g_{nn}}\frac{\partial\phi}{\partial x^n}\right)\right]. \tag{10.86a}$$

In Riemannian 3-space, Equation (10.86) can be expanded into

$$\nabla^2\phi = \frac{1}{g^{1/2}}\left\{\frac{\partial}{\partial x^i}\left(g^{1/2}\left[g^{11}\frac{\partial\phi}{\partial x^1}+g^{12}\frac{\partial\phi}{\partial x^2}+g^{13}\frac{\partial\phi}{\partial x^3}\right]\right)\right.$$
$$+\frac{\partial}{\partial x^2}\left(g^{1/2}\left[g^{21}\frac{\partial\phi}{\partial x^1}+g^{22}\frac{\partial\phi}{\partial x^2}+g^{23}\frac{\partial\phi}{\partial x^3}\right]\right)$$
$$\left.+\frac{\partial}{\partial x^3}\left(g^{1/2}\left[g^{31}\frac{\partial\phi}{\partial x^1}+g^{32}\frac{\partial\phi}{\partial x^2}+g^{33}\frac{\partial\phi}{\partial x^3}\right]\right)\right\}. \tag{10.86b}$$

In orthogonal coordinates in Riemannian 3-space, the scalar Laplacian becomes

$$\nabla^2\phi = \frac{1}{g^{1/2}}\left[\frac{\partial}{\partial x^1}\left(\frac{g^{1/2}}{g_{11}}\frac{\partial\phi}{\partial x^1}\right)+\frac{\partial}{\partial x^2}\left(\frac{g^{1/2}}{g_{22}}\frac{\partial\phi}{\partial x^2}\right)+\frac{\partial}{\partial x^3}\left(\frac{g^{1/2}}{g_{33}}\frac{\partial\phi}{\partial x^3}\right)\right]. \tag{10.86c}$$

In rectangular coordinates, $x^1=x, x^2=y, x^3=z$, Equation (10.86) becomes

$$\nabla^2\phi = \frac{\partial^2\phi}{\partial x^2}+\frac{\partial^2\phi}{\partial y^2}+\frac{\partial^2\phi}{\partial z^2}. \tag{10.86d}$$

In circular cylinder coordinates,[17] $x^1=r$, $x^2=\psi$, $x^3=z$, there is one additional term and functional coefficients appear,

$$\nabla^2\phi = \frac{\partial^2\phi}{\partial r^2}+\frac{1}{r}\frac{\partial\phi}{\partial r}+\frac{1}{r^2}\frac{\partial^2\phi}{\partial\psi^2}+\frac{\partial^2\phi}{\partial z^2}. \tag{10.86e}$$

With spherical coordinates,[14] $x^1=r$, $x^2=\theta$, $x^3=\psi$, still another term appears and the functional coefficients become more complicated:

$$\nabla^2\phi = \frac{\partial^2\phi}{\partial r^2}+\frac{2}{r}\frac{\partial\phi}{\partial r}+\frac{1}{r^2}\frac{\partial^2\phi}{\partial\theta^2}+\frac{\cot\theta}{r^2}\frac{\partial\phi}{\partial\theta}+\frac{1}{r^2\sin^2\theta}\frac{\partial^2\phi}{\partial\psi^2}. \tag{10.86f}$$

Whereas the scalar Laplacian can be defined in Riemannian n-space, the vector Laplacian[18] is defined only in Riemannian 3-space. The defining equation is

$$\text{✡}\mathbf{v} = \operatorname{grad}\operatorname{div}\mathbf{v} - \operatorname{curl}\operatorname{curl}\mathbf{v}. \tag{10.87}$$

Substitution of Equations (10.74a), (10.77b), and (10.80b) into (10.87) gives the vector Laplacian in Riemannian 3-space.

For rectangular coordinates in Euclidean 3-space,

$$\begin{aligned}\text{✡}\mathbf{v} &= \mathbf{a}_x\left(\frac{\partial^2 v_x}{\partial x^2} + \frac{\partial^2 v_x}{\partial y^2} + \frac{\partial^2 v_x}{\partial z^2}\right) + \mathbf{a}_y\left(\frac{\partial^2 v_y}{\partial x^2} + \frac{\partial^2 v_y}{\partial y^2} + \frac{\partial^2 v_y}{\partial z^2}\right) \\ &\quad + \mathbf{a}_z\left(\frac{\partial^2 v_z}{\partial x^2} + \frac{\partial^2 v_z}{\partial y^2} + \frac{\partial^2 v_z}{\partial z^2}\right) \\ &= \mathbf{a}_x \nabla^2 v_x + \mathbf{a}_y \nabla^2 v_y + \mathbf{a}_z \nabla^2 v_z. \end{aligned} \tag{10.87a}$$

Since, in rectangular coordinates, Equations (10.86d) and (10.87a) have corresponding terms, some authors have employed the reprehensible practice[19] of using the symbol ∇^2 for both scalar and vector Laplacians.

When the two Laplacians are expanded in any curvilinear coordinate system, their essential differences become apparent. Expansion of Equation (10.87) in circular cylinder coordinates gives

$$\begin{aligned}\text{✡}\mathbf{v} &= \mathbf{a}_r\left(\nabla^2 v_r - \frac{1}{r^2} v_r - \frac{2}{r^2}\frac{\partial v_\psi}{\partial \psi}\right) \\ &\quad + \mathbf{a}_\psi\left(\nabla^2 v_\psi + \frac{2}{r^2}\frac{\partial v_r}{\partial \psi} - \frac{1}{r^2} v_\psi\right) + \mathbf{a}_z \nabla^2 v_z. \end{aligned} \tag{10.87b}$$

In this coordinate system, the vector Laplacian contains four terms that do not correspond to the terms in the scalar Laplacian. If both the components v_r and v_ψ are zero, these differences disappear.

In spherical coordinates, the differences between the two Laplacians become even more marked,

$$\begin{aligned}\text{✡}\mathbf{v} &= \mathbf{a}_r\left[\nabla^2 v_r - \frac{2}{r^2} v_r - \frac{2}{r^2}\frac{\partial v_\theta}{\partial \theta} - \frac{2\cot\theta}{r^2} v_\theta - \frac{2}{r^2\sin\theta}\frac{\partial v_\psi}{\partial \psi}\right] \\ &\quad + \mathbf{a}_\theta\left[\nabla^2 v_\theta + \frac{2}{r^2}\frac{\partial v_r}{\partial \theta} - \frac{1}{r^2\sin^2\theta} v_\theta - \frac{2\cot\theta}{r^2\sin\theta}\frac{\partial v_\psi}{\partial \psi}\right] \\ &\quad + \mathbf{a}_\psi\left[\nabla^2 v_\psi + \frac{2}{r^2\sin\theta}\frac{\partial v_r}{\partial \psi} + \frac{2\cot\theta}{r^2\sin\theta}\frac{\partial v_\theta}{\partial \psi} - \frac{1}{r^2\sin^2\theta} v_\psi\right]. \end{aligned} \tag{10.87c}$$

In spherical coordinates, the vector Laplacian contains 10 terms in addition to those that occur in the scalar Laplacian.

10.12. The contracted covariant derivative of a bivalent tensor

Contracted covariant derivatives[20] were discussed in Section 7.06. In Riemannian space, since the linear connection is symmetric, there is only one kind of contracted covariant derivative of a contravariant vector. Equation (7.55) reduces to

$$\nabla_k v^k = \frac{\partial v^k}{\partial x^k} + v^l \Gamma^k_{kl}. \tag{10.88}$$

However, by Equation (10.36),

$$\Gamma^k_{kl} = \frac{1}{2g}\frac{\partial g}{\partial x^l},$$

so

$$\nabla_k v^k = \frac{\partial v^k}{\partial x^k} + \frac{v^l}{2g}\frac{\partial g}{\partial x^l} = \frac{1}{g^{1/2}}\frac{\partial}{\partial x^l}(g^{1/2}v^l) = \operatorname{div}\mathbf{v}. \tag{10.89}$$

So the contracted covariant derivative of a contravariant vector is another way of obtaining the divergence of a vector in Riemannian space.

For a contravariant bivalent tensor, the eight contracted covariant derivatives of Equation (7.57) become only two in Riemannian space,

$$\nabla_k T^{kj} = \frac{\partial T^{kj}}{\partial x^k} + T^{lj}\Gamma^k_{kl} + T^{kl}\Gamma^j_{kl} \tag{10.90a}$$

and

$$\nabla_k T^{jk} = \frac{\partial T^{jk}}{\partial x^k} + T^{lk}\Gamma^j_{kl} + T^{jl}\Gamma^k_{kl}. \tag{10.90b}$$

In the case of a bivalent mixed tensor T^k_j, the four contracted covariant derivatives of Equation (7.60) become, in Riemannian space,

$$\nabla_k T^k_j = \frac{\partial T^k_j}{\partial x^k} + T^l_j\Gamma^k_{kl} - T^k_l\Gamma^l_{kj}. \tag{10.91}$$

In rectangular coordinates, these expressions are reminiscent of those for the divergence of a vector. Since the Γ's are zero in a rectangular coordinate system in 3-space,

$$\nabla_k T^k_1 = \frac{\partial T^1_1}{\partial x^1} + \frac{\partial T^2_1}{\partial x^2} + \frac{\partial T^3_1}{\partial x^3},$$

$$\nabla_k T^k_2 = \frac{\partial T^1_2}{\partial x^1} + \frac{\partial T^2_2}{\partial x^2} + \frac{\partial T^3_2}{\partial x^3}, \tag{10.91a}$$

$$\nabla_k T^k_3 = \frac{\partial T^1_3}{\partial x^1} + \frac{\partial T^2_3}{\partial x^2} + \frac{\partial T^3_3}{\partial x^3}.$$

However, in circular cylinder coordinates, substitution of Equations (10.32b) and (10.37b) into (10.91) gives

$$\nabla_k T^k_1 = \frac{\partial T^1_1}{\partial r} + \frac{\partial T^2_1}{\partial \psi} + \frac{\partial T^3_1}{\partial z} + \frac{1}{r} T^1_1 + \frac{1}{r} T^2_2,$$

$$\nabla_k T^k_2 = \frac{\partial T^1_2}{\partial r} + \frac{\partial T^2_2}{\partial \psi} + \frac{\partial T^3_2}{\partial z} + r T^2_1 - \frac{1}{r} T^1_2, \tag{10.91b}$$

$$\nabla_k T^k_3 = \frac{\partial T^1_3}{\partial r} + \frac{\partial T^2_3}{\partial \psi} + \frac{\partial T^3_3}{\partial z}.$$

The corresponding expressions in spherical coordinates are found by substituting Equations (10.32d) and (10.37d) into (10.91):

$$\nabla_k T^k_1 = \frac{\partial T^1_1}{\partial r} + \frac{\partial T^2_1}{\partial \theta} + \frac{\partial T^3_1}{\partial \psi} + \frac{2}{r} T^1_1 + (\cot\theta) T^2_1 - \frac{1}{r} T^2_2 - \frac{1}{r} T^3_3,$$

$$\nabla_k T^k_2 = \frac{\partial T^1_2}{\partial r} + \frac{\partial T^2_2}{\partial \theta} + \frac{\partial T^3_2}{\partial \psi} + \frac{2}{r} T^1_2 + (\cot\theta) T^2_2 + r T^2_1 - \frac{1}{r} T^1_2$$
$$- (\cot\theta) T^3_3, \tag{10.91c}$$

$$\nabla_k T^k_3 = \frac{\partial T^1_3}{\partial r} + \frac{\partial T^2_3}{\partial \theta} + \frac{\partial T^3_3}{\partial \psi} + \frac{2}{r} T^1_3 + (\cot\theta) T^2_3 + r\sin^2\theta\, T^3_1$$
$$+ (\sin\theta\cos\theta) T^3_2 - \frac{1}{r} T^1_3 - (\cot\theta) T^2_3.$$

10.13. Summary

A metric is introduced:

$$(ds)^2 = g_{ij}\, dx^i\, dx^j, \tag{10.01}$$

which allows considerable increase in flexibility as compared with the previous nonmetric spaces. The metric tensors g_{ij} and g^{jk} permit

(a) change in index positions,
(b) change of valence, and
(c) change of weight.

The concepts of distance, angle, magnitude of a vector, divergence, and curl become meaningful in Riemannian space. Magnitude of a vector v^i is defined as

$$v = [g_{ij} v^i v^j]^{1/2}. \qquad (10.14)$$

The angle between vectors u^i and v^j may be expressed as

$$\cos\theta = \frac{g_{ij} u^i v^j}{uv}. \qquad (10.17)$$

Akinetors of weight w can now be changed into tensors.

In spaces with linear connection, the connection Γ^i_{jk} is arbitrary (Chapter 9). But in Riemannian space, Γ^i_{jk} is restricted to

$$\Gamma^i_{jk} = \frac{g^{il}}{2}\left[\frac{\partial g_{lj}}{\partial x^k} + \frac{\partial g_{lk}}{\partial x^j} - \frac{\partial g_{jk}}{\partial x^l}\right]. \qquad (10.31)$$

This particular form of connection is required to maintain the constancy of vector magnitude under homeomorphic displacement. It is also required to make the paths of Chapter 9 into geodesics.

The latter part of the chapter is devoted to various geometric objects in Riemannian space. The pseudoarea and pseudovolume of previous chapters now become true area and volume. Scalar and vector products can be expressed in terms of their components. The gradient of a scalar, which previously was inherently covariant, can now be expressed in contravariant form or can be written in terms of components. The generalized divergence, which was previously a nilvalent akinetor of weight +1 and applied only to a contravariant akinetor of weight +1, now becomes a scalar, which can be applied to any type of vector. In arithmetic space, generalized curl was a covariant bivalent tensor, which applied only to a covariant vector. In Riemannian space, the generalized curl can be applied to any vector and is itself a vector. The scalar Laplacian can be introduced into Riemannian n-space, but the vector Laplacian has meaning only in Riemannian 3-space. Contracted covariant derivatives of bivalent tensors in Riemannian space are also discussed.

Problems

10.01 Prove that g_{ij} and g^{jk} are tensors by employing the known transformation equations for dx^i, dx^j, and $(ds)^2$.

10.02 A student advances the following proof that g is a tensor:
(i) g_{ii} is a tensor;
(ii) g is a product of tensors, by Equation (10.03);
(iii) therefore, g is a tensor.
(a) Discuss this proof.
(b) Obtain the transformation equation for g in a Riemannian 2-space with

$$g = \begin{vmatrix} g_{11} & 0 \\ 0 & g_{22} \end{vmatrix}$$

and

$$g_{ij} = \frac{\partial x^{i'}}{\partial x^i} \frac{\partial x^{j'}}{\partial x^j} g_{i'j'}.$$

Do not use Equation (10.06) in your proof.

10.03 A coordinate system (x^1, x^2, x^3) is designated by g_{ij} with

$$g_{11} = g_{22} = \cosh^2(x^1) - \cos^2(x^2), \qquad g_{33} = 1.$$

(a) Write an expression for ds.
(b) What is the distance ds between two neighboring points on the line $x^2 = x^3 = 0$. Why is (or is not) this distance equal to dx^i?
(c) Find the distance s on the line $x^2 = a$ between points $(1, a, 0)$ and $(-1, a, 0)$.

10.04 Repeat Problem 10.03 for

$$g_{11} = g_{22} = [(x^1)^2 + (x^2)^2], \qquad g_{33} = 1.$$

10.05 Repeat Problem 10.03 for

$$g_{11} = g_{22} = (A)^2[\sinh^2(x^1) + \sin^2(x^2)],$$
$$g_{33} = (A)^2 \sinh^2(x^1) \sin^2(x^2).$$

10.06 Repeat Problem 10.03 for

$$g_{11} = g_{22} = (A)^2[\cosh^2(x^1) - \sin^2(x^2)],$$
$$g_{33} = (A)^2 \cosh^2(x^1) \sin^2(x^2).$$

10.07 Repeat Problem 10.03 for

$$g_{11} = g_{22} = (x^1)^2 + (x^2)^2, \qquad g_{33} = (x^1 x^2)^2.$$

10.08 In Euclidean 2-space, rectangular coordinates are designated as $x^{1'}, x^{2'}$. Another Cartesian system is introduced with $x^1 = x^{1'}$, and with the second axis (x^2) making an angle of α with x^1.

Determine g_{11}, g_{22}, and g and write an equation for $(ds)^2$.

10.09 Determine the metric tensor for a Euclidean 2-space with polar coordinates. Write g_{ij}, g^{jk}, g, and $(ds)^2$.

10.10 Prove Equation (10.33) for the special case of a 2×2 determinant.

10.11 Prove Equation (10.33) for the special case of a 3×3 determinant.

10.12 A vector field v^i in Euclidean 3-space is such that v^i is everywhere radial from the origin and its magnitude is inversely proportional to r.

Investigate the possibility of paths such that

$$\delta v^i/\delta s = 0.$$

10.13 A parabolic 2-space may be obtained by embedding a paraboloid of revolution in a Euclidean 3-space. Parabolic coordinates are $x^1=\mu$, $x^2=a$, $x^3=\psi$.[21] Merates of Γ^i_{jk} are zero except for

$$\Gamma^1_{11}=\frac{x^1}{(x^1)^2+(a)^2},\qquad \Gamma^1_{33}=\frac{(a)^2x^1}{(x^1)^2+(a)^2},\qquad \Gamma^3_{13}=\Gamma^3_{31}=\frac{1}{x^1}.$$

Determine the paths.

10.14

(a) By multiplication by metric tensors, obtain R_{ijkl}, R^{ijkl}, and R^{ij}_{kl} from the quadrivalent tensor R^i_{jkl}.

(b) Write the transformation equations for the three new holors. Are they tensors?

10.15 Given a bivalent tensor w^{ij}, we wish to obtain a scalar invariant w that will be a measure of w^{ij}.

(a) Write the equation for w in a Riemannian n-space.

(b) For spherical coordinates in a Euclidean 3-space, and for

$$w^{ij}=\begin{bmatrix}1 & 2 & 3\\ 4 & 5 & 6\\ 7 & 8 & 9\end{bmatrix},$$

evaluate w.

10.16

(a) Write an equation for a scalar invariant U associated with a trivalent tensor U^{ijk}.

(b) Repeat for U_{ijk}.

(c) Associate a univalent U^i with U^{ijk}.

10.17 Consider a Euclidean 2-space with rectangular coordinates $(x^{1'},x^{2'})$, and let

$$u^{1'} = ax^{1'}, \qquad v^{1'} = bx^{1'},$$

$$u^{2'} = 0, \qquad v^{2'} = cx^{2'}.$$

Show that Equation (10.17) gives the correct value of $\cos\theta$.

10.18 For a Euclidean 3-space with the metric tensor of Problem 10.03, evaluate $\cos\theta$

(a) for u^i and v^j along the same coordinate axis,

(b) along different coordinate axes, and

(c) in arbitrary directions.

10.19 Repeat Problem 10.18 for the g_{ij}'s of Problem 10.05.

10.20 A prolate spheroidal surface $x^1 = a$ is embedded in Euclidean 3-space. The surface is considered as a 2-space (x^2, x^3) with

$$g_{22} = \sinh^2 a + \sin^2(x^2), \qquad g_{33} = \sinh^2 a \sin^2(x^2).$$

(a) Determine if the 2-space is flat.

(b) Determine whether the curve $x^2 = \text{constant}$ is a geodesic.

10.21 Repeat Problem 10.20 for an oblate spheroid where coordinates of the 2-space are (x^2, x^3) and

$$g_{22} = \cosh^2 a - \sin^2(x^2), \qquad g_{33} = \cosh^2 a \sin^2(x^2).$$

10.22 Evaluate the volume of a paraboloid of revolution in Euclidean 3-space (Figure 10.04). The metric tensor is specified by

$$g_{11} = g_{22} = (x^1)^2 + (x^2)^2, \qquad g_{33} = (x^1 x^2)^2,$$

in parabolic coordinates. The paraboloid is limited by $x^1 = a$, $x^2 = b$.

10.23 Evaluate the area of the prolate spheroid of Problem 10.20.

10.24 Evaluate the area of the oblate spheroid of Problem 10.21.

10.25

(a) Obtain an expression for $\nabla^2\phi$ for the parabolic coordinates of Problem 10.22.

(b) Obtain an expression for ✡**v** in parabolic coordinates.

10.26

(a) Obtain an expression for $\nabla^2\phi$ in prolate spheroidal coordinates (x^1, x^2, x^3), where

$$g_{11} = g_{22}(a)^2[\sinh^2(x^1) + \sin^2(x^2)],$$

$$g_{33} = (a)^2 \sinh^2(x^1) \sin^2(x^2).$$

(b) Obtain an expression for ✡**v** in prolate spheroidal coordinates.

11
Euclidean space

This chapter concludes our survey of spaces. In Chapter 9, we started with an arithmetic n-space and introduced a linear connection. This was followed by Chapter 10, involving a metric. The present chapter deals with the special class of Riemannian spaces where rectangular coordinates are possible. The development for Chapters 9–11 proceeds as follows:

(a) arithmetic n-space X_n;
(b) introduction of an arbitrary linear connection Γ^i_{jk}, resulting in spaces of linear connection L_n;
(c) introduction of an arbitrary metric tensor g_{ij}, resulting in Riemannian space V_n; and
(d) consideration of the special case of (c), where $g_{i'j'}$ is the unit matrix. This is Euclidean space R_n.

The process consists in starting with the most general case and adding restrictions step by step. Note that in linearly connected spaces, the linear connection Γ^i_{jk} is completely arbitrary. It may be symmetric or not; its merates may be chosen at random. But when we take the step to Riemannian space, Γ^i_{jk} is determined uniquely by the metric tensor g_{ij}, according to Equation (10.31). The g's are arbitrary in a Riemannian space; but once the metric tensor is chosen, Γ^i_{jk} is fixed.

We now come to Euclidean space. In a sense, all properties of Euclidean space have been treated in Chapter 10, since Euclidean space is merely a special case of Riemannian space. The distinction is that the metric tensor is no longer arbitrary, as it was in Riemannian space. The metric tensor must now be obtainable by transformation from the rectangular coordinate tensor with $g_{i'j'}$ the unit matrix. Thus, an important aspect of Euclidean space theory is the determination of the metric coefficients associated with each particular set of coordinate surfaces.[1] Such calculations cannot be made in general Riemannian space, the g's being completely arbitrary.

The outline of our survey can be summarized as follows:

(a) Arithmetic space: no structure, everything arbitrary.

(b) Linearly connected space: Γ^i_{jk} arbitrary.

(c) Riemannian space: g_{ij} arbitrary, Γ^i_{jk} determined by g_{ij}.

(d) Euclidean space: g_{ij} determined by coordinate system.

The sequence from general to particular used in this book is not, of course, the historical sequence. Historically, man had Euclid's *Elements* for 2000 years before the idea dawned that Euclidean geometry is not the only geometry, that Euclidean space is not the only space. This revolutionary idea occurred to Lobachewsky. His important paper of 1826 was followed by the work of Riemann, Clifford, Klein, Ricci, Levi-Civita, and a host of others who transformed geometry into a discipline of ever-increasing generality.

Historically, then, spaces have been developed from the particular to the general. A similar sequence is employed in most textbooks. Such a procedure has certain pedagogical advantages, but it also has decided disadvantages. Theorems and definitions must be stated repeatedly in terms of increasing generality; and the student has difficulty in grasping the subject as a whole.

A more logical development (used in this book) begins with the most general case, from which all degenerate cases are deduced. This may be hard on the reader of Chapter 9; but we hope that the final result will be a more thorough understanding than would have been obtained by following the historical sequence.

11.01. Definition of Euclidean space

Riemannian space[2] is characterized by the metric

$$(ds)^2 = g_{ij}\, dx^i\, dx^j, \tag{11.01}$$

where the metric tensor g_{ij} is arbitrary. Euclidean space is a special case of the above:

Definition XVI: Euclidean space is a Riemannian space that allows the metric tensor

$$g_{i'j'} = \begin{bmatrix} 1 & 0 & 0 & \cdots & 0 \\ 0 & 1 & 0 & \cdots & 0 \\ 0 & 0 & 1 & \cdots & 0 \\ \cdot & \cdot & \cdot & \cdots & \cdot \\ 0 & 0 & 0 & \cdots & 1 \end{bmatrix}. \tag{11.02}$$

In Euclidean 3-space, Equation (11.02) reduces Equation (11.01) to

$$(ds)^2 = (dx^{1'})^2 + (dx^{2'})^2 + (dx^{3'})^2, \tag{11.03}$$

which is a formulation of the Pythagorean theorem in rectangular (orthogonal Cartesian) coordinates. Thus, Euclidean space could be defined as a Riemannian space that allows rectangular coordinates. Such a definition is sometimes useful intuitively. In 2-space, for instance, one can state immediately that a plane is Euclidean but a spherical 2-space is not Euclidean because no rectangular coordinate system can be drawn on a sphere.

A third suggested definition is that Euclidean space is a flat Riemannian space:

$$R^i_{jkl} = 0.$$

If we take a Riemannian space with $g_{i'j'}$ given by Equation (11.02), then Γ^i_{jk} and R^i_{jkl} are obviously zero. Conversely, if $R^i_{jkl} = 0$, the space is flat and one might expect intuitively that a rectangular coordinate system is possible. Indeed, it is easy to prove[3]:

Theorem XLIX. *The necessary and sufficient condition that the quadratic differential form*

$$g_{ij}\, dx^i\, dx^j$$

be reducible to a form with g_{ij} *the unit matrix is that*

$$R^i_{jkl} = 0.$$

11.02. Metric coefficients

The merates of the metric tensor g_{ij} are now evaluated in Euclidean n-space. For rectangular coordinates, Equation (11.01) reduces to

$$(ds)^2 = (dx^{1'})^2 + (dx^{2'})^2 + \cdots + (dx^{n'})^2. \tag{11.01a}$$

Since dx^i is a tensor, its general transformation is

$$dx^{i'} = \frac{\partial x^{i'}}{\partial x^i}\, dx^i.$$

Substitution into Equation (11.01a) gives

$$(ds)^2 = \left[\frac{\partial x^{1'}}{\partial x^1} dx^1 + \frac{\partial x^{1'}}{\partial x^2} dx^2 + \cdots + \frac{\partial x^{1'}}{\partial x^n} dx^n\right]^2$$
$$+ \left[\frac{\partial x^{2'}}{\partial x^1} dx^1 + \frac{\partial x^{2'}}{\partial x^2} dx^2 + \cdots + \frac{\partial x^{2'}}{\partial x^n} dx^n\right]^2 + \cdots$$
$$+ \left[\frac{\partial x^{n'}}{\partial x^1} dx^1 + \frac{\partial x^{n'}}{\partial x^2} dx^2 + \cdots + \frac{\partial x^{n'}}{\partial x^n} dx^n\right]^2$$
$$= \left[\left(\frac{\partial x^{1'}}{\partial x^1}\right)^2 + \left(\frac{\partial x^{2'}}{\partial x^1}\right)^2 + \cdots + \left(\frac{\partial x^{n'}}{\partial x^1}\right)^2\right](dx^1)^2$$
$$+ \left[\left(\frac{\partial x^{1'}}{\partial x^2}\right)^2 + \left(\frac{\partial x^{2'}}{\partial x^2}\right)^2 + \cdots + \left(\frac{\partial x^{n'}}{\partial x^2}\right)^2\right](dx^2)^2 + \cdots$$
$$+ \left[\left(\frac{\partial x^{1'}}{\partial x^n}\right)^2 + \left(\frac{\partial x^{2'}}{\partial x^n}\right)^2 + \cdots + \left(\frac{\partial x^{n'}}{\partial x^n}\right)^2\right](dx^n)^2$$
$$+ 2\left[\frac{\partial x^{1'}}{\partial x^1}\frac{\partial x^{1'}}{\partial x^2} + \frac{\partial x^{2'}}{\partial x^1}\frac{\partial x^{2'}}{\partial x^2} + \cdots + \frac{\partial x^{n'}}{\partial x^1}\frac{\partial x^{n'}}{\partial x^2}\right] dx^1 dx^2$$
$$+ 2\left[\frac{\partial x^{1'}}{\partial x^1}\frac{\partial x^{1'}}{\partial x^3} + \frac{\partial x^{2'}}{\partial x^1}\frac{\partial x^{2'}}{\partial x^3} + \cdots + \frac{\partial x^{n'}}{\partial x^1}\frac{\partial x^{n'}}{\partial x^3}\right] dx^1 dx^3 + \cdots$$
$$+ 2\left[\frac{\partial x^{1'}}{\partial x^{n-1}}\frac{\partial x^{1'}}{\partial x^n} + \frac{\partial x^{2'}}{\partial x^{n-1}}\frac{\partial x^{2'}}{\partial x^n} + \cdots + \frac{\partial x^{n'}}{\partial x^{n-1}}\frac{\partial x^{n'}}{\partial x^n}\right] dx^{n-1} dx^n. \tag{11.04}$$

But

$$(ds)^2 = g_{ij}\, dx^i\, dx^j$$
$$= g_{11}(dx^1)^2 + g_{22}(dx^2)^2 + \cdots + g_{nn}(dx^n)^2$$
$$+ 2g_{12}\, dx^1\, dx^2 + 2g_{(n-1)n}\, dx^{n-1}\, dx^n.$$

so each bracket of Equation (11.04) is obviously equal to the corresponding metric coefficient:

$$g_{ij} = \frac{\partial x^{1'}}{\partial x^i}\frac{\partial x^{1'}}{\partial x^j} + \frac{\partial x^{2'}}{\partial x^i}\frac{\partial x^{2'}}{\partial x^j} + \cdots + \frac{\partial x^{n'}}{\partial x^i}\frac{\partial x^{n'}}{\partial x^j}$$

or

$$g_{ij} = \sum_{i'=1'}^{n'} \frac{\partial x^{i'}}{\partial x^i}\frac{\partial x^{i'}}{\partial x^j}. \tag{11.05}$$

Equation (11.05) allows the unique determination of the metric tensor g_{ij} for every given coordinate system: With each coordinate system in Euclidean n-space is associated a unique metric tensor.

For the particular case of an orthogonal coordinate system, we can use the orthogonality relation of Section 10.02:

$$g_{ij}u^i v^j = 0. \tag{10.17a}$$

Equate the arbitrary vectors u^i and v^j to increments in coordinates, dx^i and dx^j, where $i \neq j$. Then orthogonality of coordinates is specified by the relation

$$g_{ij}\, dx^i\, dx^j = 0 \quad \text{for } i \neq j.$$

This eliminates all the cross products of Equation (11.05) and leaves

$$g_{ii} = \sum_{i'=1'}^{n'} \left(\frac{\partial x^{i'}}{\partial x^i} \right)^2. \tag{11.06}$$

In Euclidean 3-space, with any orthogonal coordinate system, the metric coefficients are obtained uniquely according to the simple relation

$$g_{ii} = \left(\frac{\partial x^{1'}}{\partial x^i} \right)^2 + \left(\frac{\partial x^{2'}}{\partial x^i} \right)^2 + \left(\frac{\partial x^{3'}}{\partial x^i} \right)^2. \tag{11.06a}$$

As an example, find the metric tensor for prolate spheroidal coordinates[4] in Euclidean 3-space, where

$$x^{1'} = a \sinh x^1 \sin x^2 \cos x^3,$$

$$x^{2'} = a \sinh x^1 \sin x^2 \sin x^3,$$

$$x^{3'} = a \cosh x^1 \cos x^2.$$

The surfaces $x^1 =$ constant are prolate spheroids, the surfaces $x^2 =$ constant are hyperboloids of revolution, the surfaces $x^3 =$ constant are half-planes through the axis of revolution.

Since the coordinate surfaces intersect orthogonally, we can use Equation (11.06a) and write

$$\begin{aligned} g_{11} &= (a)^2[(\cosh x^1 \sin x^2 \cos x^3)^2 + (\cosh x^1 \sin x^2 \sin x^3)^2 \\ &\qquad + (\sinh x^1 \cos x^2)^2] \\ &= (a)^2[\cosh^2 x^1 \sin^2 x^2 + \sinh^2 x^1 \cos^2 x^2] \\ &= (a)^2[\sinh^2 x^1 + \sin^2 x^2]. \end{aligned}$$

Similarly,

$$g_{22} = g_{11}, \qquad g_{33} = (a)^2 \sinh^2 x^1 \sin^2 x^2.$$

Thus, the metric tensor for prolate spheroidal coordinates is

$$g_{ij} = (a)^2 \begin{bmatrix} \sinh^2 x^1 + \sin^2 x^2 & 0 & 0 \\ 0 & \sinh^2 x^1 + \sin^2 x^2 & 0 \\ 0 & 0 & \sinh^2 x^1 \sin^2 x^2 \end{bmatrix}.$$

11.03. The linear connection

With a symmetric connection Γ^i_{jk} in Riemannian n-space, we have the general equation

$$\Gamma^i_{jk} = \frac{g^{il}}{2}\left[\frac{\partial g_{lj}}{\partial x^k} + \frac{\partial g_{lk}}{\partial x^j} - \frac{\partial g_{jk}}{\partial x^l}\right], \tag{10.31}$$

which also holds in Euclidean space.

Euclidean space always allows rectangular coordinates for which $\Gamma^i_{jk} = 0$. Thus, in Euclidean space and rectangular coordinates, the covariant derivative reduces to the partial derivative:

$$\nabla_j v^i = \frac{\partial v^i}{\partial x^j}. \tag{11.07}$$

Also, Equation (9.03) reduces to

$$\delta v^i = dv^i$$

and infinitesimal homeomorphic transport reduces to familiar parallelism.

Similarly, the differential equation for a geodesic in Cartesian coordinates in Euclidean space reduces to

$$\frac{d^2 x^i}{(ds)^2} = 0,$$

which shows that a geodesic in Euclidean space is always a straight line.

11.04. Merates and components

We now come to the very important distinction between merates of a vector and components. It is no exaggeration to say that this distinction causes more confusion and leads to more errors in tensor applications than any other aspect of the subject. In tensor theory, a vector in 3-space is written

$$v^i = (v^1, v^2, v^3), \tag{11.08}$$

where v^1, v^2, v^3 are merates. In vector analysis, a vector is written

$$\mathbf{v} = \mathbf{a}_1(v)_1 + \mathbf{a}_2(v)_2 + \mathbf{a}_3(v)_3, \tag{11.09}$$

where the $\mathbf{a}$'s are unit vectors and the v's are components.

The tendency is to glibly equate the two and to tacitly assume that merate equals component. This serious error is fostered by the fact that many tensor books write merates and components indistinguishably as v_i and call both of them "components."[5]

The distinction is evident when we consider magnitude of a vector. According to Section 10.02,

$$v = [g_{ij} v^i v^j]^{1/2}, \tag{10.14}$$

or, with an orthogonal coordinate system,

$$v = [g_{11}(v^1)^2 + g_{22}(v^2)^2 + g_{33}(v^3)^2]^{1/2}. \tag{10.14a}$$

But in vector algebra,

$$v = [((v)_1)^2 + ((v)_2)^2 + ((v)_3)^2]^{1/2}. \tag{11.10}$$

Comparison of Equations (10.14a) and (11.10) gives the relation between components and contravariant merates:

$$(v)_i = (g_{ii})^{1/2}\{v^i\}. \tag{11.11}$$

Only in rectangular coordinates are the two equal.

Also, since

$$v^i = g^{ij} v_j,$$

we can express the components of a vector in terms of covariant merates:

$$(v)_i = (g_{ii})^{1/2}\{g^{ij} v_j\}. \tag{11.12}$$

The distinction between merates and components occurs in many guises. For instance, we write an infinitesimal distance as

$$ds = [g_{11}(dx^1)^2 + g_{22}(dx^2)^2 + g_{33}(dx^3)^2]^{1/2} \tag{11.13}$$

in any orthogonal coordinate system in Euclidean 3-space. The distance ds may be regarded as the magnitude of a vector. But evidently the components of this vector are not dx^1, dx^2, dx^3: The distances along the coordinate axes must be

$$(g_{11})^{1/2}\,dx^1, \quad (g_{22})^{1/2}\,dx^2, \quad (g_{33})^{1/2}\,dx^3,$$

which is in accordance with Equation (11.11).

Similarly, an elementary area on a surface $x^1 = a$ is not expressed as

$$dx^2\,dx^3,$$

but, letting $dv^k = (0, dx^2, 0)$, $dw = (0, 0, dx^3)$, in Equation (10.59), as

$$d\mathfrak{A} = (gg^{11})^{1/2}\,dx^2\,dx^3. \tag{11.14}$$

In orthogonal coordinate systems, this can be simplified further:

$$d\mathfrak{A} = (g_{22}g_{33})^{1/2}\,dx^2\,dx^3.$$

Also, an elementary volume is not written

$$d\mathfrak{V} = dx^1\,dx^2\,dx^3$$

but as a product of distances, by letting $du^i = (dx^1, 0, 0)$, $dv^j = (0, dx^2, 0)$, $dw^k = (0, 0, dx^3)$ in Equation (10.56),

$$d\mathfrak{V} = (g)^{1/2}\,dx^1\,dx^2\,dx^3, \tag{11.15}$$

which can be expressed in orthogonal coordinates as

$$d\mathfrak{V} = (g_{11}g_{22}g_{33})^{1/2}\,dx^1\,dx^2\,dx^3. \tag{11.15a}$$

11.05. Euclidean *n*-space

In many treatments of Euclidean space, the tacit assumption is made that we are operating in a 3-space. Obviously, this restriction is not required: Euclidean space may have any number of dimensions. Thus, the equations that were developed in Chapter 10 for Riemannian *n*-space are directly applicable to Euclidean *n*-space. A few simplifications arise, however, because it is now possible to introduce rectangular coordinates. In rectangular coordinates, $\Gamma^i_{jk} = 0$ and $R^i_{jkl} = 0$, as found from Equations (10.31) and (8.09a). Since R^i_{jkl} is a tensor, it must therefore be zero in all coordinate systems (Theorem XIII). In other words, Euclidean space is flat and all the curvature tensors are zero.

In Euclidean *n*-space, an elementary (2-dimensional) area is expressed in terms of elementary vectors dv^j and dw^k:

$$d\mathfrak{A}_i = g^{1/2}\delta^{123}_{ijk}\,dv^j\,dw^k. \tag{10.57}$$

Similarly, for a 3-dimensional volume,

$$d\mathcal{V} = g^{1/2}\delta^{123}_{ijk}\, du^i\, dv^j\, dw^k. \qquad (10.56)$$

For an n-dimensional hypervolume,

$$d\mathcal{V} = g^{1/2}\,\delta^{12\cdots n}_{ij\cdots p}\, du^i\, dv^j \cdots dz^p. \qquad (10.56a)$$

The magnitude of a vector v^i is

$$v = [g_{ij}v^i v^j]^{1/2}, \qquad (10.14)$$

and the angle between vectors u^i and v^j is expressed as

$$\cos\theta = \frac{g_{ij}u^i v^j}{uv}. \qquad (10.17)$$

For orthogonality, obviously,

$$g_{ij}u^i v^j = 0. \qquad (10.17a)$$

We now come to the important combinations of derivatives. Gradient is

$$\text{grad}^i\,\phi = g^{ij}\frac{\partial\phi}{\partial x^j}, \qquad (10.73)$$

and divergence is

$$\text{div}(\mathbf{v}) = g^{-1/2}\frac{\partial}{\partial x^i}(g^{1/2}v^i). \qquad (10.75)$$

Curl is intrinsically a 3-dimensional concept, as is the vector Laplacian. But the scalar Laplacian can be applied in n-space:

$$\nabla^2\phi = g^{-1/2}\frac{\partial}{\partial x^i}\left[g^{1/2}g^{ij}\frac{\partial\phi}{\partial x^j}\right]. \qquad (10.84)$$

11.06. Skew coordinates

Orthogonal coordinate systems are so universally used that one tends to forget that nonorthogonal systems are perfectly feasible and may have important applications. The equations of Chapter 10 apply to nonorthogonal systems in Euclidean space as well as to orthogonal systems. To illustrate this fact, we introduce a simple example.

Consider Euclidean 2-space with a rectangular coordinate system $(x^{1'}, x^{2'})$. Rotation of coordinate axes through angles ζ and η (Figure 11.01) gives the relations[6]

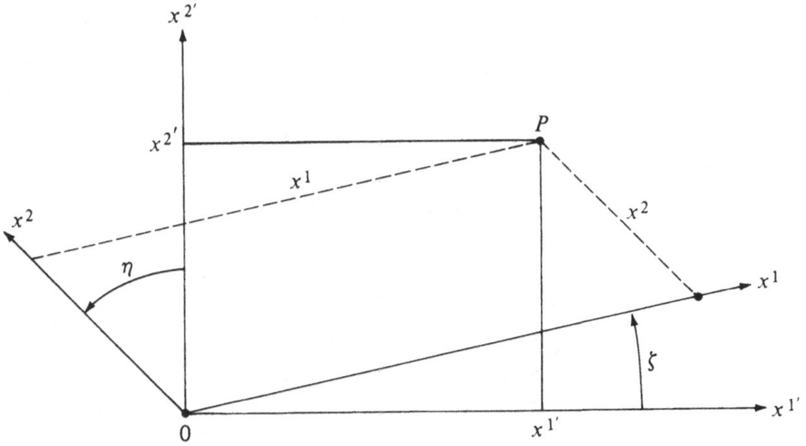

Figure 11.01. Skew coordinates in 2-space. Orthogonal coordinate system $(x^{1'}, x^{2'})$ and skew system (x^1, x^2).

$$x^{1'} = x^1 \cos\zeta - x^2 \sin\eta, \qquad x^{2'} = x^1 \sin\zeta + x^2 \cos\eta,$$

with transformation matrix

$$A_i^{i'} = \begin{bmatrix} \cos\zeta & -\sin\eta \\ \sin\zeta & \cos\eta \end{bmatrix}.$$

Metric coefficients are now evaluated by Equation (11.05):

$$g_{ij} = \sum_{i'=1'}^{n'} \frac{\partial x^{i'}}{\partial x^i} \frac{\partial x^{i'}}{\partial x^j},$$

or, for Euclidean 2-space,

$$g_{ij} = \frac{\partial x^{1'}}{\partial x^i} \frac{\partial x^{1'}}{\partial x^j} + \frac{\partial x^{2'}}{\partial x^i} \frac{\partial x^{2'}}{\partial x^j}. \tag{11.05a}$$

Thus,

$$g_{11} = \left(\frac{\partial x^{1'}}{\partial x^1}\right)^2 + \left(\frac{\partial x^{2'}}{\partial x^1}\right)^2 = \cos^2\zeta + \sin^2\zeta = 1.$$

Also,

$$g_{22} = 1, \qquad g_{12} = g_{21} = \sin(\zeta - \eta).$$

The metric tensor is therefore

$$g_{ij} = \begin{bmatrix} 1 & \sin(\zeta-\eta) \\ \sin(\zeta-\eta) & 1 \end{bmatrix}. \tag{11.16}$$

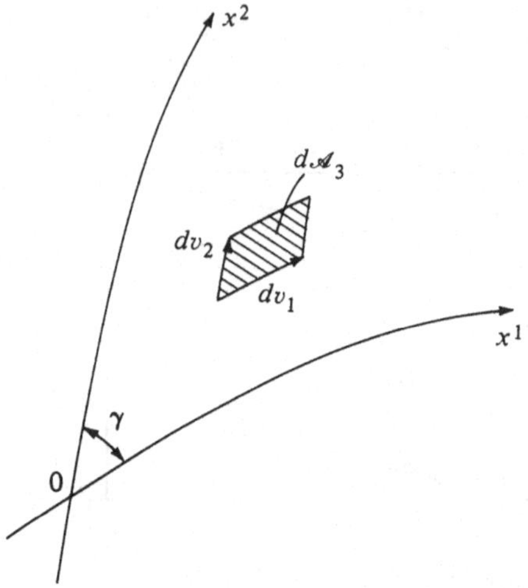

Figure 11.02. Area in a skew coordinate system:
$d\mathfrak{A}_3 = dx^1\, dx^2 \sin\gamma$.

The corresponding determinant is

$$g = 1 - \sin^2(\zeta - \eta) = \cos^2(\zeta - \eta),$$

so

$$g^{1/2} = \cos(\zeta - \eta) = \sin\gamma, \qquad (11.17)$$

$$g^{-1/2} = 1/\sin\gamma, \qquad (11.18)$$

where $\gamma = \frac{1}{2}\pi - (\zeta - \eta)$. Also,

$$g^{ij} = \frac{1}{\sin^2\gamma}\begin{bmatrix} 1 & -\cos\gamma \\ -\cos\gamma & 1 \end{bmatrix}. \qquad (11.19)$$

(1) *Area*. According to Equation (10.57),

$$d\mathfrak{A}_i = g^{1/2}\delta^{123}_{ijk}\, dv^j\, dw^k. \qquad (10.57)$$

Thus,

$$d\mathfrak{A}_3 = g^{1/2}\delta^{123}_{ijk}\, dv^j\, dw^k.$$

If we take $dv^j = (dx^1, 0, 0)$ and $dw^k = (0, dx^2, 0)$ as in Figure 11.02, then

$$d\mathfrak{A}_3 = \sin\gamma[\delta^{123}_{312}\, dv^1\, dw^2 + \delta^{123}_{321}\, dv^2\, dw^3]$$

$$= dx^1\, dx^2 \sin\gamma, \qquad (11.20)$$

which is obviously the correct area of the parallelogram.

(2) *Angle.* From Equation (10.17),

$$\cos\theta = \frac{g_{ij}u^i v^j}{uv}. \tag{10.17}$$

Suppose that we apply this equation to the determination of the angle between the axes. Then $u^i = (x^1, 0, 0)$ and $v^j = (0, x^2, 0)$. Thus,

$$\begin{aligned}\cos\theta &= (x^1x^2)^{-1}[g_{11}u^1v^1 + 2g_{12}u^1v^2 + g_{22}u^2v^2]\\ &= (x^1x^2)^{-1}[0 + 2x^1x^2\cos\gamma + 0]\\ &= \cos\gamma.\end{aligned}$$

(3) *Gradient.*[7] From Equation (10.70),

$$\text{grad}^i\,\phi = g^{ij}\frac{\partial\phi}{\partial x^j}. \tag{10.70}$$

Therefore,

$$\text{grad}^1\,\phi = g^{11}\frac{\partial\phi}{\partial x^1} + g^{12}\frac{\partial\phi}{\partial x^2} = \frac{\partial\phi}{\partial x^1} + \frac{\partial\phi}{\partial x^2}\cos\gamma,$$

$$\text{grad}^2\,\phi = \frac{\partial\phi}{\partial x^1}\cos\gamma + \frac{\partial\phi}{\partial x^2}. \tag{11.21}$$

(4) *Divergence.* According to Equation (10.61),

$$\text{div}(v^i) = g^{-1/2}\frac{\partial}{\partial x^i}(g^{1/2}v^i). \tag{10.61}$$

But $g^{1/2}$ is a constant and cancels $g^{-1/2}$. Therefore,

$$\text{div}(v^i) = \frac{\partial v^1}{\partial x^1} + \frac{\partial v^2}{\partial x^2},$$

just as in rectangular coordinates.

(5) *Scalar Laplacian.* Curl and the vector Laplacians are meaningless in 2-space, but the scalar Laplacian has significance. From Equation (10.71),

$$\nabla^2\phi = g^{-1/2}\frac{\partial}{\partial x^i}\left[g^{1/2}g^{ij}\frac{\partial\phi}{\partial x^j}\right]. \tag{10.71}$$

Here $g^{1/2}$ and $g^{-1/2}$ cancel, leaving

$$\nabla^2\phi = g^{ij}\frac{\partial^2\phi}{\partial x^i\,\partial x^j}, \tag{11.22}$$

or

$$\nabla^2\phi = g^{11}\frac{\partial^2\phi}{(\partial x^1)^2} + 2g^{12}\frac{\partial^2\phi}{\partial x^1\,\partial x^2} + g^{22}\frac{\partial^2\phi}{(\partial x^2)^2}$$

$$= \frac{\partial^2\phi}{(\partial x^1)^2} + 2\cos\gamma\frac{\partial^2\phi}{\partial x^1\,\partial x^2} + \frac{\partial^2\phi}{(\partial x^2)^2}. \tag{11.23}$$

To one familiar with Laplace's equation, the above looks very peculiar. If $\gamma = \frac{1}{2}\pi$, however, $\nabla^2\phi$ reduces to its ordinary formulation in rectangular coordinates.

11.07. Euclidean 3-space

The vast majority of Euclidean space applications deal with three dimensions and an orthogonal coordinate system. Then

$$(ds)^2 = g_{11}(dx^1)^2 + g_{22}(dx^2)^2 + g_{33}(dx^3)^2. \tag{11.24}$$

The metric tensor is

$$g_{ii} = \begin{bmatrix} g_{11} & 0 & 0 \\ 0 & g_{22} & 0 \\ 0 & 0 & g_{33} \end{bmatrix} \tag{11.25}$$

where

$$g_{ii} = \left(\frac{\partial x^{1'}}{\partial x^i}\right)^2 + \left(\frac{\partial x^{2'}}{\partial x^i}\right)^2 + \left(\frac{\partial x^{3'}}{\partial x^i}\right)^2. \tag{11.26}$$

The metric determinant is

$$g = g_{11}g_{22}g_{33}, \tag{11.27}$$

and the contravariant tensor is very simply related to the covariant tensor:

$$g^{jj} = \begin{bmatrix} 1/g_{11} & 0 & 0 \\ 0 & 1/g_{22} & 0 \\ 0 & 0 & 1/g_{33} \end{bmatrix}. \tag{11.28}$$

The magnitude of a vector v^i is

$$v = [g_{11}(v^1)^2 + g_{22}(v^2)^2 + g_{33}(v^3)^2]^{1/2}. \tag{11.29}$$

Gradient is written

$$\text{grad}\,\phi = \mathbf{a}_1\frac{1}{(g_{11})^{1/2}}\frac{\partial\phi}{\partial x^1} + \mathbf{a}_2\frac{1}{(g_{22})^{1/2}}\frac{\partial\phi}{\partial x^2} + \mathbf{a}_3\frac{1}{(g_{33})^{1/2}}\frac{\partial\phi}{\partial x^3} \tag{11.30}$$

Divergence becomes

$$\operatorname{div}\mathbf{v} = \frac{1}{g^{1/2}}\left[\frac{\partial}{\partial x^1}((g_{22}g_{33})^{1/2}(v)_1) + \frac{\partial}{\partial x^2}((g_{11}g_{33})^{1/2}(v)_2) + \frac{\partial}{\partial x^3}((g_{11}g_{22})^{1/2}(v)_3)\right]. \tag{11.31}$$

Curl can be expressed as a determinant,

$$\operatorname{curl}\mathbf{v} = \frac{1}{g^{1/2}}\begin{vmatrix} (g_{11})^{1/2}\mathbf{a}_1 & (g_{22})^{1/2}\mathbf{a}_2 & (g_{33})^{1/2}\mathbf{a}_3 \\ \dfrac{\partial}{\partial x^1} & \dfrac{\partial}{\partial x^2} & \dfrac{\partial}{\partial x^3} \\ (g_{11})^{1/2}(v)_1 & (g_{22})^{1/2}(v)_2 & (g_{33})^{1/2}(v)_3 \end{vmatrix}. \tag{11.32}$$

The scalar Laplacian is written

$$\nabla^2\phi = \frac{1}{g^{1/2}}\left[\frac{\partial}{\partial x^1}\left(\frac{g^{1/2}}{g_{11}}\frac{\partial\phi}{\partial x^1}\right) + \frac{\partial}{\partial x^2}\left(\frac{g^{1/2}}{g_{22}}\frac{\partial\phi}{\partial x^2}\right) + \frac{\partial}{\partial x^3}\left(\frac{g^{1/2}}{g_{33}}\frac{\partial\phi}{\partial x^3}\right)\right]. \tag{11.33}$$

The vector Laplacian can be written

$$\begin{aligned}
✡\mathbf{v} = \mathbf{a}_1\Bigg\{&\frac{1}{(g_{11})^{1/2}}\left[\frac{\partial}{\partial x^1}\left(\frac{1}{g^{1/2}}\left[\frac{\partial}{\partial x^1}\left(\left(\frac{g}{g_{11}}\right)^{1/2}(v)_1\right) + \frac{\partial}{\partial x^2}\left(\left(\frac{g}{g_{22}}\right)^{1/2}(v)_2\right)\right.\right.\right. \\
&\left.\left.\left.+ \frac{\partial}{\partial x^3}\left(\left(\frac{g}{g_{33}}\right)^{1/2}(v)_3\right)\right]\right)\right] \\
&+\left(\frac{g_{11}}{g}\right)^{1/2}\left[\frac{\partial}{\partial x^3}\left(\frac{g_{22}}{g^{1/2}}\left[\frac{\partial}{\partial x^3}((g_{11})^{1/2}(v)_1) - \frac{\partial}{\partial x^1}((g_{33})^{1/2}(v)_3)\right]\right)\right. \\
&- \frac{\partial}{\partial x^2}\left(\frac{g_{33}}{g^{1/2}}\left[\frac{\partial}{\partial x^1}((g_{22})^{1/2}(v)_2)\right.\right. \\
&\left.\left.\left.- \frac{\partial}{\partial x^2}((g_{11})^{1/2}(v)_1)\right]\right)\right]\Bigg\} \\
+\mathbf{a}_2\Bigg\{&\frac{1}{(g_{22})^{1/2}}\left[\frac{\partial}{\partial x^2}\left(\frac{1}{g^{1/2}}\left[\frac{\partial}{\partial x^1}\left(\left(\frac{g}{g_{11}}\right)^{1/2}(v)_1\right)\right.\right.\right. \\
&+ \frac{\partial}{\partial x^2}\left(\left(\frac{g}{g_{22}}\right)^{1/2}(v)_2\right) \\
&\left.\left.\left.+ \frac{\partial}{\partial x^2}\left(\left(\frac{g}{g_{33}}\right)^{1/2}(v)_3\right)\right]\right)\right]
\end{aligned}$$

$$
\begin{aligned}
&+\left(\frac{g_{22}}{g}\right)^{1/2}\left[\frac{\partial}{\partial x^1}\left(\frac{g_{33}}{g^{1/2}}\left[\frac{\partial}{\partial x^1}((g_{22})^{1/2}(v)_2)\right.\right.\right.\\
&\qquad\qquad\left.\left.-\frac{\partial}{\partial x^2}((g_{11})^{1/2}(v)_1)\right]\right)\\
&\qquad-\frac{\partial}{\partial x^3}\left(\frac{g_{11}}{g^{1/2}}\left[\frac{\partial}{\partial x^2}((g_{33})^{1/2}(v)_3)\right.\right.\\
&\qquad\qquad\left.\left.\left.\left.-\frac{\partial}{\partial x^3}((g_{22})^{1/2}(v)_2)\right]\right)\right]\right\}\\
&+\mathbf{a}_3\left\{\frac{1}{(g_{33})^{1/2}}\left[\frac{\partial}{\partial x^3}\left(\frac{1}{g^{1/2}}\left[\frac{\partial}{\partial x^1}\left(\left(\frac{g}{g_{11}}\right)^{1/2}(v)_1\right)\right.\right.\right.\right.\\
&\qquad\qquad+\frac{\partial}{\partial x^2}\left(\left(\frac{g}{g_{22}}\right)^{1/2}(v)_2\right)\\
&\qquad\qquad\left.\left.\left.+\frac{\partial}{\partial x^3}\left(\left(\frac{g}{g_{33}}\right)^{1/2}(v)_3\right)\right]\right)\right]\\
&+\left(\frac{g_{33}}{g}\right)^{1/2}\left[\frac{\partial}{\partial x^2}\left(\frac{g_{11}}{g^{1/2}}\left[\frac{\partial}{\partial x^2}((g_{33})^{1/2}(v)_3)\right.\right.\right.\\
&\qquad\qquad\left.\left.-\frac{\partial}{\partial x^3}((g_{22})^{1/2}(v)_2)\right]\right)\\
&\qquad-\frac{\partial}{\partial x^1}\left(\frac{g_{22}}{g^{1/2}}\left[\frac{\partial}{\partial x^3}((g_{11})^{1/2}(v)_1)\right.\right.\\
&\qquad\qquad\left.\left.\left.\left.-\frac{\partial}{\partial x^1}((g_{33})^{1/2}(v)_3)\right]\right)\right]\right\}.
\end{aligned}
\tag{11.34}
$$

11.08. Index notation versus vector notation

Vector analysis, as developed by Gibbs and Heaviside,[8] employs a simple and very effective ad hoc notation.[9] Vectors are distinguished from scalars by the use of boldface type, scalar and vector products are denoted by dot and cross, and important concepts involving derivatives are gradient, divergence, curl, and the Laplacians. One should realize, of course, that vector analysis is a very circumscribed subject compared with tensor analysis.[9] Vector theory is ordinarily limited to (a) Euclidean 3-space, (b) valences 0 and 1, and (c) orthogonal coordinate systems.

The range of the subject can be extended slightly. Some equations can be written for n-space; but vector analysis as a whole is definitely made for Euclidean 3-space. A vector product, for instance, is meaningless in 2-space or in 4-space, as is a curl or a vector Laplacian. Holors of valence 2 can also be introduced, as with Gibb's dyads,[10] but this step is not ordinarily taken. The tiny part of tensor analysis obtained by imposing the above limitations is, however, of great practical utility in geometry and in nearly all of physics.

Let us compare, therefore, the customary vector equations with the corresponding equations expressed in index notation. As noted in Section 10.04, the basic difference between the two notations is in the use of merates and components. Index notation uses merates but no unit vectors:

$$v^i = (v^1, v^2, v^3).$$

Vector notation uses unit vectors $\mathbf{a}_1, \mathbf{a}_2, \mathbf{a}_3$ and components $(v)_1, (v)_2, (v)_3$:

$$\mathbf{v} = \mathbf{a}_1(v)_1 + \mathbf{a}_2(v)_2 + \mathbf{a}_3(v)_3.$$

The relation between components and merates is

$$(v)_i = (g_{ii})^{1/2}\{v^i\}, \tag{11.11}$$

as noted in Section 11.04.

Whenever one draws a directed line segment to represent a vector, and considers this vector as the sum of other directed line segments, he is using components, not merates. For such familiar relations, then, vector notation appears to have an advantage. This occurs in the expressions for $\mathbf{v}$, $(ds)^2$, and v. In most equations, however, vector notation has no advantage. Note also that vector analysis does not always employ components: Space derivatives are always taken with respect to coordinates x^i, not with respect to the components $(x)_.$. It is interesting to see that $\nabla^2\phi$ is written exactly the same in either notation.

In handling problems of vector analysis, then, the reader has a choice between alternative notations. Neither of these notations exhibits marked advantage, so the choice is a matter of personal bias. If, however, the problem is extended to other spaces, other valences, or nonorthogonal coordinate systems, the customary vector notation is generally quite inadequate. Since index notation is so much more powerful, more general, and more precise, it will probably in time supplant vector notation even in the restricted field in which vector notation is now supreme.

11.09. Sliding vectors

In Chapter 6 it was shown that sliding vectors could be represented by bivalent alternating tensors. A vector sliding along a line (Figure 11.03)

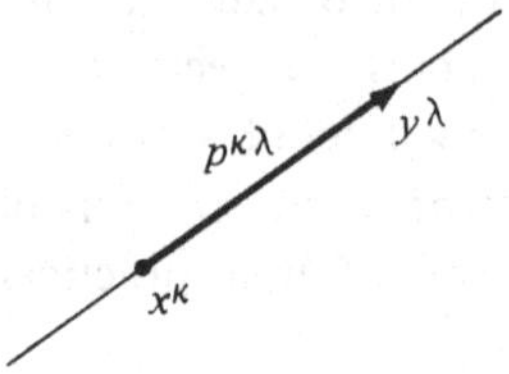

Figure 11.03. A sliding vector on a line (fixed rhabdor), $p^{\kappa\lambda}$.

was represented[11] by the alternating product of the weighted point coordinates,

$$x^{\kappa} = (1, x^1, x^2, x^3) \quad \text{and} \quad y^{\lambda} = (1, y^1, y^2, y^3).$$

Here the weights x^0 and y^0 have been arbitrarily chosen as unity. Then the fixed rhabdor of Equation (6.93) becomes

$$p^{\kappa\lambda} = 2x^{[\kappa}y^{\lambda]}, \tag{6.93}$$

or, on expanding,

$$p^{\kappa\lambda} = \begin{bmatrix} 0 & y^1 - x^1 & y^2 - x^2 & y^3 - x^3 \\ -(y^1 - x^1) & 0 & (x^1y^2 - x^2y^1) & -(x^3y^1 - x^1y^3) \\ -(y^2 - x^2) & -(x^1y^2 - x^2y^1) & 0 & (x^2y^3 - x^3y^2) \\ -(y^3 - x^3) & (x^3y^1 - x^1y^3) & -(x^2y^3 - x^3y^2) & 0 \end{bmatrix}. \tag{11.35}$$

The six distinct merates p^{01}, p^{02}, p^{03}, p^{12}, p^{13}, and p^{23} are related by the Plücker identity

$$p^{[\kappa\lambda}p^{\mu\nu]} = p^{01}p^{23} + p^{02}p^{31} + p^{13}p^{12} = 0. \tag{11.36}$$

The position of the line is defined by the merates p^{12}, p^{13}, and p^{23}. The direction of the sliding vector is from point x^{κ} to point y^{λ}. All of these statements were valid in affine space.

In Euclidean metric space, something new is added: The length of the proper rhabdor[12] can be defined as

$$p = (g_{ij}p^{0i}p^{0j})^{1/2}. \tag{11.37}$$

Since this discussion is limited to Cartesian coordinates,

$$\begin{aligned} p &= [(p^{01})^2 + (p^{02})^2 + (p^{03})^2]^{1/2} \\ &= [(y^1 - x^1)^2 + (y^2 - x^2)^2 + (y^3 - x^3)^2]^{1/2}. \end{aligned} \tag{11.38}$$

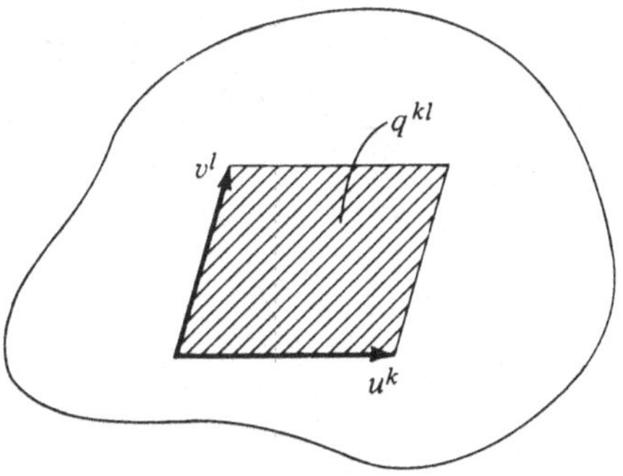

Figure 11.04. A free rhabdor q^{kl}.

Clearly, p is the distance between the two points that define the rhabdor. These points may slide along the line without changing the magnitude of the rhabdor.

In Chapter 6, we also introduced a free rhabdor, which was defined as the alternating product of two free affine vectors u^k and v^l (Figure 11.04),

$$q^{kl} = 2u^{[k}v^{l]}. \tag{6.94}$$

Expansion of Equation (6.94) gives

$$q^{kl} = \begin{bmatrix} 0 & (u^1v^2 - u^2v^1) & -(u^3v^1 - u^1v^3) \\ -(u^1v^2 - u^2v^1) & 0 & (u^2v^3 - u^3v^2) \\ (u^3v^1 - u^1v^3) & -(u^2v^3 - u^3v^2) & 0 \end{bmatrix}. \tag{11.39}$$

There are three distinct merates q^{12}, q^{31}, and q^{23}. It will immediately be noted that these merates are identical to the vector product of u^i and v^j,

$$w^k = g^{1/2} g^{kl} \delta^{123}_{lij} u^i v^j, \tag{11.40}$$

which expands into

$$w^1 = u^2v^3 - u^3v^2, \qquad w^2 = u^3v^1 - u^1v^3, \qquad w^3 = u^1v^2 - u^2v^1. \tag{11.41}$$

The magnitude of the free rhabdor[12] can be defined as the length of the associated vector w^i,

$$\begin{aligned} w &= (g_{ij} w^i w^j)^{1/2} \\ &= [(u^2v^3 - u^3v^2)^2 + (u^3v^1 - u^1v^3)^2 + (u^1v^2 - u^2v^1)^2]^{1/2}. \end{aligned} \tag{11.42}$$

In terms of the merates of the free rhabdor, its magnitude

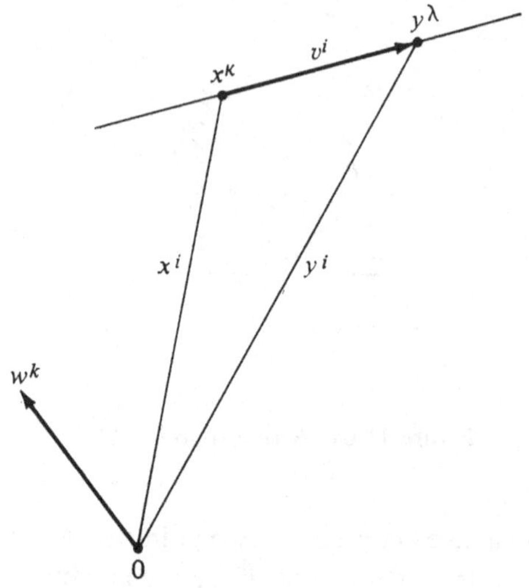

Figure 11.05. A fixed rhabdor defined in terms of proper points x^κ and y^λ.

$$q = w = [g^{ij}\delta^{123}_{ilm}q^{lm}\delta^{123}_{jnp}q^{np}]^{1/2}$$
$$= [(q^{23})^2 + (q^{31})^2 + (q^{12})^2]^{1/2}. \tag{11.43}$$

This magnitude is the area of the parallelogram formed by the vectors u^i and v^j. The same magnitude and the same q^{ij} will be associated with any member of the set of parallel planes defined by u^i and v^j. A sense of rotation from u^i to v^j is defined within the family of parallel planes. In Euclidean space, a free vector w^i that is perpendicular to the family of planes and whose length is equal to the area of the parallelogram is uniquely associated with q^{ij}.

In Chapter 6, Plücker's relation was demonstrated algebraically. In Euclidean space, it is possible to show that Plücker's equation has a simple geometric interpretation.

Consider the proper rhabdor defined in terms of points x^κ and y^λ [Equation (6.93)]. The Cartesian coordinates x^i and y^i can be visualized (Figure 11.05) as vectors from a reference point 0 to the points x^κ and y^λ. A vector v^i can be defined as the difference of y^i and x^i,

$$v^i = y^i - x^i, \tag{11.44}$$

or

$$v^1 = y^1 - x^1 = p^{01}, \quad v^2 = y^2 - x^2 = p^{02}, \quad v^3 = y^3 - x^3 = p^{03}. \tag{11.44a}$$

Another vector w^i can be defined as the vector product of x^i and y^i,

$$w^k = g^{1/2} g^{kl} \delta^{123}_{lij} x^i y^j, \tag{11.45}$$

or

$$\begin{aligned} w^1 &= x^2 y^3 - x^3 y^2 = p^{23}, \\ w^2 &= x^3 y^1 - x^1 y^3 = p^{31}, \\ w^3 &= x^1 y^2 - x^2 y^1 = p^{12}. \end{aligned} \tag{11.45a}$$

The vector v^i lies in the plane of 0, x^κ, and y^λ. The vector w^k is perpendicular to this plane. Therefore, the vectors v^i and w^k are perpendicular to each other and their scalar product is zero. But the scalar product is

$$g_{ij} v^i w^j = 0. \tag{11.46}$$

Expansion in Cartesian coordinates gives

$$v^1 w^1 + v^2 w^2 + v^3 w^3 = 0, \tag{11.46a}$$

or, in terms of the merates of the rhabdor $p^{\kappa\lambda}$,

$$p^{01} p^{23} + p^{02} p^{31} + p^{03} p^{12} = 0, \tag{11.46b}$$

which is equivalent to the Plücker relation [Equation (11.36)].

11.10. Sums of sliding vectors

If a family of several fixed and free rhabdors are added, the sum will be an alternating bivalent tensor $P^{\kappa\lambda}$ which does not generally satisfy Plücker's relation. Such a geometric figure has been called a kineor.[11] The same kineor may be obtained by adding together many different sets of fixed and free rhabdors.

Suppose that we are given the kineor $P^{\kappa\lambda}$ and that we choose one point x^κ. In Chapter 6, we showed that the kineor $P^{\kappa\lambda}$ could always be decomposed into a sum of a fixed rhabdor $p^{\kappa\lambda}$ on whose supporting line the point x^κ must lie and a free rhabdor $q^{\kappa\lambda}$ such that $q^{0i} = 0$. If we take the weight of the point $x^0 = 1$, the decomposition equations [Equation (6.95)] become

$$P^{\kappa\lambda} = p^{\kappa\lambda} + q^{\kappa\lambda} \qquad (q^0 = 0)$$

where

$$\begin{aligned} p^{0i} &= P^{0i}, \\ p^{kl} &= x^k P^{0l} - x^l P^{0k}, \\ q^{kl} &= P^{kl} - x^k P^{0l} + x^l P^{0k}. \end{aligned} \tag{11.47}$$

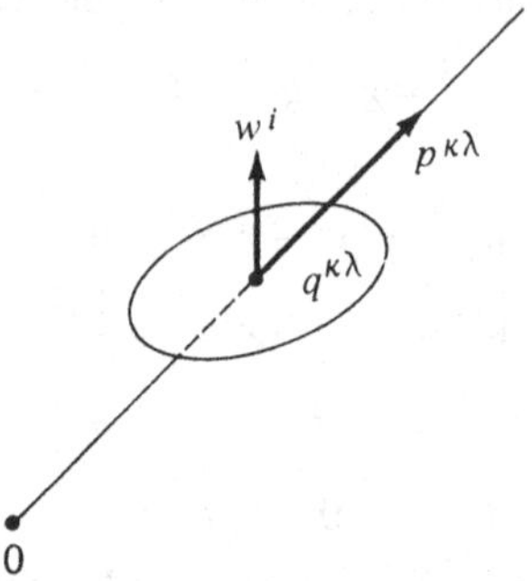

Figure 11.06. A kineor can always be decomposed into a fixed rhabdor $p^{\kappa\lambda}$ whose support passes through the origin and a free rhabdor $q^{\kappa\lambda}$.

Thus, two scalar magnitudes can be associated with each kineor.[12] By Equation (11.38),

$$p = [(P^{01})^2 + (P^{02})^2 + (P^{03})^2]^{1/2}, \tag{11.48}$$

and by Equation (11.43),

$$q = [(P^{23} - x^2P^{03} + x^3P^{02})^2 + (P^{31} - x^3P^{01} + x^1P^{03})^2 + (P^{12} - x^1P^{02} + x^2P^{01})^2]^{1/2}. \tag{11.49}$$

Note that the magnitude of the fixed rhabdor is independent of the choice of the arbitrary point x^κ. But the magnitude of the free rhabdor and the line along which the fixed rhabdor slides depend on the choice of the arbitrary point x^κ.

For example, if the origin is chosen as the arbitrary point, then

$$q = [(P^{23})^2 + (P^{31})^2 + (P^{12})^2]^{1/2}. \tag{11.49a}$$

In this example (Figure 11.06), the merates define a fixed rhabdor $p^{\kappa\lambda}$ that slides along a line passing through the origin,

$$p^{\kappa\lambda} = \begin{bmatrix} 0 & P^{01} & P^{02} & P^{03} \\ -P^{01} & 0 & 0 & 0 \\ -P^{02} & 0 & 0 & 0 \\ -P^{03} & 0 & 0 & 0 \end{bmatrix}.$$

The length of the sliding vector is p. The free rhabdor is

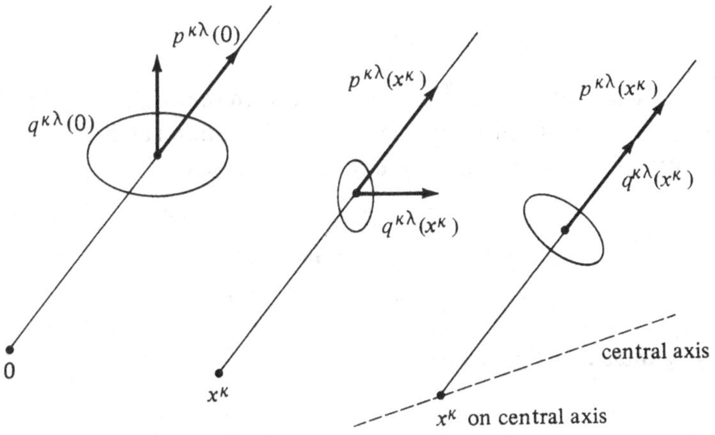

Figure 11.07. Are there any choices of x^κ such that the plane of $q^{\kappa\lambda}$ is perpendicular to the line $p^{\kappa\lambda}$?

$$q^{\kappa\lambda} = \begin{bmatrix} 0 & 0 & 0 & 0 \\ 0 & 0 & P^{12} & -P^{31} \\ 0 & -P^{12} & 0 & P^{23} \\ 0 & P^{31} & -P^{23} & 0 \end{bmatrix}.$$

The area of the parallelogram is q. However, the vector w^i, which is perpendicular to the plane direction of $q^{\kappa\lambda}$, is not generally parallel to the sliding vector.

An important question now arises. As we change the arbitrary point x^κ, the decomposition changes. Since p remains independent of the choice of x^κ, the length of the fixed rhabdor $p^{\kappa\lambda}$ remains constant (Figure 11.07). But the supporting line on which $p^{\kappa\lambda}$ slides must now pass through the arbitrarily chosen point x^κ. On the other hand, as shown in Figure 11.07, both the magnitude and direction of $q^{\kappa\lambda}$ depend on the choice of x^κ. The question is, are there any choices of x^κ such that the plane of $q^{\kappa\lambda}$ is perpendicular to the line of $p^{\kappa\lambda}$?

The condition is merely that the vector w^i, which is perpendicular to the plane of $q^{\kappa\lambda}$, be parallel to the vector v^i along the rhabdor $p^{\kappa\lambda}$ or, by Equations (11.44a), (11.45a), and (11.47),

$$\begin{aligned} \lambda P^{01} &= P^{23} - x^2 P^{03} + x^3 P^{02}, \\ \lambda P^{02} &= P^{31} - x^3 P^{02} + x^1 P^{03}, \\ \lambda P^{03} &= P^{12} - x^1 P^{02} + x^2 P^{01}. \end{aligned} \qquad (11.50)$$

Thus,

$$\lambda p = q. \tag{11.51}$$

Given $P^{\kappa\lambda}$, can points (x^1, x^2, x^3) be found such that Equation (11.50) is satisfied? Let us first rearrange the terms in Equation (11.50) so that the coefficients of the unknowns x^i are on the left side of the equation,

$$\begin{aligned} x^2P^{03} - x^3P^{02} &= P^{23} - P^{01}, \\ -x^1P^{03} \qquad\quad + x^3P^{01} &= P^{31} - P^{02}, \\ x^1P^{02} - x^2P^{01} \qquad\quad &= P^{12} - P^{03}. \end{aligned} \tag{11.52}$$

Solving these equations, we find that there is a line of points that satisfy them:

$$\begin{aligned} x^1 &= x^3\frac{P^{01}}{P^{03}} - \frac{P^{31}[(P^{01})^2 + (P^{03})^2] - P^{02}[P^{01}P^{23} + P^{03}P^{12}]}{P^{03}[(P^{01})^2 + (P^{02})^2 + (P^{03})^2]}, \\ x^2 &= x^3\frac{P^{02}}{P^{03}} - \frac{P^{23}[(P^{02})^2 + (P^{03})^2] - P^{01}[P^{02}P^{31} + P^{03}P^{12}]}{P^{03}[(P^{01})^2 + (P^{02})^2 + (P^{03})^2]}. \end{aligned} \tag{11.53}$$

This line is called the *central axis* of the kineor. However, for consistency of Equation (11.52), the parameter λ must be uniquely determined by the equation

$$\lambda = \frac{P^{01}P^{23} + P^{02}P^{31} + P^{03}P^{12}}{(P^{01})^2 + (P^{02})^2 + (P^{03})^2}. \tag{11.54}$$

Therefore, from Equations (11.51) and (11.54), we can conclude that for any point on the central axis of the kineor the ratio of the magnitudes of the fixed rhabdor p and the free rhabdor q is a constant:

$$q = \frac{P^{01}P^{23} + P^{02}P^{31} + P^{03}P^{12}}{(P^{01})^2 + (P^{02})^2 + (P^{03})^2}\, p. \tag{11.55}$$

In the special case in which $P^{\kappa\lambda}$ satisfies Plücker's equation, Equation (11.55) tells us that $P^{\kappa\lambda} = p^{\kappa\lambda}$ is a fixed rhabdor, $q^{\kappa\lambda} = 0$, and there is no free rhabdor component. Since the parameter λ is then zero, the equation of the line support of p can be found by substituting into Equation (11.52) and is

$$x^1 = x^3\left(\frac{p^{01}}{p^{03}}\right) - \left(\frac{p^{31}}{p^{03}}\right), \qquad x^2 = x^3\left(\frac{p^{02}}{p^{03}}\right) + \left(\frac{p^{23}}{p^{03}}\right). \tag{11.56}$$

This equation defines the line support of any fixed rhabdor.

11.11. Summary

Euclidean space is a special case of Riemannian space:

Definition XVI: Euclidean space is a Riemannian space that allows the metric tensor

$$g_{i'j'} = \begin{bmatrix} 1 & 0 & 0 & \cdots & 0 \\ 0 & 1 & 0 & \cdots & 0 \\ 0 & 0 & 1 & \cdots & 0 \\ \cdot & \cdot & \cdot & \cdots & \cdot \\ 0 & 0 & 0 & \cdots & 1 \end{bmatrix}. \qquad (11.02)$$

Equivalent definitions are that Euclidean space is a Riemannian space that permits a rectangular coordinate system and that Euclidean space is a Riemannian space with $R^i_{jkl} = 0$.

Theorem XLIX. *The necessary and sufficient condition that the quadratic differential form*

$$g_{ij}\, dx^i\, dx^j$$

be reducible to a form with g_{ij} *the unit matrix is that*

$$R^i_{jkl} = 0.$$

In a general Riemannian space, the metric tensor g_{ij} is arbitrary; but in Euclidean space, g_{ij} must be obtainable by transformation of the unit matrix, and

$$g_{ij} = \sum_{i'=1'}^{n'} \frac{\partial x^{i'}}{\partial x^i} \frac{\partial x^{i'}}{\partial x^j}. \qquad (11.05)$$

With an orthogonal coordinate system, Equation (11.05) reduces to

$$g_{ij} = \sum_{i'=1'}^{n'} \left(\frac{\partial x^{i'}}{\partial x^i} \right)^2. \qquad (11.06)$$

On the basis of the metric tensor g_{ij}, one can then evaluate distance, area, volume, gradient, divergence, curl, and the scalar and vector Laplacian just as in Riemannian space. Often, however, the results can be simplified because practical work in Euclidean space is usually confined to orthogonal coordinates in 3-space.

The chapter compares vector analysis in index notation and in the customary Gibbs–Heaviside notation. There are advantages to index notation, particularly if there is a possibility that one's investigations may involve higher valences or spaces other than Euclidean 3-space.

The chapter concludes with a treatment of the geometric representation of sliding vectors and their sums in Euclidean space.

Problems

11.01 Bicylinder coordinates (x^1, x^2, x^3) may be specified by the equations

$$x^{1'} = \frac{a \sinh x^1}{\cosh x^1 - \cos x^2},$$

$$x^{2'} = \frac{a \sin x^2}{\cosh x^1 - \cos x^2},$$

$$x^{3'} = x^3.$$

(a) Determine if this is an orthogonal system.
(b) Evaluate the metric tensor g_{ij}.

11.02 Repeat Problem 11.01 for Maxwell cylinder coordinates

$$x^{1'} = a(x^1 + 1 + e^{x^1} \cos x^2),$$

$$x^{2'} = a(x^2 + e^{x^1} \sin x^2),$$

$$x^{3'} = x^3.$$

11.03 Repeat Problem 11.01 for logarithmic cylinder coordinates

$$x^{1'} = \frac{a}{\pi} \ln[(x^1)^2 + (x^2)^2],$$

$$x^{2'} = \frac{2a}{\pi} \tan^{-1}\left(\frac{x^2}{x^1}\right),$$

$$x^{3'} = x^3.$$

11.04 Repeat Problem 11.01 for tangent sphere coordinates

$$x^{1'} = \frac{x^1}{(x^1)^2 + (x^2)^2} \cos x^3,$$

$$x^{2'} = \frac{x^1}{(x^1)^2 + (x^2)^2} \sin x^3,$$

$$x^{3'} = \frac{x^2}{(x^1)^2 + (x^2)^2}.$$

11.05 Repeat Problem 11.01 for bispherical coordinates

$$x^{1'} = \frac{a \sin x^2 \cos x^3}{\cosh x^1 - \cos x^2},$$

$$x^{2'} = \frac{a \sin x^2 \sin x^3}{\cosh x^1 - \cos x^2},$$

$$x^{3'} = \frac{a \sinh x^1}{\cosh x^1 - \cos x^2}.$$

11.06 In Euclidean 3-space with hyperbolic cylinder coordinates,

$$g_{11} = g_{22} = g^{1/2} = \tfrac{1}{2}[(x^1)^2 + (x^2)^2]^{-1/2}, \qquad g_{33} = 1,$$

obtain expressions for $(ds)^2$, grad ϕ, and $\nabla^2\phi$.

11.07 Repeat Problem 11.06 for inverse elliptic cylinder coordinates, where

$$g_{11} = g_{22} = g^{1/2} = \frac{a^2(\cosh^2 x^1 - \cos^2 x^2)}{(\cosh^2 x^1 - \sin^2 x^2)^2}, \qquad g_{33} = 1.$$

11.08 Repeat Problem 11.06 for toroidal coordinates, where

$$g_{11} = g_{22} = \frac{a^2}{(\cosh x^1 - \cos x^2)^2},$$

$$g_{33} = \frac{a^2 \sinh^2 x^1}{(\cosh x^1 - \cos x^2)^2}.$$

11.09 Repeat Problem 11.06 for inverse prolate spheroidal coordinates, where

$$g_{11} = g_{22} = \frac{a^2(\sinh^2 x^1 + \sin^2 x^2)}{(\cosh^2 x^1 - \sin^2 x^2)^2},$$

$$g_{33} = \frac{a^2 \sinh^2 x^1 \sin^2 x^2}{(\cosh^2 x^1 - \sin^2 x^2)^2}.$$

11.10 Repeat the derivation of Section 11.06 but for Euclidean 3-space instead of Euclidean 2-space.

11.11 In Euclidean 2-space, we wish to form a new coordinate system consisting of confocal ellipses

$$\left(\frac{x^{1'}}{a \cosh x^1}\right)^2 + \left(\frac{x^{2'}}{a \sinh x^1}\right)^2 = 1$$

and radial lines

$$\tan x^2 = x^{2'}/x^{1'}.$$

(a) Write the equations

$$x^{i'} = f^{i'}(x^i).$$

(b) Obtain the metric tensor g_{ij}.
(c) Write equations for $(ds)^2$, grad ϕ, and div $\mathbf{v}$ in the new coordinates.

11.12 Obtain an expression for the (shorter) distance along the line $x^2 = c$ between the circles $x^1 = +a$ and $x^1 = -b$ in bicylindrical coordinates (Problem 11.01) with $x^3 = 0$.

11.13 In hyperbolic cylinder coordinates (Problem 11.06) with $x^3 = 0$, find the equation for distance along the hyperbola $x^1 = a$ between $x^2 = c$ and $x^2 = d$.

11.14 The electrostatic field between semiinfinite plates of a parallel-plate capacitor can be expressed in a cylindrical coordinate system (x^1, x^2, x^3), where the equipotential surfaces are specified by $x^2 =$ constant. The electric flux lines are designated by $x^1 =$ constant, $x^3 =$ constant and the coordinate system has the metric coefficients

$$g_{11} = g_{22} = (a/\pi)^2(1 + e^{x^1}\cos x^2 + e^{2x^1}), \qquad g_{33} = 1.$$

Find the length of a flux line between plates $x^2 = -\pi$ and $x^2 = +\pi$, the line being specified by $x^1 = 0$.

11.15 Find the surface area of a toroid $x^1 = \eta_0$ using the metric coefficients of Problem 11.08.

11.16 Find the volume of the toroid of Problem 11.15.

11.17 In bispherical coordinates, $x^1 =$ constant represents a sphere, while $x^2 =$ constant is a spindle-shaped surface orthogonal to the spheres,

$$g_{11} = g_{22} = \frac{(a)^2}{(\cosh x^1 - \cos x^2)^2},$$

$$g_{33} = \frac{(a)^2 \sin^2 x^2}{(\cosh x^1 - \cos x^2)^2}.$$

Obtain an expression for the area of the surface $x^2 = \theta_0 =$ constant between the spheres $x^1 = \eta_0$ and $x^1 = -\eta_0$.

11.18 The following statement is made: The linear connection Γ^i_{jk} is zero in any coordinate system in Euclidean space. As justification, we have

(i) In rectangular coordinates, $\Gamma^{i}_{jk}=0$.
(ii) But Γ^{i}_{jk} is a geometric object. As such, it is zero in all coordinate systems if it is zero in one.
(iii) Therefore, $\Gamma^{i}_{jk}=0$ in all coordinate systems of Euclidean space.
(a) Evaluate the above argument.
(b) Obtain Γ^{i}_{jk} in a conical 2-space.

11.19 Show that the vector product

$$w^k = g^{1/2} g^{kl} \delta^{123}_{lij} u^i v^j$$

may be written

$$\mathbf{w} = \begin{vmatrix} \mathbf{a}_1 & \mathbf{a}_2 & \mathbf{a}_3 \\ (u)_1 & (u)_2 & (u)_3 \\ (v)_1 & (v)_2 & (v)_3 \end{vmatrix}$$

if the coordinate system is orthogonal.

11.20 The scalar triple product may be written

$$\mathbf{U}\boldsymbol{\cdot}\mathbf{V}\times\mathbf{W} = g^{1/2}\delta^{123}_{ijk} U^i V^j W^k.$$

(a) Write this product as a determinant in terms of merates.
(b) Repeat in terms of components.
(c) What is the geometric significance of this product?

11.21 Write the triple products

$$\mathbf{U}\times(\mathbf{V}\times\mathbf{W}) \quad \text{and} \quad (\mathbf{U}\times\mathbf{V})\times\mathbf{W}$$

in index notation. What is the geometric difference between these two triple products?

11.22 Prove that the triple product

$$U^i \otimes (V^j \otimes W^k) = T^l$$

is equal to

$$T^l = V^l (U_i W^i) - W^l (U_j V^j)$$

in Euclidean 3-space with orthogonal coordinates.

11.23 Evaluate the quadruple product

$$(U^i \otimes V^j) \odot (W^k \otimes X^l)$$

in Euclidean 3-space with orthogonal coordinates.

11.24 In Euclidean 3-space with orthogonal coordinate systems,

$$\operatorname{grad}(uv) = u \operatorname{grad} v + v \operatorname{grad} u.$$

Determine how general this relation is: Does it apply with nonorthogonal coordinates, in Riemannian n-space, in arithmetic space?

11.25 Repeat Problem 11.24 for

$$\operatorname{div}(u\mathbf{v}) = (\operatorname{grad} u)\mathbf{v} + u \operatorname{div} \mathbf{v}.$$

11.26 Repeat Problem 11.24 for

$$\operatorname{curl}(u\mathbf{v}) = (\operatorname{grad} u)\mathbf{v} + u \operatorname{curl} \mathbf{v}.$$

11.27 Repeat Problem 11.24 for

$$\operatorname{div}(\mathbf{u} \times \mathbf{v}) = \mathbf{v} \cdot \operatorname{curl} \mathbf{u} - \mathbf{u} \cdot \operatorname{curl} \mathbf{v}.$$

11.28 A covariant bivalent alternating tensor $p_{\kappa\lambda}$ was defined by Equation (6.96). Investigate the most suitable method of defining a scalar magnitude of the fixed strophor in Euclidean space. Recall Equation (6.53) relating $\tilde{p}^{\kappa\lambda}$ and $\tilde{p}_{\kappa\lambda}$. Also recall that indices can be raised or lowered in a metric space. Find a geometric meaning in terms of the planes u_κ and v_λ for your proposed magnitude of the fixed strophor.

11.29 Repeat Problem 11.28 for a free strophor expressing the geometric interpretation of the proposed scalar magnitude in terms of properties of u_k and v_l.

11.30 A fixed rhabdor is specified by the matrix

$$p^{\kappa\lambda} = \begin{bmatrix} 0 & 1 & 2 & -1 \\ -1 & 0 & 3 & 5 \\ -2 & -3 & 0 & -13 \\ 1 & -5 & 13 & 0 \end{bmatrix}.$$

(a) Find the length p of the sliding vector.
(b) Find the equation of the line on which the vector slides and sketch it.

11.31 A kineor

$$p^{\kappa\lambda} = \begin{bmatrix} 0 & 1 & 0 & 0 \\ -1 & 0 & 0 & 1 \\ 0 & 0 & 0 & 2 \\ 0 & -1 & -2 & 0 \end{bmatrix}.$$

(a) Decompose this kineor into a fixed rhabdor $p^{\kappa\lambda}$ and a free rhabdor $q^{\kappa\lambda}$ in such a way that the point $x^\kappa = (1, 1, 0, 1)$ lies on the support of the fixed rhabdor.
(b) Find the magnitude of the fixed and free rhabdors and sketch them.

11.32

(a) Find the central axis of the kineor given in Problem 11.31.

(b) Find the magnitudes p and q.

(c) Evaluate λ.

(d) Sketch the kineor for any point on the central axis.

References

Chapter 0. Historical introduction

1. J. Wallis, Letter to Collins, May 6, 1673, *Treatise of algebra* (London, 1685), p. 264; F. Cajori, "Historical notes on the graphic representation of imaginaries before the time of Wessel," *Am. Math. Mon., 19,* 167, 1912.
2. K. Weierstrass, "Zur Theorie der aus *n* Haupteinheiten gebildeten complex Grössen," *Nach. König. Ges. Wiss.* Göttingen, 395, 1884.
3. W. R. Hamilton, *Lectures on quaternions* (Dublin, 1853); W. R. Hamilton, *Elements of quaternions,* 2 vols. (Longmans, Green, London, 1899–1901); R. P. Graves, *Life of Sir William Rowan Hamilton,* Vol. 2 (Dublin University Press, Dublin, 1885), p. 435.
4. P. G. Tait, *An elementary treatise on quaternions* (Cambridge University Press, Cambridge, 1890).
5. J. C. Maxwell, *A treatise on electricity and magnetism* (Oxford University Press, Oxford, 1873).
6. B. Peirce, *Linear associative algebra* (Washington, DC, 1870); *Am. J. Math., 4,* 97, 1881.
7. J. B. Shaw, *Synopsis of linear associative algebra* (Carnegie Institution, Washington, 1907); J. A. Schouten, "Zur Klassifizierung der assoziativen Zahlensysteme," *Math. Ann., 76,* 1, 1915; H. E. Hawkes, "On hypercomplex number systems," *Am. Math. Soc. Trans., 3,* 312, 1902; G. Scheffers, "Complex Zahlensysteme," *Math. Ann., 39,* 293, 1891; L. E. Dickson, *Linear algebras* (Cambridge University Press, Cambridge, 1914), and *Algebras and their arithmetics* (Chicago University Press, Chicago, 1923).
8. A. Cayley, "Remarques sur la notation des fonctions algébriques," *J. F. Math., 50,* 282, 1855; A. Cayley, *Collected math. papers,* Vol. 2 (Cambridge University Press, Cambridge, 1889), p. 185.
9. R. A. Frazer, W. J. Duncan, and A. R. Collar, *Elementary matrices* (Cambridge University Press, Cambridge, 1938); R. E. D. Bishop, G. M. L. Gladwell, and S. Michaelson, *The matrix analysis of vibrations* (Cambridge University Press, Cambridge, 1965); B. Higman, *Applied group-theoretic and matrix methods* (Oxford University Press, Oxford, New York, 1955); L. A. Pipes, *Matrix methods for engineering* (Prentice-Hall, Englewood Cliffs, NJ, 1963).

10. H. Grassmann, *Die lineale Ausdehnungslehre, ein neuer Zweig der Mathematik* (Otto Wigand, Leipzig, 1844), *Die Ausdehnungslehre* (Enslin, Berlin, 1862); H. Grassmann, *Grassmann's gesammelte math. u. phys. Werke*, 3 vols. (B. G. Teubner, Leipzig, 1894); H. G. Forder, *The calculus of extension* (Cambridge University Press, Cambridge, 1941).
11. Letter to de Morgan, January 31, 1853.
12. J. W. Gibbs, *Elements of vector analysis* (privately printed, New Haven, 1881, pp. 17–50; 1884, pp. 50–90); *The scientific papers of J. Willard Gibbs*, Vol. 2 (Longmans, Green, London, 1906), p. 17; E. B. Wilson, *Vector analysis, founded upon the lectures of J. Willard Gibbs* (Yale University Press, New Haven, 1901).
13. O. Heaviside, *Electromagnetic theory*, 3 vols. (Electrician Printing, London, 1893); (Dover Publications, New York, 1950); See Vol. I, pp. 132–305 on vector analysis.
14. L. Brand, *Vector and tensor analysis* (Wiley, New York, 1947); H. V. Craig, *Vector and tensor analysis* (McGraw-Hill, New York, 1943); H. F. Davis, *Vector analysis* (Allyn & Bacon, Boston, 1961); R. Gans, *Vector analysis with applications to physics* (Blackie & Son, London, 1932); B. Hague, *An introduction to vector analysis for physicists and engineers* (Methuen, London, 1951); H. Lass, *Vector and tensor analysis* (McGraw-Hill, New York, 1950); E. A. Maxwell, *Coordinate geometry with vectors and tensors* (Oxford University Press, New York, 1958); P. Moon and D. E. Spencer, *Vectors* (Van Nostrand, Princeton, N.J., 1965); F. Ollendorff, *Die Welt der Vektoren* (Springer-Verlag, Wien, 1950); H. B. Phillips, *Vector analysis* (Wiley, New York, 1933); C. E. Springer, *Tensor and vector analysis* (Ronald Press, New York, 1962); A. P. Wills, *Vector analysis with an introduction to tensor analysis* (Prentice-Hall, New York, 1931; Dover Publications, New York, 1958).
15. J. G. Coffin, *Vector analysis* (Wiley, New York, 1909).
16. P. Moon and D. E. Spencer, "The meaning of the vector Laplacian," *J. Franklin Inst.*, *256*, 551, 1953, *258*, 215, 1954.
17. A. Cayley, "On the theory of linear transformations," *Cambridge Math. J.*, *4*, 193, 1845; *Collected mathematical papers*, Vol. 1, p. 80.
18. J. J. Sylvester, "On Newton's rule for the discovery of imaginary roots," *Phil. Trans.*, *154*, 579, 1864; *Collected Mathematical Papers*, Vol. 2, p. 380.
19. H. W. Turnbull, *The theory of determinants, matrices, and invariants* (Blackie & Son, London, 1929); E. Study, *Einleitung in die Theorie der Invarianten linearer Transformationen auf Grund der Vektorrechnung* (Vieweg u. Sohn, Braunschweig, 1923); R. Weitzenböck, *Invariantentheorie* (Noordhoff, Groningen, 1923); G. B. Gurevich, *Foundations of the theory of algebraic invariants* (Noordhoff, Groningen, 1964).
20. B. Riemann, "Uber die Hypothesen, welche der Geometrie zu Grunde liegen," *Werke* (Teubner, Leipzig, 1892), p. 272; English translation in D. E. Smith, *A source book in mathematics* (McGraw-Hill, New York, 1929), p. 411.
21. F. Klein, "Vergleichende Betrachtungen über neuere geometrische Forschungen," *Math. Ann.*, *43*, 63, 1893.
22. G. Ricci, "Résumé de quelques travaux sur les systèmes variable de fonctions associés a une forme différentielle quadratique," *Bull. Sci. Math.*, *16*, 167, 1892; *Opere*, Vol. 1 (Roma, 1956), p. 288.
23. G. Ricci and T. Levi-Civita, "Méthodes de calcul différentiel absolu et leurs applications," *Math. Ann.*, *54*, 125, 1901; T. Levi-Civita, *Opere matematiche*, Vol. I (Zanichelli, Bologna, 1954), p. 479.

24. A. Einstein, "Die Grundlage der Allgemeinen Relativitätstheorie," *Ann. Phys., 49,* 769, 1916; English translation in *The principle of relativity* (Dover, New York, 1923), p. 111.
25. O. Veblen, *Invariants of quadratic differential forms* (Cambridge University Press, Cambridge, 1933); T. Y. Thomas, *The differential invariants of generalized spaces* (Cambridge University Press, Cambridge, 1934); T. Levi-Civita, *The absolute differential calculus* (Blackie & Son, London, 1929); J. A. Schouten and D. J. Struik, *Einführung in die neveren Methoden der Differentialgeometric,* 2 vols. (Noordhoff, Groningen, 1935); J. A. Schouten, *Ricci-calculus* (Springer-Verlag, Berlin, 1954).
26. W. Voigt, *Die fundamental physikalischen Eigenschaften der Krystalle* (Leipzig, 1898), p. vi; *Lehrbuch der Kristallphysik* (Teubner, Leipzig, 1928).
27. J. A. Schouten, *Grundlagen der Vektor—und Affinoranalysis* (Teubner, Leipzig, 1914).
28. D. J. Struik, *Grundzüge der mehrdimensionalen Differentialgeometric in direkter Darstellung* (Springer-Verlag, Berlin, 1922).
29. J. A. Schouten, *Der Ricci-Kalkül,* 1st ed. (Springer-Verlag, Berlin, 1924); D. J. Struik, *Theory of linear connections* (Springer-Verlag, Berlin, 1934); J. A. Schouten and D. J. Struik, *Einführung in die neureren Methoden der Differentialgeometric,* 1st ed. (Noordhoff, Groningen, Vol. 1, 1935; Vol. II, 1938).
30. E. B. Christoffel, "Uber die Transformation der homogenen Differentialausdrücke zweiten Grades," *J. f. Math., 70,* 46, 241, 1869.
31. T. Levi-Civita, "Nozioni di parallelismo in una varietà qualunque e consequente specificazione geometrica della curvature Riemanniana," *Rend. Palermo, 42,* 1973, 1917.
32. Schouten, *Ricci-calculus,* p. 122; D. J. Struik, *Theory of linear connections* and T. Y. Thomas, *The differential invariants of generalized spaces* (Cambridge University Press, Cambridge, 1934).

Chapter 1. Index notation

1. P. Moon and D. E. Spencer, *Vectors* (Van Nostrand Co., Princeton, N.J., 1965), p. 1; D. E. Spencer and P. Moon, "A unified approach to hypernumbers," in R. S. Cohen, J. J. Stackel, and M. W. Wartofsky, *For Dirk Struik* (Reidel, Dordrecht, 1974).
2. M. Bôcher, *Introduction to higher algebra* (Macmillan, New York, 1927).
3. J. A. Schouten, *Der Ricci-Kalkül* (Springer-Verlag, Berlin, 1924), p. 4.
4. I. S. Sokolnikoff, *Tensor analysis* (Wiley, New York, 1964), p. 97.
5. F. D. Murnaghan, "The generalized Kronecker symbol." *Int. Math. Congress Proc.* (University of Toronto Press, Toronto, 1924), p. 928; F. D. Murnaghan, "The generalized Kronecker symbol and its application to the theory of determinants," *Am. Math. Monthly, 32,* 233, 1925; O. Veblen, *Invariants of quadratic differential forms* (Cambridge University Press, Cambridge, 1933), pp. 3, 12; J. L. Synge and A. Schild, *Tensor calculus* (University of Toronto Press, Toronto, 1949), p. 242; Sokolnikoff, *Tensor analysis,* p. 101.

Chapter 2. Holor algebra

1. F. E. Hohn, *Elementary matrix algebra* (Macmillan, New York, 1958), p. 2.
2. P. Moon and D. E. Spencer, *Vectors* (Van Nostrand, Princeton, N.J., 1965), Chap. 2.

3. O. Veblen, *Invariants of quadratic differential forms* (Cambridge University Press, Cambridge, 1933); I. S. Sokolnikoff, *Tensor analysis* (Wiley, New York, 1964); J. L. Synge and A. Schild, *Tensor calculus* (University of Toronto Press, Toronto, 1949); B. Spain, *Tensor calculus* (Oliver and Boyd, Edinburgh, 1953).
4. A. C. Aitken, *Determinants and matrices* (Interscience, New York, 1954); R. Bellman, *Introduction to matrix analysis* (McGraw-Hill, New York, 1960); G. Birkhoff and S. MacLane, *A survey of modern algebra* (Macmillan, New York, 1941); E. Bodewig, *Matrix calculus* (North-Holland, Amsterdam, 1956); M. Denis-Papin and A. Kaufman, *Cours de calcul matriciel applique* (Albin Michel, Paris, 1951); P. S. Dwyer, *Linear computations* (Wiley, New York, 1951); W. L. Ferrar, *Finite matrices* (Oxford University Press, New York, 1951); R. A. Frazer, W. J. Duncan, and A. R. Collar, *Elementary matrices and some applications to dynamics and differential equations* (Cambridge University Press, Cambridge, 1946); W. Gröbner, *Matrizenrechnung* (Oldenburg, München, 1956); P. R. Halmos, *Finite dimensional vector spaces* (Princeton University Press, Princeton, N.J., 1942); J. Heading, *Matrix theory for physicists* (Longmans, Green, London, 1958); B. Higman, *Applied group-theoretic and matrix methods* (Oxford University Press, New York, 1955); J. S. Lomont, *Applications of finite groups* (Academic Press, New York, 1959); C. C. MacDuffee, *Vectors and matrices* (Math. Assn. Am., New York, 1943), and *The theory of matrices* (Chelsea, New York, 1946); A. D. Michal, *Matrix and tensor calculus* (Wiley, New York, 1947); F. Neiss, *Determinanten und Matrizen* (Springer-Verlag, Berlin, 1955); S. Perlis, *Theory of matrices* (Addison-Wesley, Reading, Mass., 1952); W. Schmeidler, *Vorträge über Determinanten und Matrizen mit Anwendungen in Physik und Technik* (Akademie-Verlag, Berlin, 1949); O. Schreier and E. Sperner, *Einführung in die analytische Geometrie und Algebra* (Teubner, Leipzig, 1931), and *An introduction to modern algebra and matrix theory* (Chelsea, New York, 1952); H. Schwerdtfeger, *Introduction to linear algebra and the theory of matrices* (Noordhoff, Groningen, 1950); R. R. Stoll, *Linear algebra and matrix theory* (McGraw-Hill, New York, 1952); R. M. Thrall and L. Tornheim, *Vector spaces and matrices* (Wiley, New York, 1957); H. W. Turnbull, *The theory of determinants, matrices, and invariants* (Blackie & Son, London, 1929); H. W. Turnbull and A. C. Aitken, *An introduction to the theory of canonical matrices* (Blackie & Son, London, 1932); J. H. M. Wedderburn, *Lectures on matrices,* Vol. 17 (Am. Math. Soc., Colloquium Pub., 1934); P. Zurmühl, *Matrizen, eine Darstellung für Ingenieure* (Springer-Verlag, Berlin, 1958).
5. M. Bôcher, *Introduction to higher algebra* (Macmillan, New York, 1947), p. 46.
6. M. Bôcher, *Higher algebra,* p. 23.
7. M. Bôcher, *Higher algebra,* p. 36.

Chapter 3. Gamma products

1. P. Moon and D. E. Spencer, "A new mathematical representation of alternating currents," *Tensor, 14,* 110, 1963.
2. P. Moon and D. E. Spencer, *Vectors* (Van Nostrand, Princeton, N.J., 1965), p. 130.
3. F. E. Hohn, *Elementary matrix algebra* (Macmillan, New York, 1958), p. 13.
4. J. A. Schouten, *Ricci-calculus* (Springer-Verlag, Berlin, 1954), p. 14.

5. A. E. Kennelly, "Impedance," *A.I.E.E. Trans., 10,* 175, 1893; C. P. Steinmetz, "Discussion," *A.I.E.E. Trans., 10,* 227, 1893; C. P. Steinmetz, "Symbolic representation of general alternating waves and of double-frequency vector products," *A.I.E.E. Trans., 16,* 269, 1899.
6. A. E. Fitzgerald and D. E. Higginbotham, *Basic electrical engineering* (McGraw-Hill, New York, 1957), Chaps. 2 and 3; R. H. Frazier, *Elementary electric-circuit theory* (McGraw-Hill, New York, 1945), Chap. 4; O. Veblen, *Invariants of quadratic differential forms* (Cambridge University Press, Cambridge, 1933).
7. J. H. M. Manders, *Application of direct analysis to pulsating and oscillating phenomena* (Eduard Ljdo, Leiden, 1919).
8. R. A. Frazer, W. J. Duncan, and A. R. Collar, *Elementary matrices and some applications to dynamics and differential equations* (Cambridge University Press, Cambridge, 1938), p. 8.
9. G. Cayley, "Double algebra," *Math. Papers, 12,* 60, 1897.
10. E. Study, "Complexen Zahlen," *Göttingen Nach.*, 237, 1889.

Chapter 4. Tensors

1. P. Moon and D. E. Spencer, *Vectors* (Van Nostrand, Princeton, N.J., 1965), Chap. 3.
2. G. B. Gurevich, *Foundations of the theory of algebraic invariants* (Noordhoff, Groningen, 1964); O. E. Glenn, *A treatise on the theory of invariants* (Ginn, Boston, 1915); R. Weitzenböck, *Invariantentheorie* (Noordhoff, Groningen, 1923); P. Gordon, *Vorlesungen über Invariantentheorie* (Teubner, Leipzig, 1887); A. Clebsch, *Theorie der binären algebraischen Formen* (Teubner, Leipzig, 1872).
3. H. Grassmann, *Die Ausdehnungslehre* (Enslin, Berlin, 1862); H. Grassmann, *Gesammelte math. u. phys. Werke* (Teubner, Leipzig, 1894); F. Klein, *Elementary mathematics from an advanced standpoint: geometry* (Macmillan, New York, 1939), p. 21.
4. M. Bôcher, *Introduction to higher algebra* (Macmillan, New York, 1927), p. 82.
5. F. Klein, *Gesammelte mathematische Abhandlungen,* Vol. I, (Springer-Verlag, Berlin, 1921), p. 460; F. Klein, *Elementary mathematics from an advanced standpoint: geometry* (Macmillan, New York, 1939).
6. O. Veblen, *Invariants of quadratic differential forms* (Cambridge University Press, Cambridge, 1933), p. 14; M. Kucharzewski and M. Kuczma, *Basic concepts of the theory of geometric objects* (Panstwowe Wydawnictwo Naukowe, Warszawa, 1964).
7. A. Nijenhuis, *Theory of the geometric object* (Amsterdam, 1952); J. Aczel and S. Golab, *Funktionalgleichungen der Theorie der geometrischen Objekte* (Panstwowe Wydawnictwo Naukowe, Warszawa, 1960); M. Kucharzewski and M. Kuczma, *Basic concepts of the theory of geometric objects* (Panstwowe Wydawnictwo Naukowe, Warszawa, 1964).
8. E. Bauer, *Champs de vecteurs et de tenseurs* (Masson et Cie, Paris, 1955), p. 2; L. Brillouin, *Les tenseurs en méchanique et en élasticité* (Masson et Cie, Paris, 1938), p. 6; I. S. Sokolnikoff, *Tensor analysis* (Wiley, New York, 1951), p. 56; B. Spain, *Tensor calculus* (Oliver and Boyd, Edinburgh, 1953), p. 7; C. E. Weatherburn, *Riemannian geometry and the tensor calculus* (Cambridge University Press, Cambridge, 1938), p. 21.

9. P. Moon and D. E. Spencer, *Partial differential equations* (Heath, Boston, 1968), Chap. 1.
10. J. F. Nye, *Physical properties of crystals* (Oxford University Press, New York, 1957), p. 68.
11. Staff, M.I.T., *Electric circuits* (Wiley, New York, 1940), Chap. 8; G. Kron, *Tensors for circuits* (Dover, New York, 1959); L. V. Bewley, *Tensor analysis of electric circuits and machines* (Ronald Press, New York, 1961).
12. Sokolnikoff, *Tensor analysis,* p. 69; J. L. Synge and A. Schild, *Tensor calculus* (University of Toronto Press, Toronto, 1949), p. 18.
13. G. T. Kneebone, *Mathematical logic* (Van Nostrand, Princeton, N.J., 1963); O. Bird, *Syllogistic and its extensions* (Prentice-Hall, Englewood Cliffs, N.J., 1964).
14. Sokolnikoff, *Tensor analysis,* p. 66; Synge and Schild, *Tensor calculus,* p. 18.
15. Gurevich, *Foundations,* p. 170.

Chapter 5. Akinetors

1. O. Veblen, *Invariants of quadratic differential forms* (Cambridge University Press, Cambridge, 1933), p. 20; I. S. Sokolnikoff, *Tensor analysis* (Wiley, New York, 1964), p. 69; J. A. Schouten, *Ricci-calculus* (Springer-Verlag, Berlin, 1954), p. 11; and *Tensor analysis for physicists* (Oxford University Press, New York, 1954), p. 29.
2. L. Brillouin, *Tensors in mechanics and elasticity* (Academic Press, New York, 1964), Chap. 3.
3. J. L. Synge and A. Schild, *Tensor calculus* (University of Toronto Press, Toronto, 1949), Chap. 7; Veblen, *Invariants,* p. 20.
4. Brillouin, *Tensors,* p. 58.
5. Brillouin, *Tensors,* p. 84.
6. P. Moon and D. E. Spencer, Vectors (Van Nostrand, Princeton, N.J., 1965), p. 218.
7. Brillouin, *Tensors,* p. 82.
8. Moon and Spencer, *Vectors,* p. 224.
9. Synge and Schild, *Tensor calculus,* p. 246.

Chapter 6. Geometric spaces

1. M. Jammer, *Concepts of space* (Harvard University Press, Cambridge, Mass., 1954).
2. T. Y. Thomas, *The differential invariants of generalized spaces* (Cambridge University Press, Cambridge, 1934); O. Veblen, *Invariants of quadratic differential forms* (Cambridge University Press, Cambridge, 1933); O. Veblen and J. H. C. Whitehead, *The foundations of differential geometry* (Cambridge University Press, Cambridge, 1932).
3. *Commission Internationale de l'Eclairage, Compte Rendu des Seances* (Cambridge University Press, Cambridge, 1932), p. 25; W. D. Wright, "A redetermination of the mixture curves of the spectrum," *Opt. Soc. Trans., London, 31,* 201, 1929–30; J. Guild, "The colorimetric properties of the spectrum," *Roy. Soc. Phil. Trans., A230,* 149, 1932; T. Smith and J. Guild, "The C.I.E. colorimetric standards and their use," *Opt. Soc. Trans., London, 33,* 73, 1931–2; A. C. Hardy, *Handbook of colorimetry* (Technology Press, Cambridge, Mass., 1935); P. Moon and D. E. Spencer, "Analytic representation

of trichromatic data," *J. Opt. Soc. Am., 35,* 399, 1945; P. Moon and D. E. Spencer, *Lighting design* (Addison-Wesley, Reading, Mass., 1948), Chap. 11.
4. G. Kron, *Tensors for circuits* (Dover, New York, 1959).
5. T. Y. Thomas, *Differential,* p. 1; E. Cartan, *La théorie des groupes finis et continus et l'analysis situs* (Mém. Sci. Math., No. 42, Gauthier-Villars, Paris, 1930); J. A. Schouten, *The Ricci calculus* (Springer-Verlag, Berlin, 1954), p. 61.
6. Veblen, *Invariants*; Veblen and Whitehead, *Foundations*; J. A. Schouten and D. van Danzig, "Was ist Geometrie?" *Abhandlungen aus dem Seminar für Vektor und Tensoranalysis,* Vol. 2 (Moskau, 1935), p. 15.
7. F. Klein, "Vergleichende Betrachtungen über neuere geometrische Forschungen" (Erlangen, 1872); *Math. Ann., 43,* 63, 1893; *Gesammelte math. Abhandlungen,* Vol. I (Springer-Verlag, Berlin, 1921), p. 460.
8. F. Klein, *Elementary mathematics from an advanced standpoint: Geometry* (Macmillan, New York, 1939), pp. 69, 130.
9. J. A. Schouten and D. J. Struik, *Einführung in die neueren Methoden der Differentialgeometrie,* Vol. I (Noordhoff, Groningen, 1935); J. A. Schouten, *Tensor analysis for physicists* (Oxford University Press, New York, 1954).
10. Klein, *Elementary,* p. 70.
11. Klein, *Elementary,* p. 3.
12. D. L. MacAdam, "Color science and color photography," *Physics Today, 20,* 27, Jan. 1967.
13. M. Bôcher, *Introduction to higher algebra* (Macmillan, New York, 1927), p. 47.
14. T. E. French, *A manual of engineering drawing* (McGraw-Hill, New York, 1935).
15. T. A. Thomas, *Technical illustration* (McGraw-Hill, New York, 1960).
16. Bôcher, *Introduction,* p. 11; Klein, *Elementary,* p. 86.
17. Klein, *Elementary,* p. 86.
18. D. J. Struik, *Analytic and projective geometry* (Addison-Wesley, Reading, Mass., 1953), p. 43.
19. P. Moon and D. E. Spencer, "The geometry of nutrition," *J. of Nutrition, 104,* 1535, 1974.
20. J. D'Amelio, *Perspective drawing handbook* (Tudor, New York, 1964); French, *Manual.*
21. Klein, *Elementary,* p. 98; F. Morley and F. V. Morley, *Inversive geometry* (Ginn, Boston, 1933); F. S. Woods, *Higher geometry* (Ginn, Boston, 1922).
22. R. R. Lawrence, *Principles of alternating-current machinery* (McGraw-Hill, New York, 1953).
23. R. B. Adler, L. J. Chu, and R. M. Fano, *Electromagnetic energy transmission and radiation* (Wiley, New York, 1960), p. 105.
24. P. Moon and D. E. Spencer, *Field theory for engineers* (Van Nostrand, Princeton, N.J., 1961), Chap. 12; P. Moon and D. E. Spencer, *Field theory handbook* (Springer-Verlag, Berlin, 1961).
25. D. J. Struik, *Differential geometry* (Addison-Wesley, Reading, Mass., 1950); F. W. Sohon, *The stereographic projection* (Chemical, Brooklyn, N.Y., 1941).
26. D. E. Spencer, "Geometric figures in affine space," *J. Math Phys., 23,* 1, 1944; E. Study, *Geometrie der Dynamen* (Teubner, Leipzig, 1903); H. E. Timerding, *Geometrie der Kräfte* (Teubner, Leipzig, 1908); R. S. Ball, *Theory of screws* (Cambridge University Press, Cambridge, 1900).

Chapter 7. The linear connection

1. T. Levi-Civita, "Nozione di parallelismo in una varietà qualunque e conseguente specificazione geometrica della curvature Riemanniana," *Rend. Circ. Mat., Palermo, 42,* 173, 1917; T. Levi-Civita, *Opere mathematiche,* Vol. 4 (Zanichelli, Bologna, 1960), p. 1.
2. J. A. Schouten, "Die direkte Analysis zur neueren Relativitätstheorie," *Verh. Kon. Akad., Amsterdam, 12*(6), 1918.
3. H. Weyl, "Reine Infinitesimalgeometrie," *Math. Zs., 2,* 384, 1918; H. Weyl, *Raum, Zeit, Materie* (Springer-Verlag, Berlin, 1923); H. Weyl, *Mathematische Analyse des Raumproblems* (Springer, Berlin, 1923).
4. J. A. Schouten, *Ricci-calculus* (Springer-Verlag, Berlin, 1954).
5. A. S. Eddington, "A generalization of Weyl's theory of the electromagnetic and gravitational fields," *Roy. Soc. Proc., 99A,* 104, 1921; A. S. Eddington, *The mathematical theory of relativity* (Cambridge University Press, Cambridge, 1923).
6. D. J. Struik, *Theory of linear connections* (Springer-Verlag, Berlin, 1934).
7. E. B. Christoffel, "Uber die Transformation der homogenen Differentialausdrücke zweiten Grades," *Crelle's J. 70,* 46, 1869.
8. M. Cavagnero, T. Roberts, R. Semagin, and D. E. Spencer, "Tensor calculi and the linear connection," Presented to the American Mathematical Society, Pittsburgh, Pa., August 19, 1981.
9. M. Albert, D. E. Spencer, and S. Y. Uma, "Linear relations between the covariant derivatives of tensors," Presented to the American Mathematical Society, Plymouth, N.H., June 29, 1984.
10. M. Albert and D. E. Spencer, "Classification of covariant derivatives of tensors of valence *m*," Presented to American Mathematical Society, Fairfield, Conn., October 28, 1983.
11. P. J. Mann and D. E. Spencer, "The meaning of the 'divergence' of a bivalent tensor," Presented to the American Mathematical Society, Albany, N.Y., August 9, 1983.
12. M. Cavagnero, T. Roberts, R. Semagin, and D. E. Spencer, "Akinetor calculi," Presented to the American Mathematical Society, Amherst, Mass., October 26, 1985.

Chapter 8. The Riemann–Christoffel tensors

1. D. J. Struik, *Classical differential geometry* (Addison-Wesley, Cambridge, Mass., 1950).
2. M. Cavagnero, T. Roberts, R. Semagin, D. E. Spencer, and S. Y. Uma, "The Riemann–Christoffel tensors," presented to the American Mathematical Society, Worcester, Mass., April 20, 1985.
3. P. Moon and D. E. Spencer, *Partial differential equations* (Heath, Lexington, Mass., 1969).
4. J. A. Schouten and D. J. Struik, *Einführung in die neueren Methoden der Differentialgeometrie* (Noordhoff, Groningen, 1938).
5. M. Cavagnero, T. Roberts, R. Semagin, D. E. Spencer, and S. Y. Uma, "Akinetors, the Riemann–Christoffel tensors and the Ricci tensors," presented to the American Mathematical Society, Amherst, Mass., October 26, 1985.
6. P. Moon and D. E. Spencer, *Vectors* (Van Nostrand, Princeton, N.J., 1965).

7. M. K. Albert, R. Cavagnero, T. Roberts, and D. E. Spencer, "The Bianchi identities," Presented to the American Mathematical Society, Amherst, Mass., October 26, 1985.

Chapter 9. Non-Riemannian spaces

1. P. G. Bergmann, *Introduction to the theory of relativity* (Prentice-Hall, Englewood Cliffs, N.J., 1934); J. L. Synge, *Relativity, the general theory* (North Holland, Amsterdam, 1960); V. Hlavaty, *Geometry of Einstein's unified field theory* (Noordhoff, Groningen, 1957); M. A. Tonnelat, *Les théories unitaires de l'électromagnétisme et de la gravitation* (Gauthier-Villars, Paris, 1965); M. A. Tonnelat, *Einstein's unified field theory* (Gordon and Breach, New York, 1966); A. S. Petrow, *Einstein-Räume* (Akad. Verlag, Berlin, 1964).
2. D. J. Struik, *Theory of linear connections* (Springer-Verlag, Berlin, 1934), p. 15; J. A. Schouten, *Ricci-calculus* (Springer-Verlag, Berlin, 1954), p. 124; J. L. Synge and A. Schild, *Tensor calculus* (University of Toronto Press, Toronto, 1949), p. 283.
3. D. E. Spencer, "Homeomorphic displacements and the infinitesimal pentagons," Presented to the American Mathematical Society, Worcester, Mass., April 20, 1985.
4. T. Y. Thomas, *The differential invariants of generalized spaces* (Cambridge University Press, Cambridge, 1934), p. 6; L. P. Eisenhart, *Non-Riemannian geometry* (Am. Math. Soc., New York, 1927), p. 58.
5. T. Y. Thomas, *Differential invariants of generalized spaces* (Cambridge University Press, Cambridge, 1934), p. 8.
6. E. Cartan, "Sur une généralisation de la notion de courbure de Riemann et les espaces à torsion," *Acad. Sci. C.R., Paris, 174,* 593, 1922; O. Veblen and T. Y. Thomas, "The geometry of paths," *Amer. Math. Soc. Trans., 25,* 551, 1923; D. J. Struik, *Theory of linear connections* (Springer-Verlag, Berlin, 1934); J. A. Schouten, *Ricci-calculus* (Springer-Verlag, Berlin, 1954), p. 126.
7. H. Weyl, "Gravitation und Elektrizität," *Preuss. Akad. Sitz,* 465, 1918; and *Space, time, matter* (Dover Publishing, New York, 1950).
8. T. Y. Thomas, *The differential invariants of generalized spaces* (Cambridge University Press, Cambridge, 1934), p. 22.
9. L. P. Eisenhart, *Non-Riemannian geometry* (Am. Math. Soc., New York, 1927), Chap. 3; T. Y. Thomas, *The differential invariants of generalized spaces* (Cambridge University Press, Cambridge, 1934), Chap. 3; D. J. Struik, *Theory of linear connections* (Springer-Verlag, Berlin, 1934), Chap. 5; O. Veblen, *Projektive Relativitätstheorie* (Springer-Verlag, Berlin, 1933).
10. P. Moon and D. E. Spencer, *Field theory for engineers* (Van Nostrand, Princeton, N.J., 1961), p. 99.

Chapter 10. Riemannian space

1. B. Riemann, "Uber die Hypothesen, welche der Geometrie zugrunde liegen," *Ges. math. Werke* (Teubner, Leipzig, 1892; Dover Publications, New York, 1953).
2. P. Moon and D. E. Spencer, "A metric for colorspace," *J. Opt. Soc. Am., 33,* 260, 1943; P. Moon and D. E. Spencer, "A metric based on the composite color stimulus," *J. Opt. Soc. Am., 33,* 270, 1943.

3. P. Finsler, *Uber Kurven und Flächen in allgemeinen Räumen* (dissertation, Göttingen, 1918; Birkhauser, Basel, 1951); H. Rund, *The differential geometry of Finsler spaces* (Springer-Verlag, Berlin, 1939); E. Cartan, *Les espaces de Finsler* (Actualités 79, Paris, 1934); H. Busemann, *The geometry of geodesics* (Academic Press, New York, 1955).
4. P. Moon and D. E. Spencer, *Field theory for engineers* (Van Nostrand, Princeton, N.J., 1961), Chap. 3; P. Moon and D. E. Spencer, *Field theory handbook* (Springer-Verlag, Berlin, 1961).
5. Moon and Spencer, *Field theory handbook,* p. 34.
6. L. Brillouin, *Tensors in mechanics and elasticity* (Academic Press, New York, 1964), p. 150.
7. E. B. Christoffel, "Uber die Transformation der homogenen Differentialausdrücke zweiten Grades," *J. reine u. angew. Math., 70,* 46, 1869; E. B. Christoffel, *Gesammelte mathematische Abhandlungen,* Vol. I (Teubner, Berlin, 1910), p. 352; P. Moon and D. E. Spencer, *Vectors* (Van Nostrand, Princeton, N.J., 1965), p. 173.
8. J. L. Synge and A. Schild, *Tensor calculus* (University of Toronto Press, Toronto, 1949), p. 37; I. S. Sokolnikoff, *Tensor analysis* (Wiley, New York, 1951), p. 157; A. J. McConnell, *Applications of tensor analysis* (Dover Publications, New York, 1957), p. 171.
9. C. E. Weatherburn, *An introduction to Riemannian geometry* (Cambridge University Press, Cambridge, 1938); L. P. Eisenhart, *Riemannian geometry* (Princeton University Press, Princeton, N.J., 1926); E. Cartan, *Leçons sur la géométrie des espaces de Riemann* (Gauthier-Villars, Paris, 1951); T. Y. Thomas, *Differential invariants of generalized spaces* (Cambridge University Press, Cambridge, 1934).
10. J. A. Schouten and D. J. Struik, *Einführung in die neueren Methoden der Differentialgeometrie,* Vol. I (Noordhoff, Groningen, 1935), p. 118.
11. P. Moon and D. E. Spencer, *Vectors* (Van Nostrand, Princeton, N.J., 1965).
12. J. A. Schouten and D. J. Struik, *Einführung in die neueren Methoden der Differentialgeometrie,* Vol. II (Noordhoff, Groningen, 1935).
13. Moon and Spencer, *Vectors,* p. 207; Moon and Spencer, *Field theory for engineers,* p. 71.
14. Moon and Spencer, *Vectors,* p. 218; Moon and Spencer, *Field theory for engineers,* p. 43.
15. Moon and Spencer, *Vectors,* p. 223; Moon and Spencer, *Field theory for engineers,* p. 46.
16. Moon and Spencer, *Vectors,* p. 233.
17. Moon and Spencer, *Field theory handbook,* p. 12.
18. Moon and Spencer, *Field theory handbook,* Sec. V.
19. P. Moon and D. E. Spencer, "The meaning of the vector Laplacian," *J. Franklin Inst., 256,* 551, 1953.
20. P. J. Mann and D. E. Spencer, "The meaning of the 'divergence' of a bivalent tensor," presented to the American Mathematical Society, Albany, N.Y., August 9, 1983.
21. Moon and Spencer, *Field theory handbook,* p. 34.

Chapter 11. Euclidean space

1. P. Moon and D. E. Spencer, *Field theory for engineers* (Van Nostrand, Princeton, N.J., 1961), Chap. 3.

2. C. E. Weatherburn, *An introduction to Riemannian geometry* (Cambridge University Press, Cambridge, 1938).
3. L. P. Eisenhart, *Riemannian geometry* (Princeton University Press, Princeton, N.J., 1926), p. 25.
4. P. Moon and D. E. Spencer, *Field theory handbook* (Springer-Verlag, Berlin, 1961), p. 28.
5. J. L. Synge and A. Schild, *Tensor calculus* (University of Toronto Press, Toronto, 1949), p. 10; Eisenhart, *Riemannian geometry,* p. 4; J. A. Schouten, *Ricci-calculus* (Springer-Verlag, Berlin, 1954), p. 10.
6. P. Moon and D. E. Spencer, *ectors* (Van Nostrand, Princeton, N.J., 1965), p. 37.
7. Moon and Spencer, *Vectors,* Chap. 7.
8. J. W. Gibbs, *Elements of vector analysis* (privately printed, 1884); J. W. Gibbs, *The scientific papers of J. Willard Gibbs,* Vol. II (Longmans, Green, London, 1906), p. 17; O. Heaviside, *Electromagnetic theory,* Vol. I (Electrician Printing, London, 1893), p. 194 (Dover Publications, New York, 1950).
9. Moon and Spencer, *Vectors,* Chap. 1.
10. E. B. Wilson, *Vector analysis, founded upon the lectures of J. Willard Gibbs* (Yale University Press, New Haven, 1901).
11. D. E. Spencer, "Geometric figures in affine space," *J. Math. Phys., 23,* 1, 1944.
12. D. E. Spencer, "The tensor interpretation of the figures of Study's *Geometrie der Dynamen,*" *J. Math. Phys., 23,* 103, 1944.

Index